ASTRONOMY AND
ASTROPHYSICS LIBRARY

ASTRONOMY AND
ASTROPHYSICS LIBRARY

Series Editors: M. Harwit, R. Kippenhahn, J.-P. Zahn

Tools of Radio Astronomy
K. Rohlfs

Physics of the Galaxy and Interstellar Matter
H. Scheffler and H. Elsässer

Galactic and Extragalactic Radio Astronomy, 2nd Edition
G.L. Verschuur and K.I. Kellermann, Editors

Astrophysical Concepts, 2nd Edition
M. Harwit

Observational Astrophysics
P. Léna

Supernovae
A.G. Petschek

Albert G. Petschek

Editor

Supernovae

With 111 Illustrations

Springer-Verlag
New York Berlin Heidelberg
London Paris Tokyo Hong Kong

Albert G. Petschek
Department of Physics
New Mexico Institute of Mining and Technology
Socorro, NM 87801,
U.S.A.

Series Editors

Martin Harwit
National Air and Space Museum
Smithsonian Institution
7th St. and Independence Ave. S.W.
Washington, DC 20560
U.S.A.

Rudolf Kippenhahn
Max-Planck-Institut für
    Physik und Astrophysik
Institut für Astrophysik
Karl-Schwarzschild-Straße 1
D-8046 Garching,
F.R.G.

Jean-Paul Zahn
Université Paul Sabatier
Observatoires du Pic-du-Midi
    et de Toulouse
14, Avenue Edouard-Belin
F-31400 Toulouse
France

Cover photograph: Cassiopeia A (Cas A; 3c 461). Computer-enhanced radiograph of Cassiopeia A, supernova remnant. Observed by P.E. Angerhofer, R. Braun, S.F. Gull, R.A. Perley, and R.J. Tuffs, 1983. Courtesy NRAO/AUI.

Library of Congress Cataloging in Publication Data
Supernovae / Albert Petschek, editor.
        p.   cm. — (Astronomy and astrophysics library)
    Includes bibliographical references.

    ISBN-13: 978-1-4612-7951-8   e-ISBN-13: 978-1-4612-3286-5
    DOI: 10.1007/978-1-4612-3286-5

    1. Supernovae.   I. Petschek, Albert.   II. Series.
QB843.S95S97 1990
523.8'446—dc20   90-30473
                                                          CIP

Printed on acid-free paper.

Typset by Asco Trade Typesetting Ltd., Hong Kong

9 8 7 6 5 4 3 2 1

Contents

Introduction

For millennia mankind has watched as the heavens move in their stately progression from night to night and from year to year, presaging with their changes the changing seasons. The sun, the moon, and the planets move in what appears to be an unchanging firmament, except occasionally when a new "star" appears. Among the new stars there are comets, novae, and finally supernovae, the subject of this book. Superstitious mankind regarded these events as significant portents and recorded them carefully so that we have records of supernovae that may reach back as far as 1300 B.C. (Clark and Stephenson, 1977; Murdin and Murdin, 1985). The Cygnus Loop, believed to be a 15,000-year-old supernova remnant at a distance of only 800 pc (Chevalier and Seward, 1988), must have awed our ancestors. Tycho's supernova of 1572, at a distance of 2500 pc, had a magnitude of −4.0, comparable to Venus at its brightest, and Kepler's supernova of 1604 had a magnitude of −3 or so. Thus the Cygnus Loop supernova might have had a magnitude of −6 or so, and should have been readily visible in daytime. A supernova in Vela, about 8000 B.C. was comparably close, as was SN 1006, whose magnitude may have been −9. While most of the supernova records come from the Old World, the supernova of 1054 is recorded in at least one petroglyph in the American West.

Nowadays, we watch the heavens not only with our eyes but also with telescopes. The number of supernovae observed each year just about uses up the alphabet, 1988A through 1988Z, for example. All of these are in galaxies other than the Milky Way. In fact, we cannot be certain of any naked-eye supernova, or any other supernova, in the Milky Way since that of Kepler, and thus none has been investigated with a telescope. Besides telescopes, which make possible the observation of more distant supernovae, we now have spectroscopes which allow us to determine the elements present in the supernovae; radiotelescopes which allow us to find remnants of supernovae which went unobserved, such as, for example, the one in Saggitarius near the galactic center; satellites which give access to yet other regions of the spectrum; and even several large underground detectors intended perhaps for particle physics, but which can detect the elusive neutrino. Undoubtedly, the best observed supernova, so far, is 1987A, a naked-eye supernova in the Large Magellanic Cloud. It was close, and was discovered early in its light curve, and could be investigated with all our modern equipment. Because the neutrinos from the original collapse were detected, the time of collapse is known precisely.

Not only have we greatly improved means of observation, but we also have greatly improved means of calculation and a better understanding of the underlying physics. With all of this new material, it is out of the question to try to bring the reader up to date on all of it. Thus this is a book about supernovae, not all about supernovae, nor even about all the new things since the last book about supernovae. It was the intent to write a general book, and not to emphasize the fascinating new data from 1987A. However, 1987A kept intruding, so that considerable material about it is presented as well.

In the general plan of the book we have, first, a chapter on the classification of supernovae. Supernovae are classified largely according to their spectra, but also to some extent according to their light curves. Separate chapters discuss both the spectra and the light curves. Unfortunately, there is no universal agreement on the proper nomenclature. At one time, a variety of types of supernovae was perceived to exist. Nowadays, it is more common to consider only two fundamental types, depending on whether the spectra show hydrogen lines or not, and then subclassify those types according to a more detailed examination of the spectra or according to their light curves. Since the existence or absence of hydrogen in the spectrum exhausts the possibilities, this latter scheme leaves no room for any type beyond I (no hydrogen) and II (hydrogen).

The authors writing here agree to disagree about classification. The extensive table of supernovae classifications by Branch, in the chapter on Spectra of Supernovae, relies only on the spectra for classification, and thus has only Type I and Type II. Spectra that do not fit the normal Type II are labeled IIpec. Harkness and Wheeler (Classification of Supernovae) and Sramek and Weiler (Radio Supernovae), on the other hand, believe IIpec is too general a classification, and prefer "Type V" for some of them. (None of the contributors wishes to refer to any supernova by the old classifications, Type III or Type IV, thus leaving a gap in the numbering used.) The light curves are commonly invoked in the classification of Type II supernovae, some of whose light curves show a characteristic plateau while others are more linear. A classification by spectrum does not allow for this distinction. Finally, Woosely writing on the mechanism of Type I supernovae suggests "Type I$\frac{1}{2}$" for supernovae which show hydrogen in their spectra, but explode by a mechanism resembling that of Type I's. As editor, I was unable even to persuade him to use consistent roman numeration, which would be "Type I S" (S for semi, Menninger (1969)). Woosley thought that this minor bit of erudition would lead to confusion with some subtype of Type I's. Since this book is about a subject in a continuing state of flux, the reader should expect, and will have to bear with, this sort of difference of opinion.

After the chapters on classification, spectra, and light curves, further chapters deal with other parts of the electromagnetic spectrum emitted by supernovae. First, there is a chapter on the radio emission by supernovae. Then a chapter on the interaction of supernovae with the circumstellar material, which is responsible for much of the radio emission, but of course also affects other things such as the behavior of the supernova remnants, which can be observed not only in the radio but also in the visible and the X-ray region. Chapter 6 deals with X- and γ-ray emission. Chapter 7 deals with the neutrinos in which core-collapse supernovae get rid of the overwhelming majority of the prodigious energy they have available. Supernova neutrinos were first observed in 1987A, and came out just as the calculations predicted, a marvelous check on the theory, except perhaps for the poor statistics, and the chance that the agreement is illusory, as suggested by Dar (1989). Kristian *et al.*

(1989, see also Chapter 3) detected a pulsation of 1987A in the visible on one day. Although the detection was highly significant, repeated attempts to detect the pulsation since have failed. Dar presents evidence that the neutrinos were pulsed at the same frequency as the optical light. If this were true, our view of the neutrino diffusion processes in supernovae and our view of the energy budget would require extensive revision. An even nearer supernova while the neutrino detectors are turned on would be most welcome.

There is a lot of theory in the chapter on neutrinos, because until lately there have been no experiments. The three final chapters deal more explicitly with theory. The first of these discusses the theory of Type I supernovae, and in a speculative section suggests that the burning rate of a deflagration depends on the Taylor instability of the burned material underlying the unburned material. Chapter 8 deals with the direct mechanism of supernovae, that is, with supernovae whose optics and hydro-dynamics are produced directly by the collapse, rather than being mediated by the pulse of neutrinos. For this mechanism to work, rather special equations of state are required. This chapter deals with these explicitly. The final chapter deals with the indirect mechanism in Type II supernovae. In this mechanism, the outgoing shock is reinvigorated by the deposition of energy by the neutrinos.

So far as is known, there is no metallic palladium in supernova precursors, so that the current controversy about cold fusion is not relevant, but of course if the experiments turn out to be correct and the theory of picnonuclear reactions must be revised, supernova theory would probably change as well.

As editor, I am hugely indebted to Hans Bethe who made several suggestions regarding the list of authors, and to Stirling Colgate who has long been fascinated with explosions, the bigger the better, and who introduced me to the whole topic.

References

Chevalier, R.A. and Seward, F.D. 1988, Supernovae and supernova remnants, in *Multiwavelength Astrophysics*, ed. F.A. Cordova (Cambridge: Cambridge University Press).

Clark, D.H. and Stephenson, F.R. 1977, *The Historical Supernovae* (Oxford: Pergamon Press).

Dar, A. 1989, preprint TECH-PR-89-17 (submitted to *Phys. Rev. Lett.*).

Kristian, J. and 14 others, 1989, *Nature*, **338**, 234.

Meanninger, K.W. 1969, *Number Words and Number Symbols* (Cambridge, MA: The MIT Press). Translated from the German by P. Broneer.

Murdin, P. and Murdin, L. 1985, *Supernovae* (Cambridge: Cambridge University Press).

List of Contributors

EDWARD A. BARON, Department of Physics, State University of New York, Stony Brook, NY 11794, U.S.A.

DAVID BRANCH, Department of Physics and Astronomy, University of Oklahoma, Norman, OK 73019, U.S.A.

ADAM S. BURROWS, Departments of Physics and Astronomy, University of Arizona, Tucson, AZ 85721, U.S.A.

ROGER A. CHEVALIER, Department of Astronomy, University of Virginia, Charlottesville, VA 22903, U.S.A.

JERRY COOPERSTEIN, Department of Physics, Brookhaven National Laboratory, Upton, NY 11973, U.S.A.

ROBERT P. HARKNESS, Department of Astronomy, University of Texas, Austin, TX 78712, U.S.A.

ROBERT P. KIRSHNER, Harvard-Smithsonian Center for Astrophysics, 60 Garden Street, Cambridge, MA 02138, U.S.A.

RONALD W. MAYLE, Lawrence Livermore National Laboratory, Livermore, CA 94550, U.S.A.

ALBERT G. PETSCHEK, Department of Physics, New Mexico Institute of Mining and Technology, Socorro, NM 87801, U.S.A.
and
Mail Stop D434, Los Alamos National Laboratory, Los Alamos, NM 87545, U.S.A.

RICHARD A. SRAMEK, VLA/NRAO, P. O. Box O, Socorro, NM 87801, U.S.A.

PETER G. SUTHERLAND, Physics Department, McMaster University, Hamilton, Ontario L8S 4M1, Canada.

KURT W. WEILER, Radio and Infrared Astronomy Branch, Naval Research Laboratory, Washington, D.C. 20375–5000, U.S.A.

J. CRAIG WHEELER, Department of Astronomy, University of Texas, Austin, TX 78712, U.S.A.

S.E. WOOSLEY, Board of Studies in Astronomy and Astrophysics, Lick Observatory, University of California, Santa Cruz, CA 95064, U.S.A.

1. Classification of Supernovae

ROBERT P. HARKNESS and J. CRAIG WHEELER

1.1. Introduction

The fundamental classification scheme for supernovae is based on their spectra. As we learn more, the scheme becomes more complex, and it may be that every supernova is unique in some aspect. There have certainly been a number of interesting surprises in the last few years for a field that is nearly 50 years old. In this chapter we give an updated summary of supernova spectral classifications, emphasizing recently introduced subtypes and the importance of late-time spectra. Other recent discussions of supernova classifications may be found in the chapter by Branch in this volume and by Filippenko (1989). Porter and Filippenko (1987) have summarized the observed properties of SN Ib events.

1.1.1. Type I and Type II at Maximum Light

Figure 1.1 shows the basic classification scheme. The class of a supernova has traditionally depended on the spectrum near maximum light. The basic differentiating property was whether or not the spectra showed evidence of hydrogen. If so, the event was classified as Type II, if not, Type I (Minkowski, 1939, 1940). Among the SN II a further differentiation has been proposed based on the shape of the light curve. Those with a pronounced plateau have been termed Type II plateau (SN II-P), and those with a nearly linear decline in magnitude with time from peak are termed Type II linear (SN II-L) (Barbon, Ciatti, and Rosino, 1973). With the advent of SN 1987A we now must add a third subclass. The light curve of SN 1987A was very different due to the blue compact nature of the progenitor, although it shows some similarity to SN II-P on the exponential tail (see Dopita, 1988; Arnett et al., 1989, and references therein). SN 1909A (Young and Branch, 1988), 1923A, 1948B, and 1965L (Schmitz and Gaskell, 1988) may have been of the same light curve class. In addition, SN 1987A has given a new spectral behavior. Traditional SN II-L and P events have shown a nearly pure continuum near maximum light, a week or two after the explosion. SN 1987A, in contrast, showed strong P Cygni features of the Balmer lines from the first day. Fritz Zwicky created "Type V" for an event (SN 1961V) that had narrow hydrogen absorption lines. These are a variant on Type II, but there have now been one or two others, so we retain this somewhat archaic terminology as a shorthand for this potential subclass.

There has been a suggestion for 25 years that Type I supernovae should be subclassified according to features of the spectra (Bertola, 1962), and we have finally learned to do that successfully in the last 5 years (Elias et al., 1985; Branch, 1986; Harkness et al., 1987; Porter and Filippenko, 1987). One key is the presence or absence of the strong Si II absorption feature at 6150 Å. Classical Type Ia events

SUPERNOVA CLASSIFICATION

MAXIMUM LIGHT SPECTRA

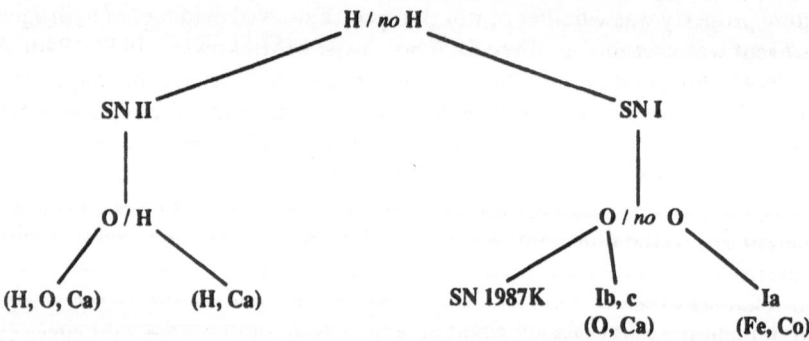

Figure 1.1. The basic classification scheme for supernovae based on spectral features at early and late times.

show this strong Silicon P Cygni feature. Others do not, and they have come to be known as a separate subclass. The events that fail to show the strong Silicon feature near maximum light can be further differentiated by the presence or absence of strong lines of He I (Wheeler, Harkness, Barker *et al.*, 1987). The events that show no silicon near maximum light, but do show He I, especially He I $\lambda 5876$ are identified as Type Ib. There are other events that fail to show either hydrogen or silicon near maximum light, but show only weak evidence for helium. Wheeler and Harkness (1986) proposed the category SN Ic for these events. Wheeler *et al.* (1987) then presented atmosphere models that suggested that SN Ib and the candidates for Type

Ic might be similar except for the relative concentrations of helium and oxygen in the envelope. Nevertheless, these events can be differentiated from SN Ib events like SN 1983N and 1984L, and so from purely taxonomical considerations we will refer to these events as Type Ic which, at the very least, is less cumbersome than referring to helium-rich and helium-poor SN Ib. These SN Ic events are probably physically closely related to the SN Ib. As we will see, growing physical understanding and gradations of properties will ultimately supplant the purely observationally derived classification schemes.

1.1.2. Late-Time Nebular Spectra

Recent years have seen a great deal of work and new insight into the nature of spectra in the later phases of supernovae, roughly 6 months after maximum when the photosphere recedes, densities drop, and the spectra become dominated by nebular emission lines. Most events preserve the distinction of SN I or SN II as defined by the early-time spectra. In the nebular phase, events with hydrogen lines near maximum light also show strong Balmer line emission in the nebular phase. Some events show strong lines of [O I], particularly the $\lambda6300, 6364$ doublet as well as hydrogen and permitted and forbidden Ca II lines. Others are dominated by the hydrogen and calcium emission with weaker evidence for oxygen. Whether this represents a fundamental difference in the physical properties and whether it is correlated with the SN II-L and II-P subclassification scheme is not clear at this time.

The Type I events have proved to show very distinct differences in their nebular spectra. In this phase SN Ia show strong overlapping emission lines of [Fe II], [Fe III] and probably [Co III] with Ca II in absorption but no obvious evidence for oxygen (Axelrod, 1980). SN Ib events show strong [O I] $\lambda6300, 6364$ as well as calcium and Mg I $\lambda4562$ emission (Gaskell *et al.*, 1986). It is not clear whether the SN Ic events show identifiable differences from the SN Ib in the nebular phase.

1.1.3. Other Aspects

There are other properties by which we may differentiate supernovae. Some give rise to detectable radio emission as discussed by Sramek and Weiler in this volume. Radio emission has been observed from some SN II-P, SN II-L, and SN Ib events and from SN 1987A. The radio emission probably arises in the shock between the supernova ejecta and a surrounding circumstellar nebula, perhaps ejected in a wind by the progenitor.

Most SN II show an ultraviolet spectrum which is basically Planckian, whereas SN I, of both subtypes a and b, show a pronounced ultraviolet deficit and rather similar spectrum. Curiously, SN 1987A displayed an ultraviolet spectrum very similar to that of SN Ia and Ib. We will argue below that the ultraviolet spectrum may also contain clues to the interaction with a circumstellar nebula and be related to the relative compactness of the progenitor.

Elias *et al.* (1985) studied the infrared light curves of a number of SN I with broad-band photometry and proposed the category of SN Ib on the basis of a pronounced difference from the classical SN Ia events. The distinguishing property is that SN Ia show a post-maximum dip in the J and H light curves at about 20 days after optical maximum and then a secondary peak at about 30 days. Elias *et al.* (1981) proposed that the dip is due to some absorption that sets in at 20 days and then disappears. The SN Ib events (including SN 1983I which we now differentiate as a SN Ic) showed no such post-maximum dip. Graham (1986) proposed that the dip was due to silicon which shows prominently in the optical spectra of SN Ia, but not in SN Ib. One problem with this suggestion is that the dip is broad band and cannot be due to a single spectral line. The origin of the dip in the infrared spectra of SN Ia awaits a proper physical explanation.

1.2. Observed Time Dependence

1.2.1. The First Few Months

1.2.1.1. *SN II-L*
Figure 1.2 shows the development of spectra in a bright SN II-L, SN 1979C (Branch *et al.*, 1981). Near maximum, the spectrum is nearly a continuum. The International

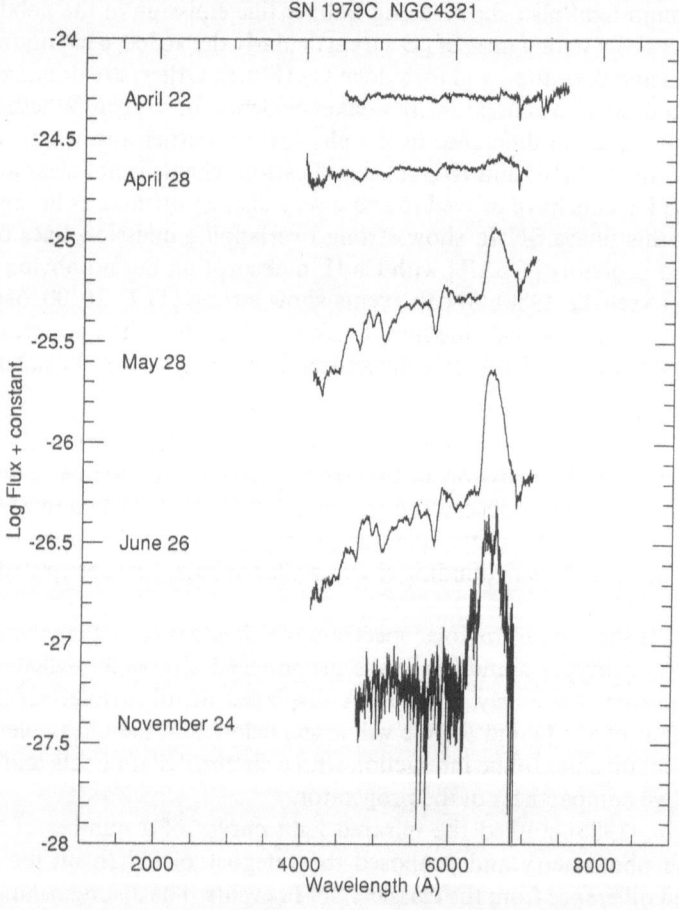

Figure 1.2. The development in time of spectra of the bright SN II-L, SN 1979C (Branch *et al.*, 1981). The spectrum near maximum is almost a pure continuum. A month after maximum, Hα is prominently in emission with no blue absorption component. Seven months after maximum the line of [O I] λ6300, 6364 becomes observable.

Ultraviolet explorer (IUE) spectra in the ultraviolet of both SN 1979C and SN 1980K, another bright SN II-L, showed an ultraviolet excess which Fransson (1986) attributes to Compton scattering of hard radiation from the shock by electrons in the circumstellar nebula. About a month after maximum, the emission line of Hα begins to become prominent along with multiple features of Fe II. Note that there is no blue absorption component to the Hα line. Seven months after maximum (November 24) the line of [O I] λ6300, 6364 becomes observable. The onset of this line has been thought to signify the exposure of oxygen-rich core material, but recent work on SN 1987A has shown that the oxygen content in the solar abundance material in the hydrogen envelope can be sufficient to produce strong metal line emission, so this conclusion is not mandated (Swartz, Harkness, and Wheeler, 1989).

1.2.1.2. SN II-P

The sample of well-studied events is too small to determine if there are systematic differences in the spectra of SN II-L and SN II-P. Figure 1.3 shows a series of spectra

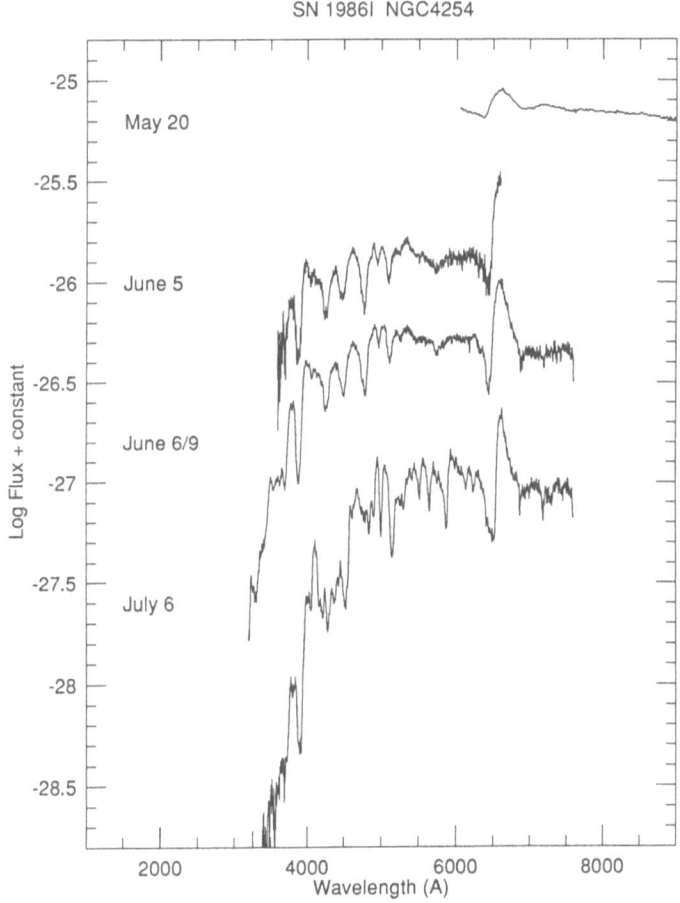

Figure 1.3. The development in time of spectra of the SN II-P event SN 1986I (Pennypacker *et al.*, 1989). The first spectrum shows Hα, but otherwise a featureless continuum. Two weeks later the strong Hα emission is well established with a noticeable blueshifted P Cygni absorption component.

from the SN II-P event SN 1986I (Pennypacker *et al.*, 1989). This event shows some sign of Hα in the first spectrum, but otherwise a featureless continuum extending to the red. Two weeks later the strong Hα emission is well established along with the Fe II features. This early onset of the Hα feature may be significant, or it may just be that a somewhat different epoch is being sampled compared to the first spectrum in Figure 1.2. In this event the Hα does display a noticeable blueshifted P Cygni absorption component. Once again, there is insufficient data to say whether this feature successfully differentiates an SN II-P from an SN II-L, and, if so, why.

The spectra of SN 1987A from 2 days to about 3 weeks are shown in Figure 1.4 (Phillips *et al.*, 1988). At 2 days the photospheric temperature was still too high for metals such as Ca II and Fe II to be apparent, but Ca II H + K was strong by 3 days. (The steep decline in the spectra below 3800 Å for February 26 and later is real.) The Balmer series was well developed even at this early time. No other supernova has shown this development, but no other event has been observed at a corresponding stage in its development.

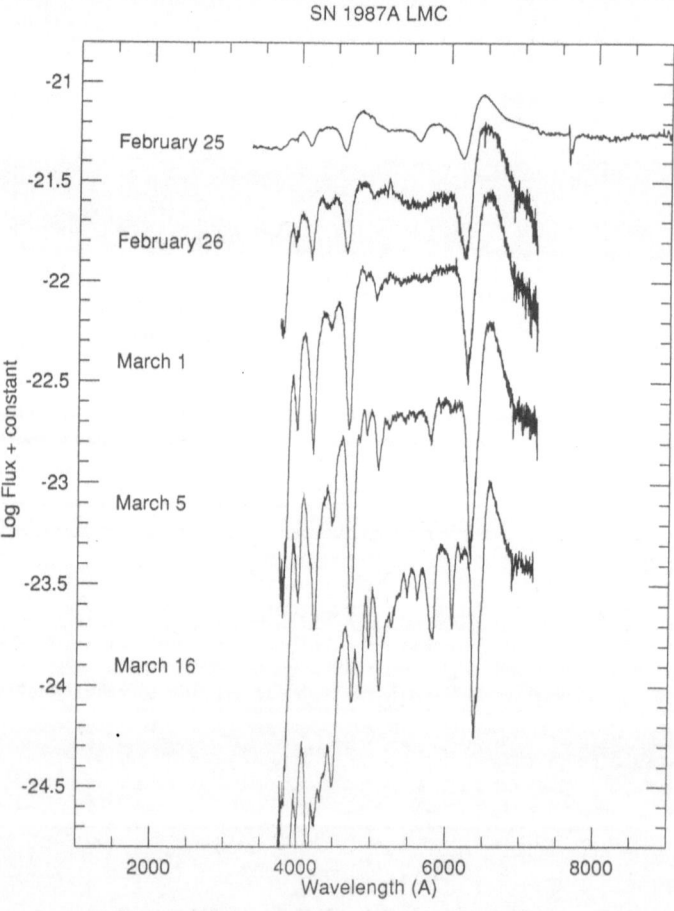

Figure 1.4. Spectra of SN 1987A from 2 days to nearly 3 weeks after explosion (Phillips *et al.*, 1988).

After 3 weeks the spectrum of SN 1987A changed only slowly. At 3 weeks the spectrum resembles the Type II-P supernova SN 1986I and also SN 1985H.

1.2.1.3. *SN Ia*

The time evolution of the spectra of SN 1981B, a classical SN Ia (Branch *et al.*, 1983) is given in Figure 1.5. The optical spectrum at maximum light has been modeled with some success as a blend of P Cygni profiles of intermediate to iron-peak heavy elements (Branch *et al.*, 1985; Harkness, 1986, 1987). The principal absorption components from red to blue in the first spectrum of Figure 1.5 are Ca II, O I, Si II, S II, Fe II, Ca II, and perhaps, at the very blue end of the spectrum, Co II. The observed spectrum makes a transition at some point so that by 2 weeks after maximum (the second spectrum in Figure 1.5), the spectrum is dominated by Fe II. The other species have become too cool to be sufficiently excited with the

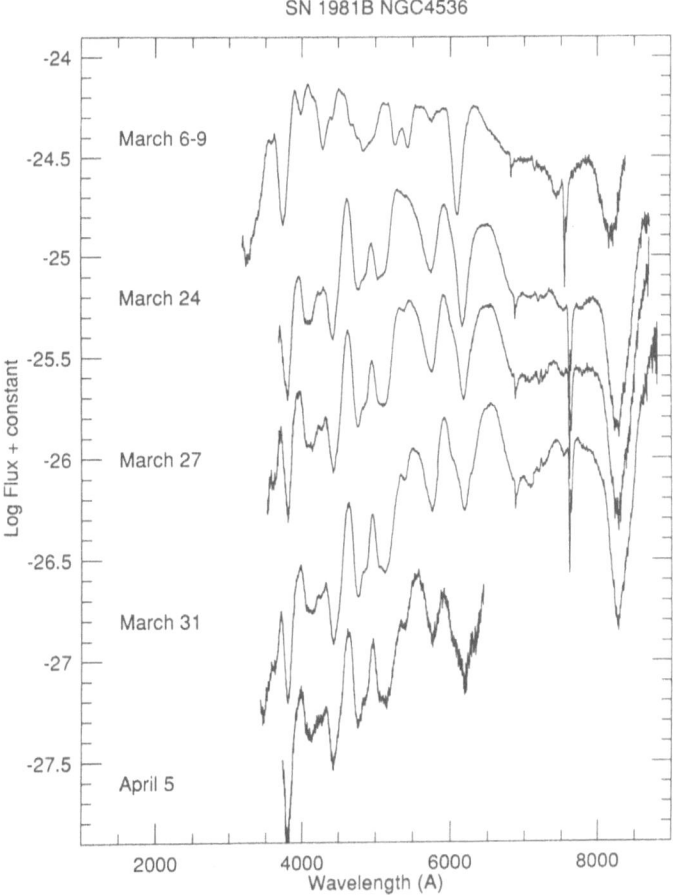

Figure 1.5. The development in time of spectra of SN 1981B, a classical SN Ia (Branch *et al.*, 1983). The principal absorption components from red to blue in the first spectrum are Ca II, O I, Si II, S II, Fe II, Ca II, and perhaps, at the very blue end of the spectrum, Co II. Two weeks after maximum, the spectrum is dominated by Fe II.

exception of the calcium infrared triplet which grows in strength. Models suggest that this transition could occur in a few days shortly after maximum light as the photosphere recedes into the iron-rich ejecta. After this the spectrum changes little even through models predict a transition from a dense state dominated by permitted Fe II to a more rarefied state dominated by forbidden Fe II. In the final spectrum the feature at 5900 Å has been identified as Na D (Branch *et al.*, 1985), but models at even later times identify this feature as CO III (Axelrod, 1980; see below).

Branch (1987) has argued that the spectrum of SN 1984A in NGC 3227 obtained prior to maximum light showed sufficiently high Doppler shift as to be inconsistent with other SN Ia. He thus argues that this is definite spectral evidence that SN Ia

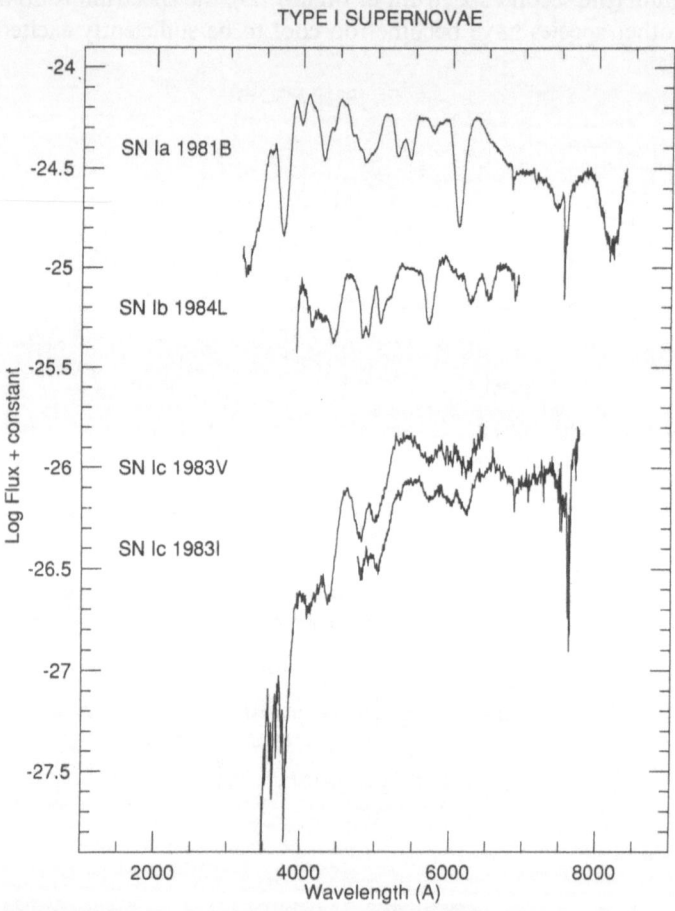

Figure 1.6. The differences among the perceived categories of Type I supernovae are illustrated. The top spectrum is the earliest spectrum from the SN Ia in Figure 1.5. The second spectrum corresponds to nearly maximum light in the SN Ib 1984L. Neither of these spectra shows evidence for hydrogen. The first spectrum shows the strong absorption at 6150 Å attributed to Si II which is absent in the second. The second shows the strong absorption at 5700 Å attributed to He I $\lambda5876$ which is absent in the first. The lower two spectra in Figure 1.6 are examples of SN Ic, showing no hydrogen, no silicon, and only weak evidence for helium.

are not completely homogeneous. Pearce, Colgate, and Petschek (1988) have responded that this higher than normal photospheric velocity may just be due to the early epoch of observation, but Branch, Drucker, and Jeffery (1988) argue against this possibility. SN 1986G in Centaurus A (NGC 5128) was heavily reddened, E $(B - V) = 0.9$, but well studied (Phillips et al., 1987; Frogel et al., 1987). The light curve dropped from maximum distinctly faster than that of SN 1981B. The spectrum showed subtle, but definable differences from SN 1981B, and resembled spectra of SN 1971I which also had a similar light curve (Barbon, Ciatti, and Rosino, 1973). The photospheric velocity was always systematically slower than that of SN 1981B at a given phase. The question of whether or not there are significant differences in peak magnitude between SN 1986G and other SN Ia is obscured by observational uncertainties in distance and reddening. SN 1987N was another classic SN Ia discovered prior to maximum (Schneider et al., 1987). This supernova showed distinctly faster photospheric velocities than SN 1981B. SN 1987A has taught us that the photospheric velocity can recede very rapidly in the early epochs of a compact progenitor, as Pearce, Colgate, and Petschek (1988) point out, but the reality of genuine spectral and light curve differences among SN Ia seems to have been confirmed.

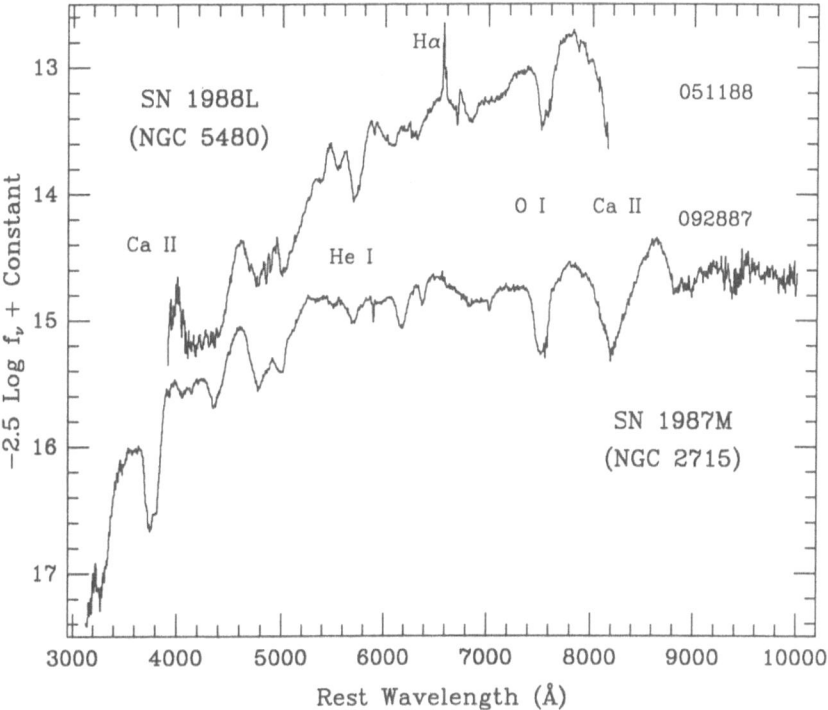

Figure 1.7. Two recent examples of SN Ic (Filippenko, 1988). Note the bump at 6600 Å surmounted by the Hα emission from the host H II region in SN 1988L which could be Hα and the strong O I and Ca II absorption features and the enhanced Ca II hydrogen and sodium emission feature in SN 1988L. SN 1988L also shows a more distinct He I λ5876 feature than other SN Ic.

1.2.1.4. *SN Ib and Ic*

Figure 1.6 shows the differences among the perceived categories of Type I supernovae. The top spectrum is the earliest spectrum from the SN Ia in Figure 1.5. The second spectrum corresponds to nearly maximum light in the SN Ib 1984L. Neither of these spectra shows evidence of hydrogen. The first spectrum shows the strong absorption at 6150 Å attributed to Si II which is absent in the second. The second shows the strong absorption at 5700 Å attributed to He I λ5876 which in turn is absent in the first. The lower two spectra in Figure 1.6 are examples of SN Ic. These spectra are not from maximum light, but about 25 days after maximum (Wheeler *et al.*, 1987), so some caution should be exercised, given the rapid transition of the spectra in SN Ia in such an interval. These spectra may show weak evidence for He I λ5876, and hence have a spectral connection to SN Ib, but Ca II, Fe II, and O I are the only clear line identifications. Figure 1.7 shows two more recent examples

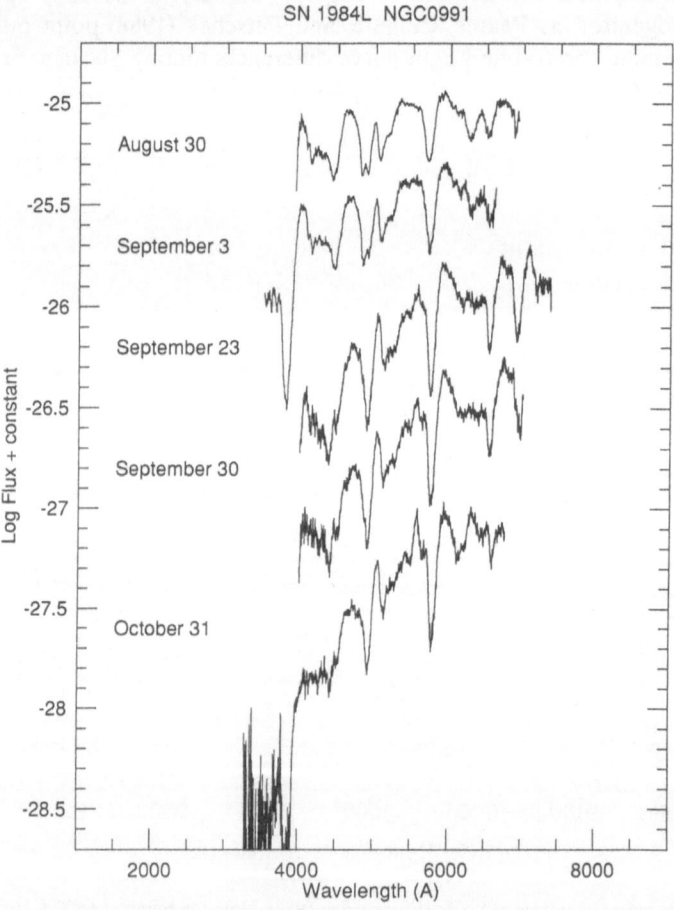

Figure 1.8. The development in time of spectra of the prototypical SN Ib event SN 1984L. The strong He I features make the identification unambiguous 2 weeks after maximum. The last spectrum in Figure 1.8 shows evidence for emission of O I λ5577 and [O I] λ6300, 6364 signifying the onset of the nebular phase.

of SN Ic, SN 1987M and SN 1988L. Filippenko (1988) has suggested that the bump at 6600 Å surmounted by the Hα emission from the host H II region in SN 1983I (Figure 1.6) and SN 1988L (Figure 1.7) could be Hα in the supernova which would imply some generic link to SN II. Note in Figure 1.7 the strong O I and Ca II absorption features. SN 1988L also shows a more distinct absorption at about 5700 Å than other SN Ic. This could be stronger He I λ5876, but there is no evidence for stronger He I 6678 or 7026, so it may represent enhanced Na I. In any case this stronger feature indicates a gradation in some property that belies strict attempts to subclassify the spectra.

The time evolution of the prototypical SN Ib event SN 1984L is given in Figure 1.8. Note the strengthening of the He I features that make the identification unambiguous 1 or 2 weeks after maximum. The last spectrum in Figure 1.8 shows evidence for emission of [O I] λ5577 and [O I] λ6300, 6364 that signifies the onset of the nebular phase.

1.2.2. Six Months Later

1.2.2.1. *SN II*

Figure 1.9 shows an example of an SN II-L in the nebular phase about 6 months after maximum. This particular example, SN 1980K in NGC 6946, has strong Hα and Hβ emission features and the [O I] λ6300, 6364 component is clearly visible. Otherwise, it is typical of many SN II at this epoch. Figure 1.10 shows the spectrum of SN 1987A about 6 months after maximum (Terndrup *et al.*, 1989). This spectrum is distinguished by the strong P Cygni components to both Hα and Hβ, in addition to the strong [O I] and Ca II features.

1.2.2.2. *SN I*

The nebular spectrum of SN Ia 1981B about 4 months after maximum is given in Figure 1.11. The spectrum is dominated by [Fe II], [Fe III], and [Co III]. Figure 1.12 shows the nebular spectrum of the first SN Ib found in that phase, SN 1985F,

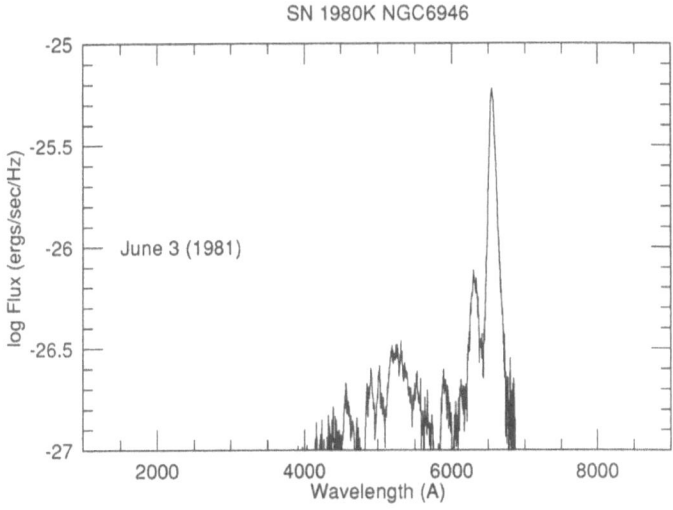

Figure 1.9. The spectrum of SN II 1980K in the nebular phase about 6 months after maximum.

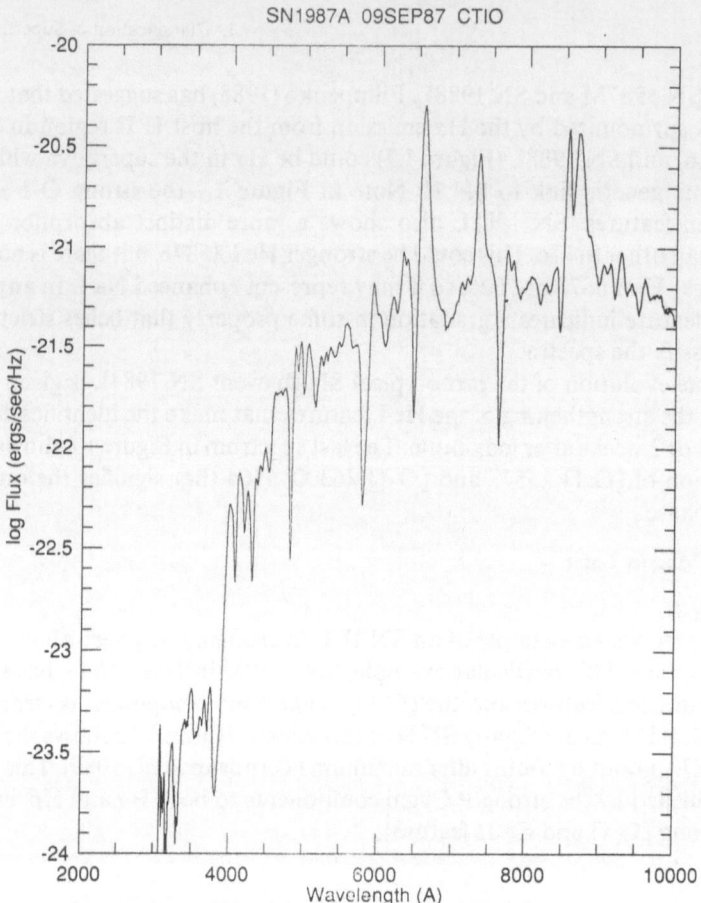

Figure 1.10. The spectrum of SN 1987A about 6 months after maximum (Terndrup *et al.*, 1989). Note the strong P Cygni components to both Hα and Hβ, in addition to the strong O I and Ca II features.

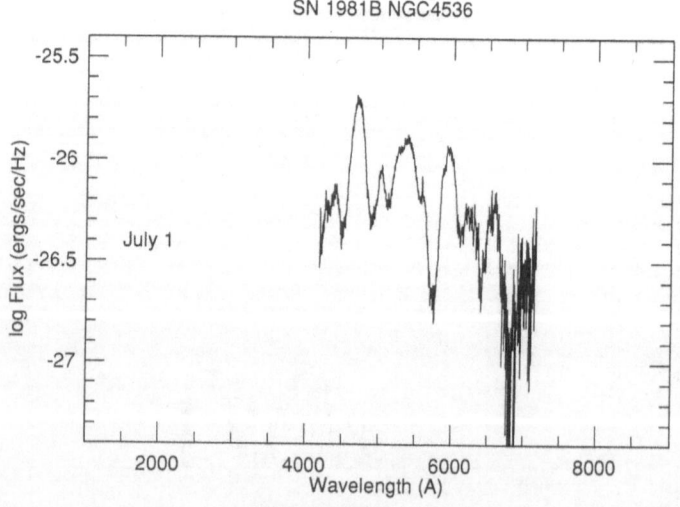

Figure 1.11. The nebular spectrum of SN Ia 1981B about 4 months after maximum. The spectrum is dominated by [Fe II], [Fe III], and [Co III].

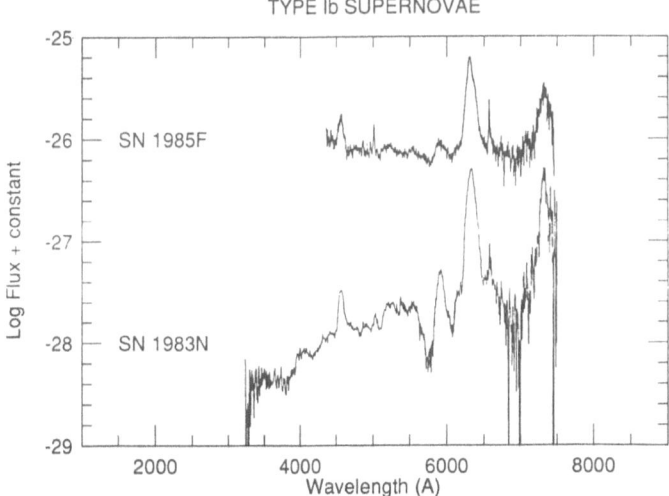

Figure 1.12. The nebular spectrum of the first SN Ib found in that phase, SN 1985F, and the very similar spectrum of the SN Ib 1983N.

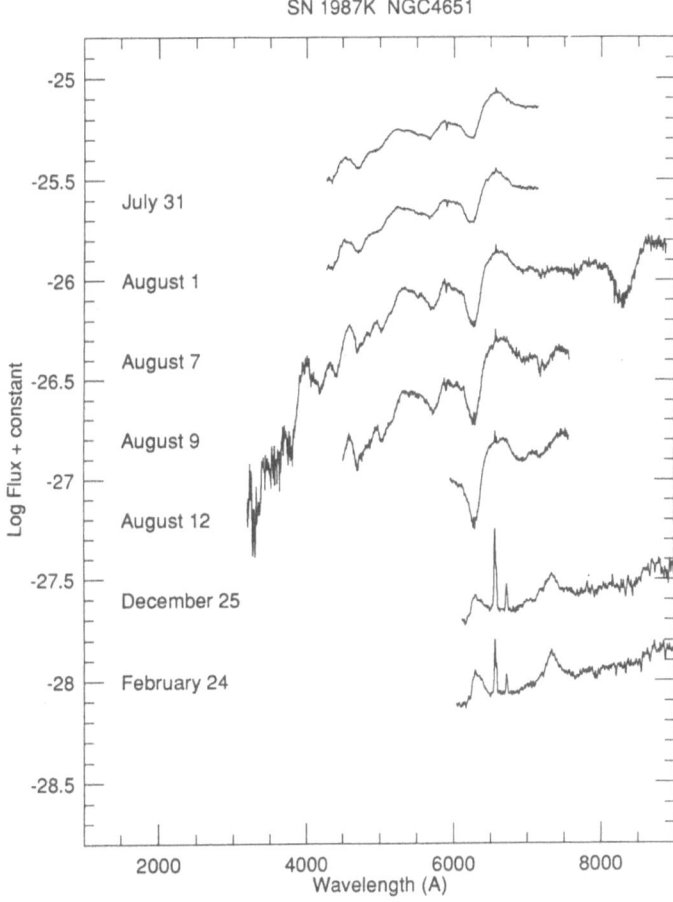

Figure 1.13. The development in time of spectra of SN 1987K (Filippenko, 1988). This event seems to show strong Balmer features in the first 2 weeks characteristic of an SN II. In the nebular phase there is no evidence of hydrogen and the spectrum is indistinguishable from an SN Ib and Ic.

Table 1.1. Identifying spectral features of supernova classes.

Class	Event	Major features	
		Near maximum	About 6 months
II-P	1986I	H, Fe	H, Ca
II-L	1979C	H, Fe	H
	1987A	H, Fe, Ba, Sc	H, Ca
	1987K	H, Fe, Ca	O, Ca
Ib	1983N, 1984L	He, Fe, Ca	O, Mg
Ic	1983I, V, 1987M	O, Ca, Fe	O, Ca, Mg
Ia	1981B	O, Mg, Si, S, Ca, Fe	Fe, Co

and the very similar spectrum of SN 1983N which strongly suggested the identification of SN 1985F as an SN Ib (Gaskell *et al.*, 1986).

A major challenge to the classification scheme of supernovae has emerged with SN 1987K, studied in detail by Filippenko (1988). As illustrated in Figure 1.13, SN 1987K showed a strong P Cygni feature at around 6300 Å which may be Hα so the event was identified as a SN II. The feature at 4700 Å could be Hβ; however, in the spectrum of August 7 it is no longer clear that it is not Fe II. This spectrum is very similar to an SN Ic spectrum minus the O I λ7773 line, but with an additional strong line at 6300 Å. Four months later SN 1987K was in the nebular phase, and there was no evidence of hydrogen. This remarkable transition, if true, tends to support the notion that SN Ib are more closely related to SN II than to SN Ia. The identification of hydrogen near maximum light must be closely scrutinized. The suggestion is that SN 1987K had only a small mass hydrogen envelope which contributed to the formation of the spectrum near maximum light, but was too dilute to contribute later, a hypothesis that will be explored below. The behavior of SN 1987K thus reopens the question of the spectral behavior and type of SN 1985F near maximum light since no spectrum was obtained then.

Table 1.1 summarizes some of the distinguishing spectral characteristics that delineate the various categories of supernovae near maximum light and in the later phase.

1.3. Theory

To classify supernovae properly we must understand their physics. There are several key parameters that determine the nature of a supernova explosion and the development of its spectrum. The energy and the mass tend to enter as a single parameter, E/M, which determines the velocity profile. This parameter, along with the mean opacity, determines the width of the early peak in the light curve. The radius of the progenitor determines the magnitude of adiabatic losses in the expansion. If the progenitor is very small then radioactive heating will dominate the radiation of shock deposited energy near maximum light. The progenitor radius also has a strong effect on the formation of the ultraviolet spectrum. The amount of ^{56}Ni mass ejected has a major effect on the peak luminosity through reheating after adiabatic expansion, and accounts for the tail luminosity in at least some

Table 1.2. Key properties of supernova explosions.

Property	Ia	Ib	II-P	II-L	SN 1987A
Initial radius R_0 (cm)	$\lesssim 10^9$	$\gtrsim 10^{10}$	$\gtrsim 10^{14}$	$\gtrsim 10^{14}$	$\sim 3 \times 10^{12}$
56 Ni $(M/M_\odot)$	> 0.6	0.15?	0.075?	0.6?	0.075
Envelope mass $(M/M_\odot)$	None	He (< 5)	H ($\gg 1$)	H (?)	H/He (~ 10)
Winds/circumstellar nebula	Negligible?	Yes	Yes	Yes	Yes (small)
Neutron star	No	Yes?	Yes?	No?	Yes (finally?)

events. The envelope composition affects the post-maximum light curve, especially by setting the electron density and hence the electron scattering component of the opacity. Winds prior to explosion affect the radio emission but also the density profile in the ejecta which in turn affects the maximum velocities in the ejecta and the formation of the spectrum, especially in the ultraviolet. Finally, some supernovae are expected to leave neutron star remnants whereas others are expected to be completely disrupted. Table 1.2 gives a representation of these parameters for selected events.

1.3.1. SN Ia

1.3.1.1. *Early Times*
The models for SN Ia presume the explosion of a carbon–oxygen white dwarf. A deflagration model naturally gives some unburned intermediate mass elements on the outside of a nickel–iron core. Some artificial homogenization of the composition improves the spectral fit, but the models give a fair representation of the spectrum and the light curve. More detailed spectral data near maximum light are needed to confirm the presence of Co II (below 3500 Å) and the rapid transition to the iron dominated phase.

Figure 1.14 shows a comparison of observed and theoretical spectra from the ultraviolet through the near infrared for SN Ia. The observed spectrum is from SN 1981B near maximum light. The theoretical model is based on the deflagration model W7 of Nomoto, Thielemann, and Yokoi (1984). The model gives a good representation of the optical and ultraviolet spectrum based on features due to neutral and once-ionized intermediate mass elements. The dominant physical process involved in the formation of the ultraviolet spectrum is resonant scattering from many strong lines of Fe II with some contribution from Mg II $\lambda 2797$. The peak at 2200 Å is not an emission feature, but the affect of a natural lack of strong Fe II lines in that wavelength range (Harkness, 1986, 1987; Wheeler and Harkness, 1986).

1.3.1.2. *Late Times*
The models for the late-time spectra of SN Ia by Axelrod (1980) have given a good account of the spectrum and provide evidence for the decay of the [Co III] feature at 5900 Å. These models suggest that deflagration models provide a more satisfactory fit than do detonation models (Woosley, Axelrod, and Weaver, 1984). The models published to date are based solely on forbidden emission lines. They do not account for the strong calcium absorption lines observed, nor have they attempted

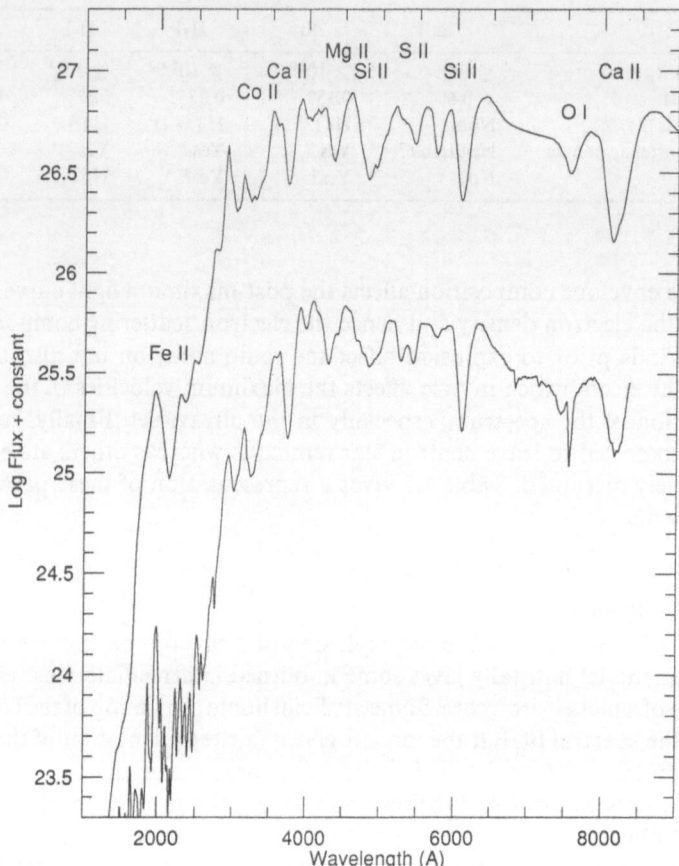

Figure 1.14. The observed spectrum is from SN Ia 1981B near maximum light from the ultraviolet through the near infrared is compared to the theoretical spectrum based on the deflagration model W7 of Nomoto, Thielemann, and Yokoi (1984).

to model the effect at late times of the intermediate mass elements such as C, O, Mg, Si, and S, which are intrinsic to the deflagration model and which show so distinctly in the maximum light spectra.

1.3.2. SN Ib and Ic

1.3.2.1. *Early Times*

Model atmospheres for SN Ib and Ic events based on evolutionary models have not yet been computed. The most successful models for SN Ib are based on the notion that they are the evolved cores of massive stars although the published spectral models depend on simple power-law representations of the density. These models do suggest a large helium abundance despite uncertainties regarding non-LTE (local thermodynamic equilibrium) effects. The SN Ic may be similar, but with a greater loss of helium giving a composition enhanced in oxygen at the expense of helium.

Models based on power-law density profiles have been presented for SN Ib by Harkness *et al.* (1987) and for SN Ic by Wheeler, Harkness, Barker *et al.* (1987). These models assume that the density scales like r^{-7} with $v \propto r$ and the radii chosen to correspond to expansion for 20 days with a fiducial velocity of 5000 km s^{-1}. The total mass of the models is formally very large, $\sim 30\ M_\odot$ but more realistic models would have flatter density profiles at smaller radii, and hence less mass. Two homogeneous composition mixes were considered, a "helium-rich" mix with $X_{He} = 0.9$, $X_C = 0.01$, and $X_O = 0.09$, and an "oxygen-rich" mix with $X_{He} = 0.1$, $X_C = 0.1$, and $X_O = 0.8$. The ionization and excitation is assumed to be in LTE. The He I level populations were multiplied by a departure coefficient to mimic the effect of departure of these levels from LTE.

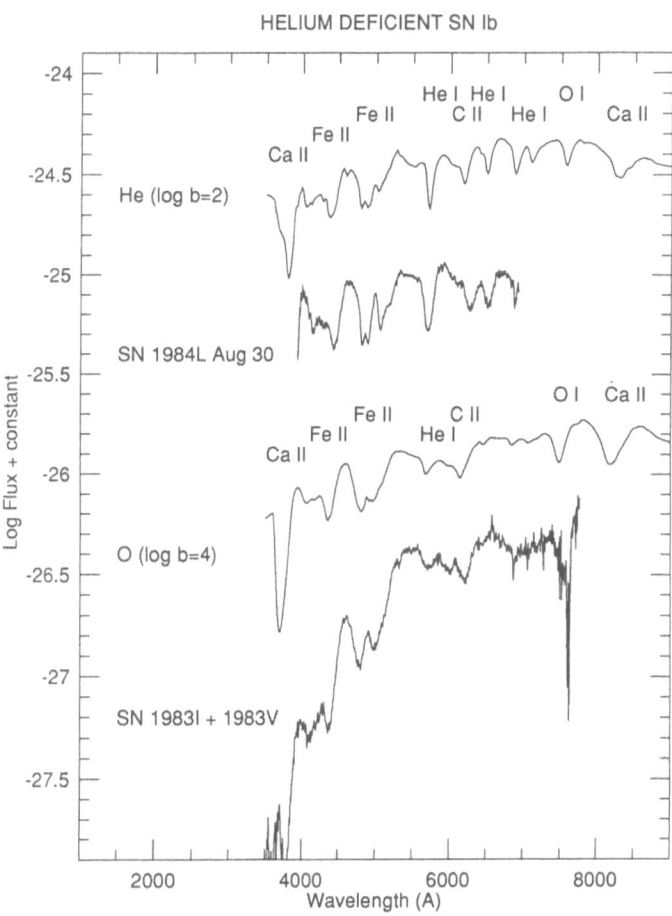

Figure 1.15. The spectra of SN Ib 1984L and the combined spectra of SN Ic events 1983I and 1983V are compared to theoretical atmosphere models. The upper model corresponds to a helium-rich mix with a departure coefficient of $b = 100$ and reproduces many of the basic features of the observed spectrum, especially the lines of He I. In the lower spectra the helium lines remain weak with the smaller abundance in the oxygen-rich mix despite the larger departure coefficient, $b = 10^4$, and the model gives a reasonable representation of the SN Ic spectra.

Figure 1.15 shows the comparison of such models with the spectra of SN Ib 1984L and the combined spectra of SN Ic, events 1983I and 1983V. The upper model corresponds to the helium-rich mix with a departure coefficient of $b = 100$ and reproduces many of the basic features of the observed spectrum, especially the lines of He I. In the lower spectrum, that of an oxygen-rich atmosphere, the helium lines are indistinct despite a larger departure coefficient, $b = 10^4$, because the electron scattering opacity is much larger than in the helium-rich atmosphere (Harkness et al., 1987). These atmosphere models encourage the view that there is a close relation between SN Ib and SN Ic events, namely a smaller helium abundance in the latter, despite the distinct spectral differences.

Figure 1.16 shows the complete optical ultraviolet spectrum from the model corresponding to the helium-rich composition in Figure 1.15 together with the

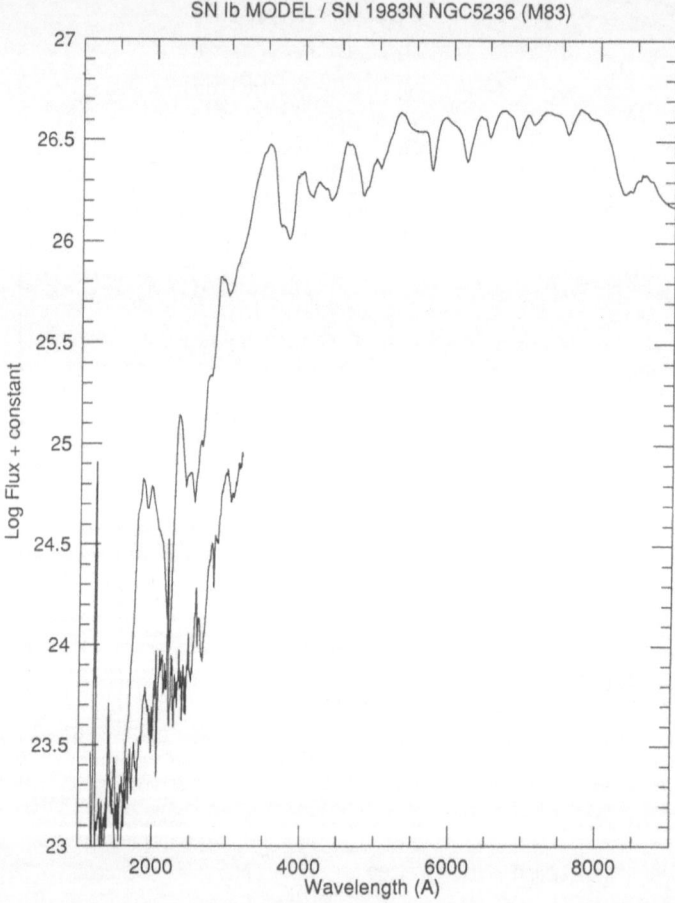

SN Ib MODEL / SN 1983N NGC5236 (M83)

Figure 1.16. The complete optical ultraviolet spectrum is given for the model corresponding to the helium-rich composition in Figure 1.15. The observed ultraviolet spectrum of SN Ib 1983N shows many of the features of the model spectrum. The theoretical ultraviolet spectrum is very similar to that of models of SN Ia despite the fundamental differences in the underlying models and the distinct differences in the optical portion of the spectra.

observed ultraviolet spectrum of SN Ib 1983N. The ultraviolet spectrum is similar
to that of models of SN Ia despite the rather large differences in the underlying
models and the distinct differences in the optical portion of the spectra. In particular,
the SN Ib atmosphere model shows the same ultraviolet deficit due to the presence
of strong Fe II lines, and the peak at about 2000 Å due to their absence. In the case
of the Ib models, the iron that gives rise to the ultraviolet deficit is just that due to
the solar abundance of iron in the helium-rich envelope. A large abundance of iron
is not required to get such a spectrum.

1.3.2.2. *Late Times*
Fransson and Chevalier (1989) and Axelrod (1988) have discussed the physics of the
formation of the late-time emission line spectrum of SN Ib. They adopt a model
based on the core of a massive star, but do not account for a number of the questions
raised by studies of SN 1987A. In particular, they assume that the composition
stratification predicted in the computer models is maintained and that the radioactive matter is confined in the innermost zones. Models of the light curves and optical,
X-ray, and γ-ray spectra of SN 1987A suggest that the composition is strongly mixed
and that the radioactive matter may reside rather far out in the hydrogen-rich
envelope. If this has happened in a star which otherwise conformed to our expectations for the evolution and explosion of a massive star, then we must also ask
whether such phenomena are within the range of possibility for SN Ib events, the
progenitors of which are still debated. In addition, Fransson and Chevalier assume
that the strong oxygen and calcium lines arise in the heavy element core. Swartz,
Harkness, and Wheeler (1989) show that the small abundances of these elements in
the outer envelope can be sufficient to account for the strong lines in the spectrum
of SN 1987A. The question then also must be addressed as to whether the strong
oxygen and calcium lines in SN Ib can arise in the small amount of oxygen and
calcium in the helium envelope suspected for SN Ib events, even if heavy elements
from the core are not mixed up into the envelope. Begelman and Sarazin (1986)
estimated that SN 1985F contained a minimum of 5 $M_\odot$ of oxygen. This estimate
needs to be revised with models which account for the stratification of composition
and ionization structure and variations in the γ-ray deposition. Only with such
studies will we be able to understand the varying line strengths among the SN Ib
and Ic events.

1.3.3. SN II
The basic model for SN II-P is the core-collapse explosion of a massive star within
an extended red supergiant envelope. SN II-L may just represent a continuation of
that theme with perhaps a smaller envelope mass. An alternative view is that SN
II-L may represent a different explosion mechanism and be more closely related to
SN Ia (Doggett and Branch, 1985; Wheeler, Harkness, and Cappellaro, 1987). The
fact that the SN II-L events SN 1979C and 1980K showed normal hydrogen-rich
nebular spectra suggests that they did not have very small hydrogen envelopes. If
they had, they might be expected to make the transition manifested by SN 1987K
where the hydrogen seems to have been present and then disappeared. To date there
has been relatively little in the way of detailed modeling of the early spectra of SN
II. SN 1987A is the distinct exception.

1.3.3.1. *Early Times in SN 1987A*

The ultraviolet spectra of SN 1987A presented one of the most interesting aspects of its early spectral evolution. The two SN II-L 1979C and 1980K showed a nearly Planckian or even an excess of flux in the ultraviolet. This is thought to be due to Compton scattering in the circumstellar matter (Fransson, 1986). SN 1987A, on the other hand, showed a very similar ultraviolet deficit to SN Ia and Ib events. Theoretical study of the formation of the ultraviolet spectrum in SN 1987A has shown that the compact nature of the progenitor and the velocity structure play an important role. Figure 1.17 shows a series of spectra based on evolutionary/

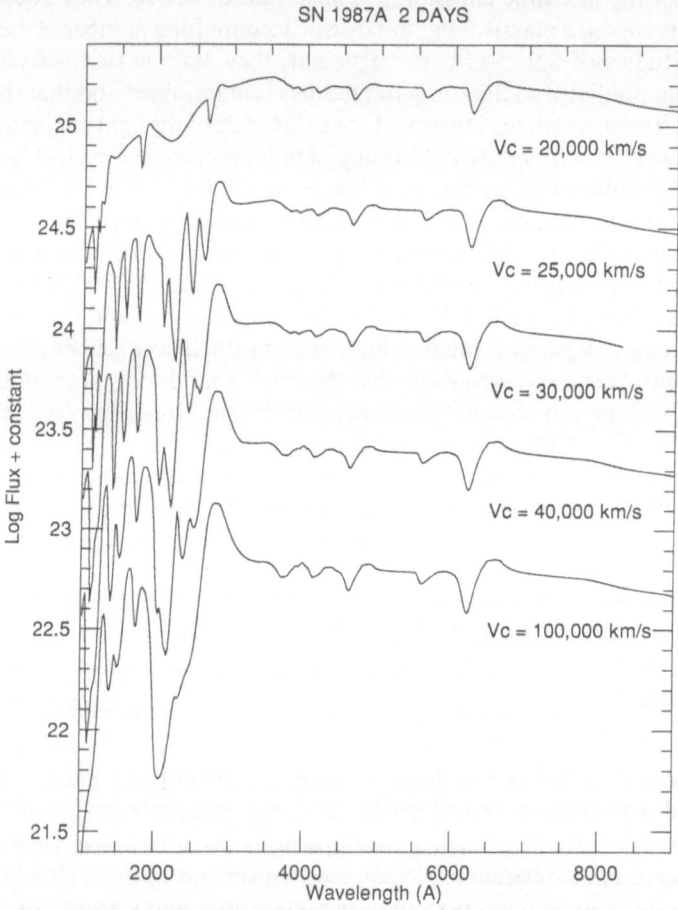

Figure 1.17. A series of model spectra for SN 1987A 2 days after the explosion assuming a heavy element core of 6 $M_\odot$ and an envelope of 10 $M_\odot$ composed of equal amounts of hydrogen and helium with $Z = 0.005$. The outermost layers are truncated at different velocities. If the velocity is truncated at a velocity as low as 20,000 km s^{-1} insufficient Fe II forms. If the outermost matter expands as rapidly as 100,000 km s^{-1} there is a surfeit of Fe II and the resulting features in the ultraviolet are too strong. The best representation of the observed ultraviolet spectrum at about 2 days is for the model with the velocity truncated at about 25,000–30,000 km s^{-1}.

dynamical models for SN 1987A. The particular model had a heavy element core of 6 $M_\odot$ and an envelope of 10 $M_\odot$ composed of equal amounts of hydrogen and helium by mass with $Z = 0.005$. The epoch corresponds to 2 days after the explosion. The series of spectra in Figure 1.17 show the effect of removing matter with expansion velocity greater than a certain cutoff, so that it does not contribute to the spectrum (Harkness and Wheeler, 1989). If matter is removed for velocities greater than 20,000 km s^{-1} the model does not sample enough cool, high-velocity matter. If the model includes matter expanding as rapidly as 100,000 km s^{-1}, it samples too much material cooled by the rapid adiabatic expansion and Fe II dominates the ultraviolet spectrum. The closest representation of the observed ultraviolet spectrum at about 2 days is for the model with the velocity truncated at about 25,000–30,000 km s^{-1} (Figure 1.18). This is just the range of maximum velocity predicted by models of the radio emission in which the ejecta are slowed by collision

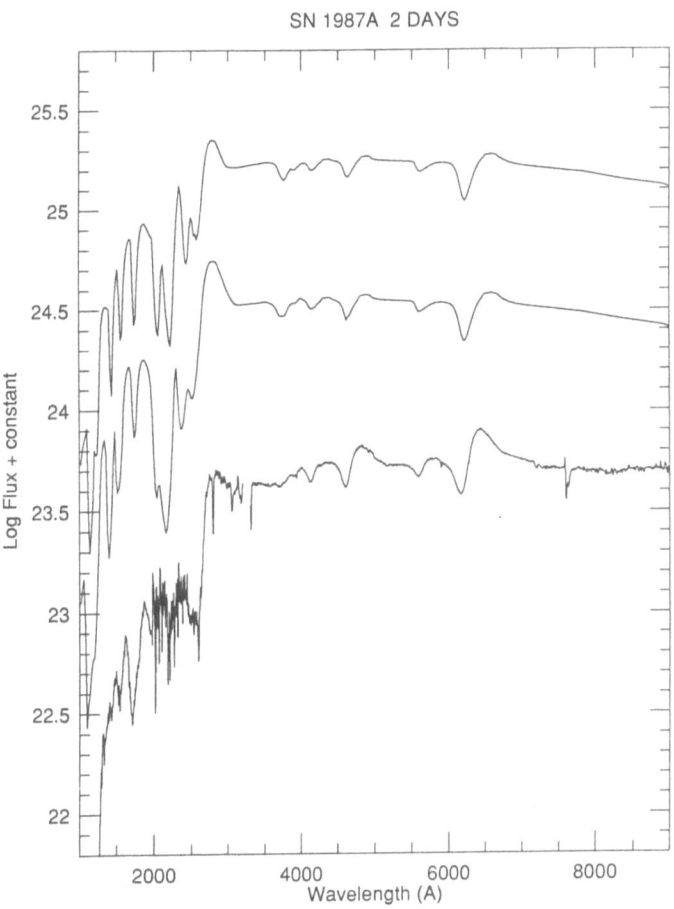

Figure 1.18. The spectra 1.52 days after the explosion for the model of Figure 1.17 truncated at 25,000 and 30,000 km s^{-1} is shown in comparison with the ultraviolet and optical spectrum of SN 1987A obtained at that epoch.

with a circumstellar wind (Chevalier, 1984). The atmosphere models of Figure 1.17 were computed assuming LTE and ionization equilibrium. We caution that relaxing these assumptions, especially in the high-velocity matter, might lead to a different interpretation of the formation of the ultraviolet spectrum.

These calculations imply some possible lessons for SN Ia and Ib and other SN II as well. In the computed model W7, used for the atmosphere calculations in Figure 1.14, the maximum velocity was about 25,000 km s^{-1} due to the finite zoning. Trial calculations of SN Ia spectra extending the atmosphere to higher velocities show the Fe II and Mg II features to be too strong in the ultraviolet, akin to the last spectrum in Figure 1.17. If departures from ionization or local thermodynamic equilibrium do not significantly alter these models, this suggests that the arbitrary truncation in the models may have some basis in physical reality. This may imply that SN Ia events also explode in some sort of circumstellar nebula, perhaps left over from mass transfer in a binary system. Establishing the existence of such a circumstellar nebula in SN Ia by study of the ultraviolet spectra would be a major step forward. The SN Ib have also demonstrated radio emission, and thus collision with a wind and the truncation of the velocity profile might be expected and its effect also manifested in the ultraviolet spectrum. On the other hand, SN 1979C and 1980K showed radio emission and did not show the ultraviolet deficit that characterizes the SN Ia, b and SN 1987A. Thus the ultraviolet spectrum does seem to be of importance in classifying supernovae, but it is apparently a measure of the compactness of the progenitor and the density of circumstellar material more than of the composition of the ejecta.

1.3.3.2. *Late Times in SN 1987A*

The late-time spectra of SN 1987A have been studied by Fransson and Chevalier (1987), Axelrod (1988), and by Swartz, Harkness, and Wheeler (1989). Fransson and Chevalier and Axelrod neglect the hydrogen envelope, and, as for their models of SN Ib, assume that the emission lines of oxygen and calcium arise in the heavy element core. Swartz *et al.* include the hydrogen-rich envelope and explore the effects of composition mixing and the placement of the photosphere. They find that in order to excite the Balmer emission lines, substantial γ-ray energy deposition must occur in the hydrogen-rich envelope. This may require some ^{56}Co to be physically present in the envelope. This is consistent with studies of the early observation of X-rays and γ-rays from the supernova (Nomoto, Shigeyama, and Hashimoto, 1988; Pinto and Woosley, 1988). While definitive statements are not possible, Swartz *et al.* find that an adequate model for SN 1987A at about 200 days is one in which the composition, except for the ^{56}Co, is unmixed and the heavy element core is obscured by a quasi-continuum (modeled by placing a photosphere at the base of the hydrogen-rich envelope). Figure 1.19 shows the observed spectrum at 200 days in SN 1987A and three theoretical models based on no mixing, moderate mixing, and heavy mixing from the core to the envelope. The first model gives a reasonable representation of the observations while the last model gives emission lines that are too strong and broad. These models suggest that there is adequate oxygen and calcium in the hydrogen-rich envelope to account for the observed emission line strength for conditions where the Balmer lines are adequately excited even with no

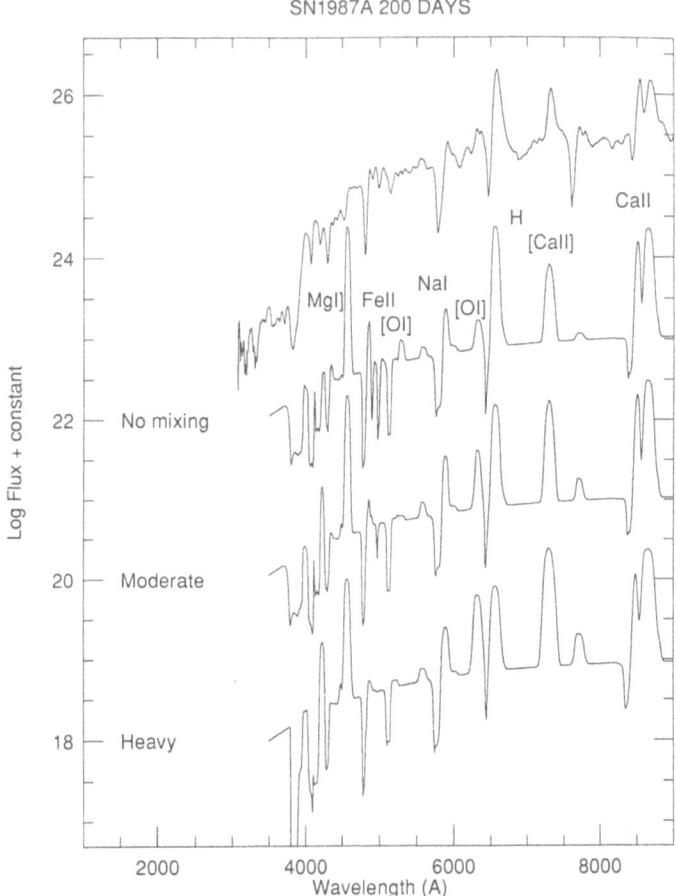

Figure 1.19. The observed spectrum of SN 1987A at 200 days is given with three theoretical models based on no mixing, moderate mixing, and heavy mixing from the core to the envelope.

mixing from the inner layers. The models also suggest that a quasi-continuum may form below the hydrogen envelope so that the inner heavy element core is shielded from direct view, at least in the optical. Models in which the core region contributes to the emission lines tend to give emission lines with narrow peaks arising from the slow-moving high concentration of heavy elements in the core. Such narrow peaks are not observed in the spectra of SN 1987A. As mentioned above, some of these aspects may also pertain to spectra of SN Ib and Ic in the nebular phase.

As yet little work has been done on the late-time phase of SN II aside from that on SN 1987A by Fransson and Chevalier (1988) and by Swartz, Harkness, and Wheeler (1989). This work must be extended to explore the effect of envelopes of varying mass, of models with detonating or deflagrating cores surrounded by hydrogen envelopes and other configurations to understand the physics behind the various classes of SN II.

1.4. SN 1987K

It thus remains an interesting question to determine where SN 1987K fits into the classification and physical scheme of supernovae. It was apparently a Type II near maximum from the taxonomical definition that such events show hydrogen in the spectrum. It was indistinguishable from a Type Ib (or c) at later times and definitely neither an SN II nor an SN Ia. The late-time spectrum suggests an origin in a massive star, given the preponderance of circumstantial evidence that SN Ib events arise in the cores of massive stars. The primary evidence for that is the association with H II regions and the mass of the ejecta as determined from the light curve to be several solar masses (Wheeler and Levreault, 1985; Schaeffer, Cassé, and Cahen, 1987; Ensman and Woosley, 1988; Nomoto, Shigeyama, and Hashimoto, 1988).

The appearance of hydrogen near maximum light and its subsequent disappearance at later times could require that the amount of hydrogen was small and that it became too dispersed to be detectable at later times. As we noted earlier, the suggestion that SN II-L are distinguished from SN II-P only by the small size of the hydrogen envelope gives a qualitatively similar picture, and yet SN II-L show appreciable evidence for hydrogen in their later phases so they must be substantially different than SN 1987K. The question of the appearance and then disappearance of hydrogen must be approached carefully. Studies of SN 1987A have shown that to see the hydrogen at late times there must be substantial radioactive heating by direct mixing of ^{56}Co into the envelope or by inhomogeneities in the envelope which allow greater deposition of γ-rays from the interior than given by spherically symmetric models. The failure to detect hydrogen may then be determined by the deposition of energy in the envelope, and not only by amount of hydrogen.

SN 1987A has also re-emphasized a question not yet adequately addressed. That is the amount of H an SN Ib (or an SN Ia, for that matter), could possess in a thin outer envelope that would escape detection. To answer such a question and explore the possible nature of SN 1987K we have constructed a model with a hydrogen envelope of $\sim 0.1\ M_\odot$ on top of a core of about 5.3 $M_\odot$ of helium and heavier elements. This model is similar to those put forward for SN 1987A in which strong mass loss from a star of $\sim 20\ M_\odot$ left a small amount of hydrogen at the time of the explosion (Wood and Faulkner, 1987). Such models do not work for SN 1987A because they give insufficient tamping of the inner core, but they are not astrophysically unreasonable. The inner 1.4 $M_\odot$ is assumed to remain behind as a neutron star, and the outer 4 $M_\odot$ is ejected by imposing a high velocity in the zones beyond the core to simulate a shock of chosen energy of $\sim 10^{51}$ ergs. The outer radius of the star was $\sim 3 \times 10^{12}$ cm, appropriate to SN 1987A, but not necessarily to SN 1987K.

The light curve of this model was calculated for various values of the energy. Some models included a careful treatment of the recombination, although this did not seem to have a significant qualitative effect. A bolometric light curve has been presented for the SN Ib 1983N by Panagia (1985) and a partial optical light curve exists for SN 1987K (Filippenko, 1988; although Evans (private communication) believes his visual observations suggest errors in the reduction of the CCD data). A model with 2×10^{51} ergs is shown in Figure 1.20. This model fit the peak of the bolometric light curve of SN 1983N fairly well and the light curve of SN 1987K

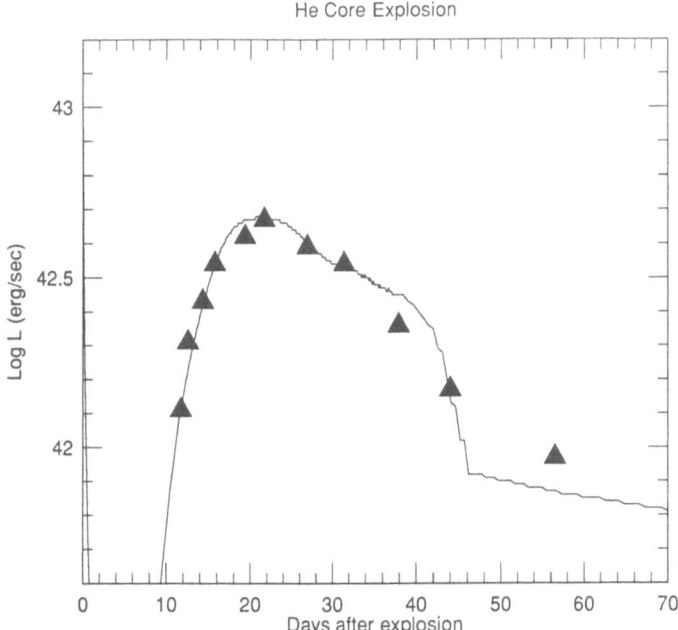

Figure 1.20. The light curve of a model with an original helium core of 5.3 $M_\odot$ plus a hydrogen envelope of 0.1 $M_\odot$ in which 1.4 $M_\odot$ is left as a neutron star and an explosion is initiated with an energy of 2×10^{51} ergs to eject the outer 4 $M_\odot$. The model resembles the light curves of SN Ib 1983N and SN 1987K.

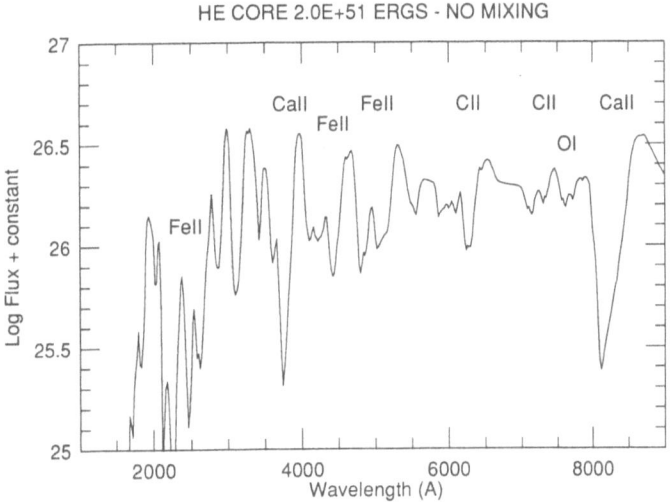

Figure 1.21. A model spectrum corresponding to the model of Figure 1.20 at maximum light, 20 days.

within the rather large uncertainty. A model with an energy of 4×10^{51} ergs gave too narrow a light curve, and one with 10^{51} ergs, a light curve that was too broad. These conclusions are similar to those of Schaeffer, Cassé, and Cahen (1987), Ensman and Woosley (1988) and Nomoto, Shigeyama, and Hashimoto (1988) for models of SN Ib which omit the thin outer layer of hydrogen.

Figure 1.21 shows the spectrum of the model whose light curve is shown in Figure 1.20. The spectrum was computed at 20 days, corresponding to the epoch of maximum light in the theoretical light curve. The scattering photosphere is in the oxygen-rich layers at the base of the helium layer. The Balmer series is completely obscured by strong metal lines formed by the solar abundance of heavy elements in the hydrogen and helium layers in the theoretical spectrum. Homogenizing the composition and thus mixing the hydrogen with the heavier elements from the inner core resulted in little change. The conditions are too cool and too low in density and the resulting spectrum bears little resemblance to that observed. A related experiment was performed with the W7 model for a SN Ia. In LTE, 0.1 $M_\odot$ of hydrogen was not detectable in the theoretical spectrum either in an outer layer, or homogenized with the heavy elements lying above 11,000 km s^{-1}.

There are two lessons to be learned here. One is that the most obvious explanation for the SN 1987K phenomenon does not work in its first manifestation. There may be some simple variation, varying the mass and radius of the hydrogen envelope, that brings better agreement, but it is certainly not obvious that this is so. The problem of making the hydrogen disappear remains to be solved. The second lesson is that while a decent light curve is necessary to constrain the nature of a supernova explosion, it is not sufficient. Clearly, something rather substantial must be changed to find a model with a light curve that matches the observations as well, but does a much better job of reproducing the spectrum near maximum light. Only after such a task is accomplished will we have a better view of where SN 1987K fits into the classification scheme of supernovae. The related question of how much hydrogen could be present in SN Ia and SN Ib events is still open, and very important.

1.5. Conclusions

We began by remarking that nearly all supernovae display some peculiarity if examined closely enough. In many cases the differences are not subtle, but we do not know whether the differences arise in subtle or major differences in the physical properties. Table 1.3 gives a collection of some supernovae that serve to define a category of one. For example, Figure 1.22 gives the spectrum of SN 1987B which tended to be overshadowed by its more famous antecedent, but which displayed one of the most bizarre spectra we have seen. The spectrum has strong, narrow Balmer emission lines. It resembles the spectrum of a dwarf nova, but the line peaks fall at the rest wavelength of the host galaxy, so there is no question that the object is extragalactic.

For some time there has been a tendency to confuse the physical phenomena of core collapse and thermonuclear explosion with the observed categories of SN II and SN I. With the elucidation of the categories of SN Ib and Ic, the discovery of

Table 1.3. Peculiar supernovae.

Event	Galaxy	Type	Peculiarity
SN 1885A	NGC 224 M31	I?	Very rapid light curve[1,2]
SN 1954A	NGC 4214	Ib?	Faster than average light curve, strong helium lines[3]
SN 1961V	NGC 1058	V?	Very massive star explosion?[4,5]
SN 1971I	NGC 5055	Ia	Fast light curve, spectra variations[6]
SN 1980I	Intergalactic?	?	Mildly peculiar? Si II present plus 6680 absorption[7]
SN 1983K	NGC 4699	II	"Wolf–Rayet" features in pre-maximum spectra[8]
SN 1984E	NGC 3169	II	Narrow (~ 360 km s^{-1}) strong Hα line[9]
SN 1985A	NGC 2748	Ia	Superposed narrow Hα; falls in H II region?[10]
SN 1985H	NGC 3359	II	Strong barium line; related to SN 1987A?[11]
SN 1986A	NGC 3367	Ia	Superposed narrow Hα; falls in H II region?[12]
SN 1986G	NGC 5128 Centaurus A	Ia	Fast light curve, spectra variations[13,14]
SN 1986J	NGC 891	V?	Another SN V?[15]
SN 1987A	LMC	II	Unique light curve, spectra variations[16]
SN 1987B	NGC 5850	II?	Exceedingly narrow Balmer emission lines,[17] unique spectrum resembles dwarf nova
SN 1987F	NGC 4615	II	Peculiar spectra variations, late-time permitted emission lines, spectra resembles Seyfert I[18]
SN 1988I	Anonymous	II	Spectra resembles Seyfert I[18]

[1] de Vaucouleurs, G. and Corwin, H.G. Jr. 1985, *Ap. J.*, **295**, 287.
[2] Chevalier, R.A. 1988, *Ap. J. (Letters)*, **331**, L109.
[3] McLaughlin, D.B. 1963, *Pub. A.S.P.*, **75**, 133.
[4] Doggett, J.B. and Branch, D. 1985, *Ap. J.*, **90**, 11.
[5] Utrobin, V.P. 1984, *Ap. and Space Sci.*, **98**, 115.
[6] Kirshner, R.P., Oke, J.B., Penston, M.V., and Searle, L. 1973, *Ap. J.*, **185**, 303.
[7] Smith, H.A. 1981, *Ap. J.*, **86**, 998.
[8] Niemela, V.S., Ruiz, M.T. and Phillips, M.M. 1985, *Ap. J.*, **289**, 52.
[9] Gaskell, C.M. 1984, *I.A.U.C.*, 3936.
[10] Wagner, G. and McMahan, R.K. 1987, *Ap. J.*, **93**, 287.
[11] Wheeler, J.C., Harkness, R.P., and Barkat, Z. 1988, Lecture Notes in Physics, Vol. 305, p. 305.
[12] Dekker, M., D'Odorico, S., and Arsenault, R. 1988, *Astr. and Ap.*, **189**, 353.
[13] Frogel, J.A. *et al.* 1987, *Ap. J. (Letters)*, **315**, L129.
[14] Phillips, M.M. *et al.* 1987, *Pub. P.A.S.P.*, **99**, 592.
[15] Rupen, M.P., van Gorkom, J.H., Knapp, G.R., Gunn, J.E., and Schneider, D.P. 1987, *Ap. J.*, **94**, 61.
[16] Arnett, W.D., Bahcall, J.N., Kirshner, R.P., and Woosley, S.E. 1989, *Ann. Rev. Astr. Ap.*, **27**, 629.
[17] *I.A.U.C.*, 4321, 4325, 4329, 4386.
[18] Filippenko, A.V. 1989, *Astron. J.*, **97**, 726.

SN 1987K, and questions of whether SN II-L might represent thermonuclear explosions, it is clear that we must be careful not to use these terms casually and interchangeably. As the spectral classes of supernovae proliferate, however, it may yet prove best to go back to a fundamental classification scheme wherein the most basic categories are core collapse and thermonuclear explosion, but we cannot yet rigorously assign most supernovae into such classes. For the time being we are forced to maintain the phenomenological categories despite the encroaching fuzziness of the boundaries.

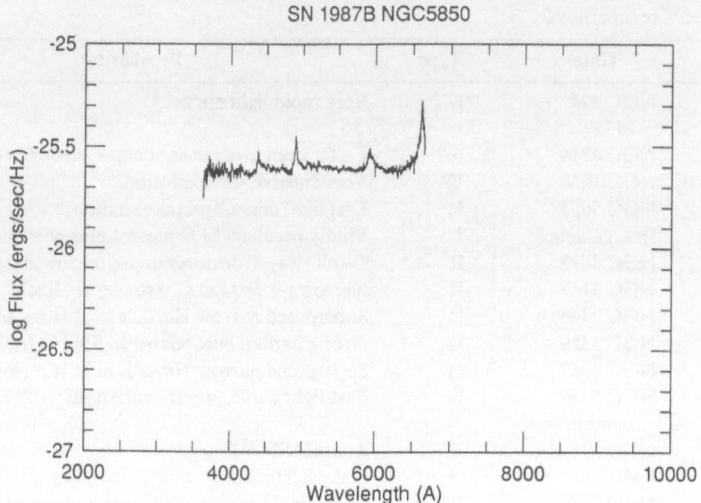

Figure 1.22. The spectrum of SN 1987B showing unprecedented strong, narrow Balmer emission lines.

Acknowledgments

We are grateful to David Branch who continued to encourage us to write this chapter and for teaching us so much about supernova spectra. We also thank Mike Dopita, Mark Phillips, and Nick Suntzeff for providing spectra of SN 1987A; Alex Filippenko for his spectacular spectra of other interesting events and permission to use some of them prior to publication; Carl Pennypacker for his efforts to discover early supernovae and for permission to show the spectra of his first "baby," SN 1986I; Doug Swartz for many valuable discussions of nebular spectra and for permission to use his nebular models of SN 1987A prior to publication; Anita Cochran and Ed Barker for their long-standing willingness to obtain supernova spectra at McDonald Observatory, and in particular for permission to show the spectrum of SN 1987B prior to publication; and Tim Jones for help in preparing Figure 1.1. This research is supported in part by NSF Grant 8717166, by grants from the R.A. Welch Foundation and Cray Research, Inc., and by the services of the University of Texas System Center for High Performance Computing.

References

Arnett, W.D., Bahcall, J.N., Kirshner, R.P. and Woosley, S.E. 1989, *Ann Rev. Astr. and Ap.*, **27**, 629.

Axelrod, T.A. 1980, in *Type I Supernovae*, ed. J.C. Wheeler (Austin: University of Texas), p. 80.

Axelrod, T.A. 1988, in *Proceedings of the IAU Colloquium No. 108, Atmospheric Diagnostics of Stellar Evolution*, ed. K. Nomoto, Lecture Notes in Physics, Vol. 305, 375.

Barbon, R., Ciatti, F., and Rosino, L. 1973, *Astr. and Ap.*, **29**, 57.

Begelman, M.C. and Sarazin, C.L. 1986, *Ap. J. (Letters)*, **302**, L59.

Bertola, F. 1962, *Asiago Contr.*, No. 128.

Branch, D. 1986, *Ap. J. (Letters)*, **300**, L51.

Branch, D. 1987, *Ap. J. (Letters)*, **316**, L81.

Branch, D., Doggett, J.B., Nomoto, K., and Thielemann, F.-K. 1985, *Ap. J.*, **294**, 619.

Branch, D., Drucker, W., and Jeffery, D.J. 1988, *Ap. J. (Letters)*, **330**, L117.

Branch, D., Falk, S.W., McCall, M.L., Rybski, P., Uomoto, A.K., and Wills, B.J. 1983, *Ap. J.*, **270**, 123.

Branch, D., Lacy, C.H., McCall, M.L., Sutherland, P.G., Uomoto, A.K., Wheeler, J.C., and Wills, B.J. 1981, *Ap. J.*, **244**, 780.

Chevalier, R.A. 1984, *Ap. J. (Letters)*, **285**, L63.

Doggett, J.B. and Branch, D. 1985, *Astron. J.*, **90**, 2303.

Dopita M.A. 1988, *Space Sci. Rev.*, **46**, 225.

Elias, J.H., Frogel, J.A., Hackwell, J.A., and Persson, S.E. 1981, *Ap. J. (Letters)*, **251**, L13.

Elias, J.H., Mathews, K., Neugebauer, G., and Persson, S.E. 1985, *Ap. J.*, **296**, 379.

Ensman, L.M. and Woosley, S.E. 1988, *Ap. J.*, **333**, 754.

Filippenko, A.V. 1988, *Astron. J.*, **96**, 1941.

Filippenko, A.V. 1989, *Proceedings of the Astronomical Society of Australia*, **7**, 412.

Fransson, C. 1986, in *Radiation Hydrodynamics in Stars and Compact Objects*, eds. D.H. Mihalas and K.-H. Winkler (Berlin: Springer-Verlag), p. 141.

Fransson, C. and Chevalier, R.A. 1987, *Ap. J. (Letters)*, **322**, L15.

Fransson, C. and Chevalier, R.A. 1988, *Ap. J.*, **343**, 323.

Frogel, J.A., Gregory, B., Kawara, K., Laney, D., Phillips, M.M., Terndrup, D., Vrba, F., and Whitford, A.E. 1987, *Ap. J. (Letters)*, **315**, L129.

Gaskell, C.M., Cappellaro, E., Dinerstein, H.L., Garnett, D., Harkness, R.P., and Wheeler, J.C., 1986, *Ap. J. (Letters)*, **306**, L77.

Graham, J.R. 1986, *M.N.R.A.S.*, **220**, 27p.

Harkness, R.P. 1986, in *Radiation Hydrodynamics in Stars and Compact Objects*, eds. D.H. Mihalas and K.-H. Winkler (Berlin: Springer-Verlag), p. 183.

Harkness, R.P. 1987, in *Relativistic Astrophysics*, ed. M.P. Ulmer (Singapore: World Scientific), p. 413.

Harkness, R.P. and Wheeler, J.C. 1989, *Proceedings of the Astronomical Society of Australia*, **7**, 431.

Harkness, R.P. and 10 others, 1987, *Ap. J.*, **317**, 355.

Minkowski, R. 1939, *Ap. J.*, **89**, 156.

Minkowski, R. 1940, *Pub. A.S.P.*, **52**, 206.

Nomoto, K., Shigeyama, T., and Hashimoto, M. 1988, in *Proceedings of the IAU Colloquium No. 108, Atmospheric Diagnostics of Stellar Evolution: Chemical Peculiarity, Mass Loss and Explosion*, ed. K. Nomoto, Lecture Notes in Physics, Vol. 305, p. 319.

Nomoto, K., Thielemann, F.-K., and Yokoi, K. 1984, *Ap. J.*, **286**, 644.

Panagia, N. 1985, Lecture Notes in Physics, Vol. 224, p. 14.

Pearce, E. C., Colgate, S.A., and Petschek, A.G. 1988, *Ap. J. (Letters)*, **325**, L33.

Pennypacker C.R. and 23 others, 1989, *Astron. J.*, **97**, 186.

Phillips, M.M., Heathcote, S.R., Hamuy, M., and Navarrete, M. 1988, *Astron. J.*, **95**, 1087.

Phillips, M.M. and 27 others, 1987, *Pub. A.S.P.*, **99**, 592.

Pinto, P.A. and Woosley, S.E. 1988, *Ap. J.*, **329**, 820.

Porter, A.C. and Filippenko, A.V. 1987, *Astron. J.*, **93**, 1372.

Schaeffer, R., Cassé, M., and Cahen, S. 1987, *Ap. J. (Letters)*, **316**, L31.

Schmitz, M.F. and Gaskell, C.M. 1988, in *Supernova 1987A in the Large Magellanic Cloud*, eds. M. Kafatos and A. Michalitsianos (Cambridge: Cambridge University Press), p. 112.

Schneider, D.P., Mould, J.R., Porter, A.C., Schmidt, M., Bothun, G.D., and Gunn, J.E. 1987, *Pub. A.S.P.*, **99**, 1167.

Swartz, D.A., Harkness, R.P., and Wheeler, J.C., 1989, *Nature*, **337**, 439.

Terndrup, D.M., Elias, J.H., Gregory, B., Heathcote, S.R., Phillips, M.M., Suntzeff, N.B., and Williams, R.E. 1989, *Proceedings of the Astronomical Society of Australia*, in press.

Wheeler, J.C. and Harkness, R.P. 1986, in *Galaxy Distances and Deviations from Universal Expansion*, eds. B.F. Madore and R.B. Tully (Dordrecht: Reidel), p. 45.

Wheeler, J.C., Harkness, R.P., Barker, E.S., Cochran, A.L., and Wills, D. 1987, *Ap. J. (Letters)*, **313**, L69.

Wheeler, J.C., Harkness, R.P., and Cappellaro, E. 1987, in *Relativistic Astrophysics*, ed. M.P. Ulmer (Singapore: World Scientific), p. 403.

Wheeler, J.C. and Levreault, R. 1985, *Ap. J. (Letters)*, **294**, L17.

Wood, P.R. and Faulkner, D.J. 1987, *Proceedings of the Astronomical Society of Australia*, **7**, 75.

Woosley, S.E., Axelrod T.S., and Weaver T.A. 1984, in *Stellar Nucleosynthesis*, eds. C. Chiosi and A. Renzini (Dordrecht: Reidel), p. 263.

Young, T.R. and Branch, D. 1988, *Nature*, **333**, 305.

2. Spectra of Supernovae

DAVID BRANCH

2.1. Introduction

Ever since the spectrum of a supernova was first observed in 1885, astronomers have known that a correct reading of the spectrum would reveal otherwise unattainable information about the physical conditions and composition of the radiation source. But the spectra of supernovae resisted interpretation for a very long time. The outstanding obstacle was the great characteristic width of the spectral features—150–300 Å—corresponding to Doppler broadening velocities of 5000–10,000 km s^{-1}. Because only a dozen or so overlapping spectral features occupy the whole optical spectrum at any one time, line identifications proved to be so difficult that very little was known about Type II spectra, and even less about Type I, until around 1970. In the hope that a brief history of the subject will be interesting (if not of great utility), I attempt in Section 2.2 to outline the development of supernova spectrum interpretations, and to cite in one place most of the significant papers. Section 2.3 then presents a list of classifications of all supernovae whose spectra I have seen, and a (nearly) complete spectrum bibliography.

We now know that during the first months after an explosion the ejected matter remains optically thick; "P Cygni" profiles of spectral lines formed in the outer layers are superimposed on a thermal continuum emitted from a *photosphere*, and spectrum formation is analogous to that which occurs in an expanding stellar atmosphere. As the ejected matter becomes optically thin to continuum photons via expansion and cooling, the supernova gradually becomes a self-excited *nebula*, with emission lines dominating the spectrum. In Section 2.4, the basic theory of spectrum formation during the photospheric and nebular phases, and methods for calculating synthetic spectra that were developed before the time of SN 1987A, are reviewed in qualitative terms. As summarized in Section 2.5, it was only during the early 1980s that the classical mysteries of supernova spectra were fully resolved—definite line identifications were established and a good semiquantitative picture of spectrum formation was achieved. Now the spectra of supernova 1987A are stimulating the development of more detailed, quantitative analyses. The techniques now being tested on SN 1987A are likely to lead in a few years to reliable information on the composition of the matter ejected by supernovae of all types—information which will have a strong impact on attempts to relate the various kinds of supernovae to their stellar progenitors and explosion mechanisms. The first applications of the new quantitative spectroscopy, and the prospects for further work, are discussed briefly in Section 2.6.

2.2. Historical Overview

2.2.1. Early Observations and Interpretations

The visual spectrum of SN 1885A (S Andromedae) in Messier 31 was examined directly by many observers, through spectroscopes. The first published representation of a supernova spectrum was that of Sherman (1888), who presented the blue spectrum in the *Monthly Notices of the Royal Astronomical Society* under the heading "A Discussion of Bright Lines in Stellar Spectra." Among the interpretations was that of Backhouse (1888), who declared "the whole spectrum of star and nebula is very much like that of an ordinary comet." In their centennial review of S And, de Vaucouleurs and Corwin (1985) summarized the contemporary descriptions of the spectrum and presented their own impression of how the visible spectrum must have appeared.

SN 1895B (Z Centauri) in NGC 5253 had its blue spectrum photographed, but not immediately published. Together with S And, Z Cen was assigned by Cannon (1916) to spectral class R (carbon stars). A microphotometer tracing of the Z Cen spectrum eventually was published by Johnson (1936), who suggested that the spectrum consisted of emission lines on a continuum. Payne-Gaposchkin (1936a) was more specific: the spectrum of Z Cen was like that of an ordinary nova in the late "4640" stage (featuring high-excitation emission lines such as those of N III), but with abnormally broad lines. S And was similar, but "to make definite line identifications would be to push the material too far" (Payne-Gaposchkin, 1936b).

SN 1917A in NGC 6946 also had its spectrum photographed, but never published. Humason (1936) did publish Mount Wilson blue spectra of SNe 1926A in NGC 4303 and 1936A in NGC 4273. He identified Balmer emission lines in the two "remarkably similar" supernovae, and concluded that their spectra "confirm the prediction by Baade and Zwicky that extremely wide emission lines are to expected in the spectra of supernovae, indicating that gaseous shells are ejected at great speed." Baade (1936) followed with a spectrum of SN 1936A obtained 100 days after maximum light that consisted essentially of a single wide band centered on 4671 Å. Baade suggested that the feature was produced by N III, and he concluded "the new spectrum leaves no doubt that the interpretation of the earlier spectra as a superposition of wide emission lines is correct."

Popper (1937) obtained spectra of SN 1937C at the Lick Observatory, and introduced the notion that there might be more than one kind of supernova by remarking that SNe 1937C and 1895B lacked two deep minima (later recognized to be Hβ and Hγ absorptions) that appeared in Humason's spectra of SN 1936A.

The classic paper of Minkowski (1939) on SNe 1937C in IC 4182 and 1937D in NGC 1003 was the first to present long series of spectra, both blue and red, and a thorough discussion. Minkowski thought that the blue and red spectra were of independent origin, and that the gradual redshifting of the blue bands was either a reflection of the deepening gravitational potential near the surface of a collapsing star (as had been suggested by Fritz Zwicky) or a decreasing blueshift of forbidden absorption lines. Except for relatively narrow [O I] 6300, 6360 lines that first appeared 6 months after maximum light, no spectral features could be definitely identified. After discussing the spectra of S And, Z Cen, and SNe 1917A, 1926A, and 1936A, Minkowski wrote "with the possible exception of S And, the spectra of all supernovae are similar and differ only in minor details." Since hydrogen lines were not seen in SNe 1937C and 1937D, "former identifications ... with lines of N III and H cannot be maintained." Despite the presence of low-excitation [O I] lines, which suggested "the possibility of an increased abundance of oxygen as a result of nuclear reactions," Minkowski thought the absence of hydrogen and helium lines implied a very high degree of ionization. So, although he discussed the possibilities of He II and O IV in absorption and Ca V in emission, he concluded that "these bands are probably emitted under conditions which are so peculiar that an identification with lines of known origin can hardly be expected."

Nevertheless, Whipple and Payne-Gaposchkin (1941; summarized by Payne-Gaposchkin and Whipple (1940)), tried. After agreeing with Zwicky (1940) that the [O I] lines in SN 1937C originated in interstellar matter stimulated by the supernova, they turned their attention to the unidentified features. They calculated synthetic spectra consisting of broad, symmetrical emission lines, primarily of Fe II, N III, N II, He I, He II, C II, and Fe III, superimposed on a continuum. Comparison with the spectra of SNe 1937C and 1937D was judged to be generally satisfactory. A deep minimum observed near 6140 Å was recognized to be an absorption feature, and the redshifting of spectral features with time was "now explained as a fortuitous effect arising from changes in the emission spectra with increasing temperature." Their conclusion that Fe II lines (in emission) played an important role in the spectra of SN 1937C, although correct, was to have little influence on subsequent discussions.

Minkowski (1940) abandoned the view that all supernovae have similar spectra when he observed SN 1940B in NGC 4725 and discovered that "the spectrum of this supernova is entirely different from that of any other nova or supernova previously observed. It is continuous ...." Soon after, Humason and Minkowski (1941) found that the red spectrum of SN 1941A in NGC 4559 "does not resemble that of any other supernova." Minkowski (1941) then made the distinction between a homogeneous group of Type I supernovae like 1937C and 1937D that lack hydrogen lines completely, and a less homogeneous group of Type II, including 1936A, 1940B, and 1941A, whose initially continuous spectra develop conspicuous Balmer lines and other structures soon after maximum light. He remarked that the synthetic spectra of Whipple and Payne-Gaposchkin "disagree in many details with the observed spectra of Type I" but "agree better with spectra of Type II and provide a very satisfactory confirmation of the identifications which, in this case, are already suggested by the pronounced similarity to the spectra of ordinary novae." In his review article on supernovae for the *Publications of the Astronomical Society of the Pacific*, Hubble (1941) wrote of Type I "the spectra are radically different from those of any other known phenomena, and none of the major features has yet been identified," but of Type II he declared "all the major features in the photographic region have been identified, and the story is reasonably clear. It seems to be the story of a nova on a gigantic scale." The confident statements of Minkowski and Hubble about Type II do not seem to have been supported by the facts; not a single microphotometer tracing of a Type II spectrum had yet been published, and the only specific identifications that had been proposed in print were with the hydrogen lines and N III.

Between 1941 and the time of Zwicky's (1965) classic review in the *Stellar Structure* volume of the Stars and Stellar Systems Series, many more spectra were observed, particularly by Francesco Bertola and colleagues at the Asiago Observatory and by Zwicky and Jesse Greenstein at Palomar, but very little progress was made with interpretation. After reviewing the observations of Type I spectra, Zwicky could only write: "Curiously enough, in spite of extremely arduous work by observers, experimentalists, and theoretical experts, the origin and the meaning of the spectra of Type I have remained completely mysterious." Type II spectra still were thought to be analogous to the spectra of novae during their high-excitation emission-line phases. Zwicky's article included a section on spectra by Greenstein, who identified spectral features of the Type II SN 1959D in NGC 7331 with lines of H, He I, He II, C III, and C IV (Figure 2.1). The high-excitation interpretation of Type II spectra continued to have adherents (e.g., Ciatti, Rosino, and Bertola, 1971; Schild, 1985) until, but not after, SN 1987A.

Zwicky (1965) also introduced Type III, based on SN 1961I in NGC 4303, Type IV (SN 1961F in NGC 3003), and Type V (SN 1961V in NGC 1058). All three prototypes appeared in 1961, and all three had hydrogen lines (some of Zwicky's unpublished spectra of SN 1961F show the Balmer lines more clearly than the published ones). Greenstein pointed out that SN 1961I and the Type II SN 1959D were "remarkably similar from the spectroscopic point of view."

Around 1970, the continuing mystery of the Type I spectra stimulated a number of interesting interpretations. Morrison and Sartori (1969, and subsequent papers) proposed that the spectral features are formed not in the ejected matter, but in circumstellar or interstellar gas ionized by a very powerful initial ultraviolet pulse from the supernova. Gordon (1972, and subsequent papers) argued that helium is the most abundant element, and that the spectra consist of overlapping emission lines of He I, He II, and coronal ions such as [Fe XIV], with superimposed absorptions of He I and Fe II that form in the outermost layers. Seddon's (1969) cautious suggestion that the spectral features might be intrinsically broad absorptions similar to the interstellar diffuse bands was taken up by Huffman (1970) and Manning (1970), who made specific identifications with absorption bands of iron ions in crystals such as garnet silicates, and Graham and Duley (1971) who proposed bands of metal atoms trapped in hydrocarbons.

2.2.2. The Thermal Interpretation of the Photospheric Phase

The modern recognition that supernovae of both types initially emit thermal continuous spectra, with superimposed *low-excitation* emissions and blueshifted absorptions, can be traced back to the 1960s. The light curve and broad-band colors of the Type II SN 1959D were discussed by Arp (1961) in terms of a blackbody continuum from an opaque, expanding, cooling envelope. The relatively narrow lines (2000 km s^{-1}) in the spectra of SN 1961V (Zwicky's Type V) enabled Bertola (1963b) to make the first identifications of individual Fe II emission lines (Figure 2.2). Branch and Greenstein (1971) calculated very simple synthetic spectra for SN 1961V that showed the importance of blueshifted absorption components

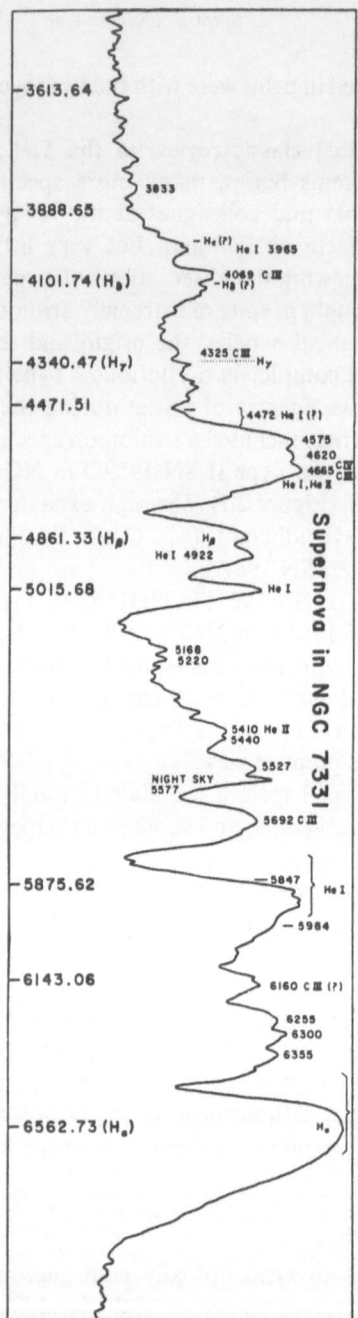

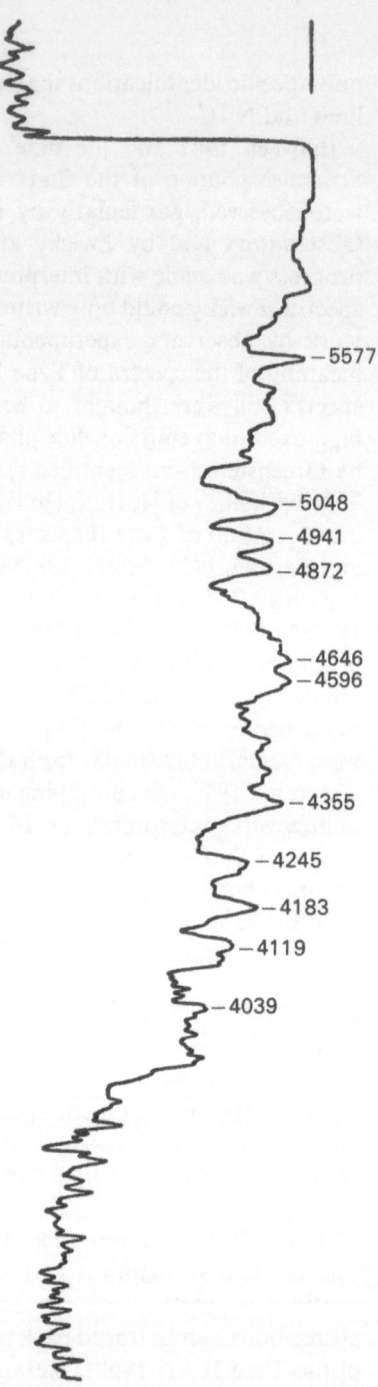

Figure 2.1. Spectrum of the Type II SN 1959D in NGC 7331, obtained with the 200-inch telescope of the Palomar Observatory by J.L. Greenstein on September 1, 1959. The wavelength scale at the bottom was established by a comparison spectrum of H, He I, and Ne I. Identifications of emission lines at the top illustrate the high-excitation interpretation of Type II spectra. Reproduced from Zwicky (1965), with permission.

Figure 2.2. Spectrum of the peculiar Type II SN 1961V in NGC 1058, obtained with the 122-cm telescope of the Asiago Observatory by F. Bertola on November 2, 1961. The majority of the relatively narrow emission peaks were identified by Bertola as lines of H and Fe II. Reproduced from Bertola (1963b), with permission.

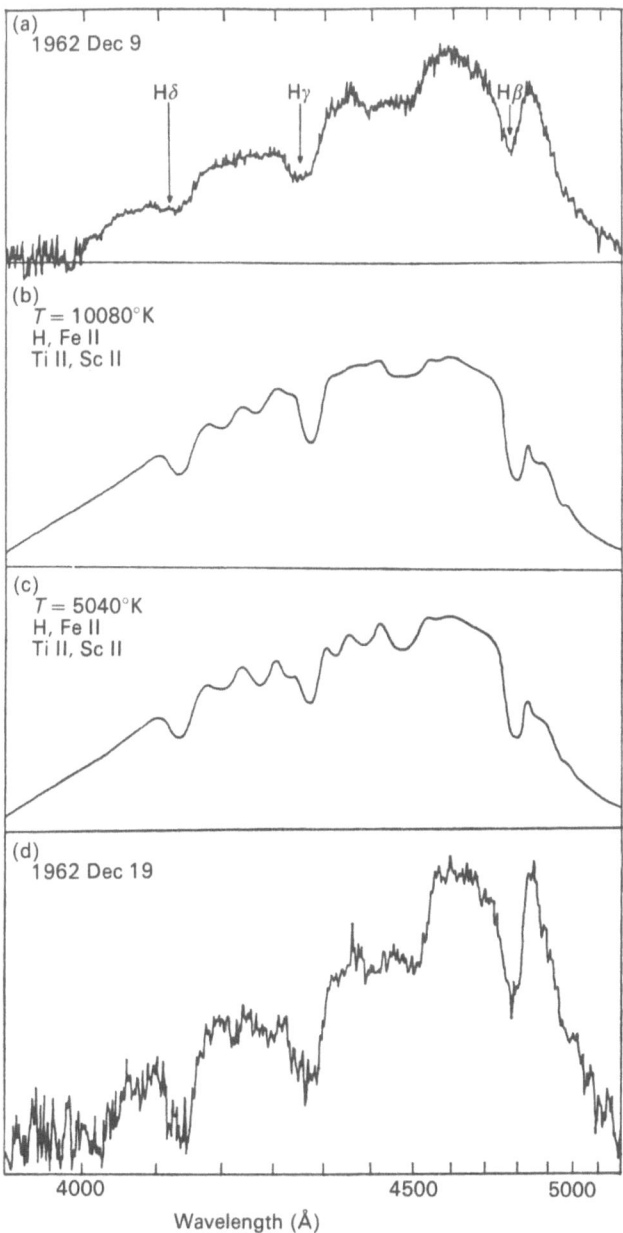

Figure 2.3. Top and bottom panels: spectra of the Type II SN 1962M in NGC 1313, obtained with the 74-inch telescope of the Radcliffe Observatory by P.W. Hill on December 9 and 19, 1962. Middle panels: synthetic spectra, consisting of blueshifted absorption lines of H, Fe II, Ti II, and Sc II, superimposed on a continuum. Reproduced from Patchett and Branch (1972), with permission.

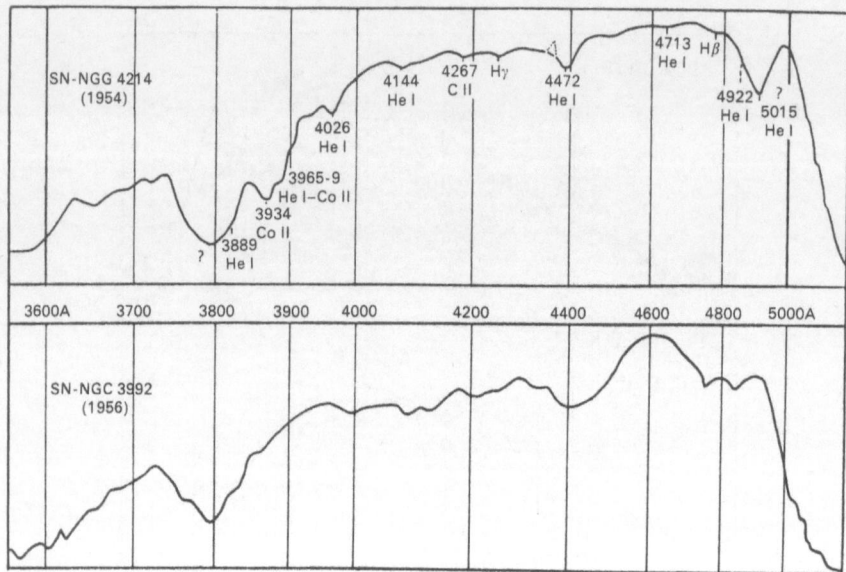

Figure 2.4. Spectra of the Type I SNe 1954A in NGC 4214 and 1956A in NGC 3992, obtained with the 36-inch telescope of the Lick Observatory by N.U. Mayall on June 24, 1954, and April 4, 1956. Most of the identifications of absorption features are with lines of He I. Reproduced from McLaughlin (1963), with permission, courtesy of the Publications of the Astronomical Society of the Pacific.

associated with each of the emissions, and replaced Bertola's identifications of He I and C III with lower-excitation lines of Ca II, Na I, Ti II, and Sc II. Patchett and Branch (1972) extended the interpretation to the broader-lined spectra of ordinary Type II's (Figure 2.3), and argued that the correct analogy with novae is with the very early, low-excitation "principal" spectrum (McLaughlin, 1937) rather than with the later high-excitation emission-line phase.

On the Type I front, McLaughlin (1963), the foremost authority on the spectra of novae, introduced a departure from the classical view that the spectra are principally overlapping emission lines. He suggested that Lick Observatory spectra of SN 1954A in NGC 4214, which were "in some measure peculiar," contained strong absorption features of He I (Figure 2.4). Minkowski (1963) quickly demurred, but McLaughlin's He I identifications were later supported by a study of Mount Wilson and Palomar spectra of SN 1954A (Branch, 1972). It is clear now that McLaughlin's He I line identifications were correct, and that SN 1954A belongs to the observationally rare subclass recently designated Type Ib. The first paper to propose most of the presently accepted low-excitation line identifications for classical Type I (now Type Ia) was that of Pskovskii (1969); he identified He I, Si II, Fe II, Mg II, Ca II, and S II, all in absorption (Figure 2.5), and credited McLaughlin with putting him on the right track. Mustel (1971, and subsequent papers) criticized Pskovskii severely, and correctly replaced Pskovskii's He I by Na I, but many of his other identifications were either like Pskovskii's or referred to weak features that now appear to have been produced by observational noise. The simple synthetic-

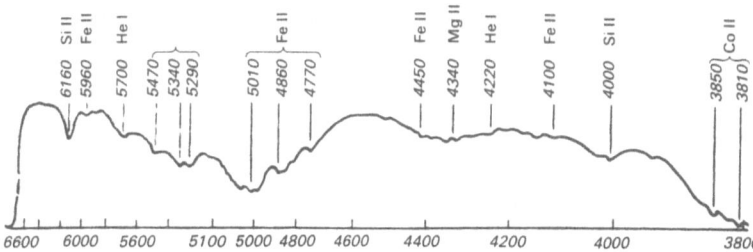

Figure 2.5. Spectrum of the Type I SN 1937C in IC 4182, obtained with the 60-inch telescope of the Mount Wilson Observatory by R. Minkowski on September 1, 1937. Most of the identifications of absorption features are with lines of singly ionized heavy elements. Reproduced from Pskovskii (1969), with permission.

spectrum technique that had been used for SN 1961V and Type II was applied to Type I by Branch and Patchett (1973), who reached conclusions similar to those of Pskovskii, and introduced a method analagous to Baade's (1926) method of distance-determination for variable stars, in an attempt to make an astrophysical, supernova-based, determination of the extragalactic distance scale.

2.2.3. From Photographic Plates to Linear Detectors

Shortly after the thermal interpretation of supernova spectra was introduced, Greenstein and Minkowski (1973) concluded the era of photographic supernova spectroscopy by presenting previously unpublished Palomar data. At the same time, and the same place, the modern era of linear-detector spectroscopy began. Kirshner *et al.* (1973a) used a multichannel scanner to obtain absolute low-resolution (20–180 Å) spectral energy distributions, including a comprehensive set of observations spanning 230 days of the evolution of the bright Type I SN 1972E in NGC 5253. The spectral scans confirmed that in both Type I and Type II a thermal continuum carries the bulk of the radiated energy, and that the characteristic line profile consists of an unshifted emission flanked by a violet-shifted absorption (a P Cygni profile). The review article on supernova spectra by Oke and Searle (1974) in the *Annual Reviews of Astronomy and Astrophysics* provided a list of supernova spectral types, through SN 1972E, and a description of the spectral evolution of both types, based on both the photographic spectra that had accumulated over decades and on the new scanner data. Kirshner and Kwan (1974, 1975) calculated a line profile produced by scattering in a homologously expanding atmosphere, applied Baade's method to the Type II SNe 1969L in NGC 1058 and 1970G in M101, and showed that the spectral line strengths in Type II supernovae are roughly consistent with ordinary cosmic abundances in the line-forming regions of the ejected matter.

The resolution of the first Palomar spectral scans was not high enough to fully resolve the line profiles, but within a decade optical spectra were being obtained at sufficient resolution (about 10 Å), especially at the McDonald Observatory. High-quality McDonald spectra of the Type II SN 1979C in M100 (Branch *et al.*, 1981) and the Type I SN 1981B (Branch *et al.*, 1983) were compared to synthetic spectra that, unlike those used previously, included the essential radiative transfer effects. The first supernova spectra outside the optical region were obtained with the

International Ultraviolet Explorer (IUE) satellite (Panagia *et al.*, 1980), and an atlas displaying ultraviolet spectra of six supernovae was published by Benvenuti *et al.* (1982). Limited spectroscopic information in the near-infrared was first obtained for the Type II SN 1980K in NGC 6946 (Dwek *et al.*, 1983) and the Type I SN 1983N in M83 (Graham *et al.*, 1986). Good coverage of near-infrared spectra finally was obtained for the Type I SN 1986G in Cen A (Frogel *et al.*, 1987; Graham *et al.*, 1987).

By the time supernovae enter their nebular emission-line phases they are faint, so for a long time spectra of the nebular phases were seldom obtained. The Palomar scans (Kirshner *et al.*, 1973a; Kirshner and Kwan, 1975) showed that after hundreds of days Type II spectra develop strong emission lines of [O I] and [Ca II]. Kirshner and Oke (1975) followed the Type I SN 1972E for almost 2 years and found the nebular spectrum of Type I to be entirely different from that of Type II; they interpreted the observed features as a complex blend of emission lines, reasoned that forbidden lines were likely to be stronger than permitted lines, and suggested specifically that blended [Fe II] lines dominated the spectrum. The [O I] lines that Minkowski had seen in SN 1937C after 6 months never appeared in SN 1972E.

It was not until the 1980s that the necessity to split Type I into two subclasses became clear. Supernovae 1983N in M83 (Panagia *et al.*, 1985) and 1984L in NGC 991 (Wheeler and Levreault, 1985) were observed to be spectroscopically distinct from normal Type I's. These two, like SNe 1962L in NGC 1073 (Rosino, 1963; Rousseau, Prevot, and Bardin, 1963; Bertola, 1964) and 1964L in NGC 3938 (Bertola, Mammano, and Perinotto, 1965), lacked the conspicuous red absorption feature that characterizes normal Type I during the first month after maximum. Harkness *et al.* (1987) established the presence of strong He I lines in SNe 1983N and 1984L. Filippenko and Sargent (1985) found what first appeared to be a unique nebular spectrum—SN 1985F showed strong lines of [O I] and [Ca II], like Type II, but no trace of hydrogen—but within the year Kirshner (unpublished) and Gaskell *et al.* (1986) discovered that SNe 1983N and 1984L developed similar nebular spectra. The designation Type Ib was first applied to SNe 1983N, 1984L, 1985F, and a few others by Elias *et al.* (1985).

2.3. Individual Supernovae: Spectral Types and Bibliography

Minkowski's simple rule that supernovae with hydrogen lines are Type II and those without hydrogen are Type I is a good basis for spectroscopic classification because it is based on a simple *empirical* criterion that would appear to exhaust the possibilities (but see below). In this scheme, only two main types are needed; subtypes, such as Ia and Ib, can be introduced as required, and peculiarities of individual supernovae can be acknowledged by using IIpec and Ipec. In this view, the prototypes of Zwicky's Types III, IV, and V become IIpec, because they had hydrogen lines; if the Types III and IV were maintained, then we would be in for type proliferation, because many supernovae are individually peculiar in one way or another. The division of Type II into II-L ("linear") and II-P ("plateau"), based on light-curve shape, is outside the scope of this chapter; systematic spectroscopic differences between II-P and II-L have not yet been established.

Table 2.1 presents spectral classes for all supernovae whose spectra I have seen.

Table 2.1. Supernova spectral types.

SN	Galaxy	Type	References
1885A	N224	I	de Vaucouleurs and Corwin, 1985, and references therein.
1895A	N5253	I	Johnson, 1936
1926A	N4273	II	Humason, 1936
1936A	N4303	II	Baade, 1936; Humason, 1936
1937C	I4182	Ia	Popper, 1937; Minkowski, 1939; Greenstein and Minkowski, 1973; Lick
1937D	N1003	Ia	Minkowski, 1939; Zwicky, 1958, 1965; Lick
1939A	N4636	I	Lick
1940A	N5907	II	Zwicky, 1958, 1965
1948B	N6946	II	Mayall, 1948; Lick
1954A	N4214	Ib	Wellman, 1954; McLaughlin, 1963; Branch, 1972; Greenstein and Minkowski, 1973; Greenstein; Lick
1954B	N5668	I	Greenstein; Lick
1956A	N3992	I	McLaughlin, 1963; Zwicky and Karpowicz, 1964; Lick
1957A	N2841	I	Zwicky and Karpowicz, 1965; Branch and Doggett, 1985; Lick
1957B	N4374	I	McLaughlin, 1963; Greenstein and Minkowski, 1973; Branch and Doggett, 1985; Greenstein; Lick
1959C	Anon	I	Greenstein and Zwicky, 1962; Lick
1959D	N7331	II	Zwicky, 1965; Greenstein and Minkowski, 1973; Greenstein; Lick
1960F	N4496A	Ia	Rosino and Bertola, 1961; Vorontosov-Velyaminov and Savel'eva, 1961; Bloch et al., 1964; Zwicky; Greenstein; Lick
1960H	N4096	Ia	Greenstein and Minkowski, 1973; Greenstein; Lick
1960N	Anon	I	Greenstein
1960R	N4382	I	Bertola, 1962; Zwicky, 1965; Greenstein and Minkowski, 1973; Zwicky; Greenstein
1961D	Anon	I	Zwicky, 1965
1961F	N3003	II	Zwicky, 1965; Zwicky
1961H	N4564	Ia	Bertola, 1962; Zwicky
1961I	N4303	II	Zwicky, 1965; Greenstein and Minkowski, 1973; Greenstein; Lick; Zwicky
1961P	Anon	Ia	Bertola, 1962
1961U	N3938	II	Bertola, 1963a; Zwicky, 1965; Greenstein; Zwicky
1961V	N1058	IIpec	Bertola, 1963b; Zwicky, 1964b; Branch and Greenstein, 1971; Greenstein and Minkowski, 1973; Greenstein; Zwicky
1962A	Anon	I	Zwicky and Barbon, 1967
1962E	Inter	?	Rudnicki and Zwicky, 1967
1962J	N6835	Ia	Bertola, 1965; Zwicky
1962L	N1073	Ib	Rosino, 1963; Rousseau et al., 1963; Bertola, 1964
1962M	N1313	II	Hill, 1965; Patchett and Branch, 1972
1963I	N4178	I	Bertola, 1965; Lick
1963J	N3913	Ia	Chincarini and Margoni, 1964; Zwicky, 1964a; Bertola, 1965
1963P	N1084	Ia	Bertola et al., 1965
1964E	Anon	I	Lick
1964F	N4303	II	Zwicky
1964H	N7292	II	Greenstein
1964L	N3938	Ib	Bertola et al., 1965; Zwicky
1965H	N4666	II	Dibai et al., 1967
1965I	N4753	Ia	Aller and Ross, 1965; Dibai et al., 1967; Ciatti and Barbon, 1971
1965N	N3074	II	Rudnicki, 1966
1966B	N4688	II	Gates et al., 1967
1966J	N3198	Ib	Chalonge and Burnichon, 1968; Chincarini and Perinotto, 1968; Borzov et al., 1969; Lick

Table 2.1 (*continued*)

SN	Galaxy	Type	References
1967C	N3389	Ia	Andrillat, 1967; Rubin and Ford, 1967; Ford and Rubin, 1968; Borzov *et al.*, 1969; Greenstein and Minkowski, 1973
1967H	N4254	?	Fairall, 1975
1968E	N2713	Ia	Ford and Rubin, 1968; Lick
1968I	N4981	I	Lick
1968L	N5236	II	Wood and Andrews, 1974
1969C	N3811	Ia	Bertola and Ciatti, 1971; Kirshner *et al.*, 1973a
1969H	N4725	I	Kirshner *et al.*, 1973a
1969L	N1058	II	Ciatti *et al.*, 1971; Kirshner *et al.*, 1973a; Kirshner and Kwan, 1974
1970G	M101	II	Barbon *et al.*, 1973a; Kirshner *et al.*, 1973a; Kirshner and Kwan, 1974; Pronik *et al.*, 1977
1970J	N7619	Ia	Assousa *et al.*, 1976
1971G	N4165	I	Lick
1971I	N5055	Ia	Kikuchi, 1971; Barbon *et al.*, 1973b; Kirshner *et al.*, 1973a; Iye *et al.*, 1975
1971L	N6384	Ia	Kirshner *et al.*, 1973a; Assousa *et al.*, 1976
1971U	Anon	I	Assousa *et al.*, 1976
1972E	N5253	Ia	Sistero and Castore Sistero, 1972; Ciatti, 1973; Greenstein and Minkowski, 1973; Kirshner *et al.*, 1973a; Kirshner *et al.*, 1973b; McCarthy and Araya, 1973; Bolton *et al.*, 1974; McCarthy, 1974; Kirshner and Oke, 1975; Abdulwahab and Morrison, 1978; Branch and Tull, 1979; Meyerott, 1980; Branch, 1984
1973N	N7495	Ia	Ciatti and Rosino, 1977
1973R	N3627	II	Kirshner and Kwan, 1975; Ciatti and Rosino, 1977
1974G	N4414	Ia	Andrillat, 1975; Iye *et al.*, 1975; Patchett and Wood, 1976; Ciatti and Rosino, 1977; Wyckoff and Wehinger, 1977
1974J	N7343	Ia	Ciatti and Rosino, 1977
1975A	N2207	Ia	Kirshner *et al.*, 1976
1975B	Anon	I	Kirshner *et al.*, 1976
1975G	Anon	Ia	Ciatti and Rosino, 1978
1975N	N7723	Ia	Ciatti and Rosino, 1978
1976B	N4402	I	de Vaucouleurs *et al.*, 1981
1976D	N5427	Ia	A. J. Longmore and J. R. Graham, unpublished
1978A	M43223	IIpec	Elliot *et al.*, 1978
1978B	U6335	II	McDonald Observatory, unpublished
1978E	Anon	Ia	Barbon *et al.*, 1982b; McDonald Observatory, unpublished
1978G	I5201	II	Benvenuti *et al.*, 1982 (UV)
1979B	N3913	Ia	Barbon *et al.*, 1982b
1979C	M100	II	Panagia *et al.*, 1980 (UV); Penston and Blades, 1980; Branch *et al.*, 1981; Barbon *et al.*, 1982b; Benvenuti *et al.*, 1982 (UV); Fransson *et al.*, 1984 (UV)
1980D	N3733	II	Kirshner, 1985
1980I	Inter	Ia	Smith, 1981
1980K	N6946	II	Prabhu, 1981; Barbieri *et al.*, 1982; Barbon *et al.*, 1982a; Benvenuti *et al.*, 1982 (UV); Dwek *et al.*, 1983 (IR); Uomoto and Kirshner, 1986; Harkness and Wheeler, 1989
1980N	N1316	Ia	Prabhu, 1981; Benvenuti *et al.*, 1982 (UV)
1981B	N4536	Ia	Barbon *et al.*, 1982a; Benvenuti *et al.*, 1982 (UV); Branch *et al.*, 1982; Branch *et al.*, 1983; Branch, 1984; Branch and Venkatakrishna, 1986 (UV)
1982B	N2268	Ia	Benvenuti *et al.*, 1982 (UV); Ciatti *et al.*, 1988
1982V	U02174	I	Wegner and McMahan, 1987

Table 2.1 (*continued*)

SN	Galaxy	Type	References
1982W	N5485	Ia	Barbon *et al.*, 1989a
1983G	N4753	Ia	Harris *et al.*, 1983; Prabhu, 1983; McCall *et al.*, 1984 (POL)
1983I	N4051	Ib	Wheeler *et al.*, 1987a
1983K	N4699	II	Niemela *et al.*, 1985
1983L	N7038	I	Schild, 1985
1983N	N5236	Ib	Richter and Sadler, 1983; Panagia *et al.*, 1985 (UV); Gaskell *et al.*, 1986; Graham *et al.*, 1986 (IR); Harkness *et al.*, 1987.
1983P	N5746	I	C. Willmer, unpublished
1983R	I1731	Ia	Filippenko and Sargent, 1986; Barbon *et al.*, 1989a
1983S	N1448	II	Hempe, 1986
1983U	N3227	Ia	de Robertis and Pinto, 1985; Wheeler, 1985; Barbon *et al.*, 1989a
1983V	N1365	Ib	Wheeler *et al.*, 1987a; Branch, 1988; R. Cannon unpublished, P. O. Lindblad, unpublished
1983W	N3625	Ia	Wheeler, 1985
1984A	N4419	Ia	Wegner and McMahan, 1987; Barbon *et al.*, 1989b
1984E	N3169	II	Dopita *et al.*, 1984; Henry and Branch, 1987
1984L	N991	Ib	Wheeler and Levreault, 1985; Harkness *et al.*, 1987
1984N	N7184	pec	J.B. Oke, unpublished
1985A	N2748	Ia	Wegner and McMahan, 1987
1985B	N4045A	Ia	Wegner and McMahan, 1987
1985F	N4618	Ib	Filippenko and Sargent, 1985, 1986; Gaskell *et al.*, 1986; Filippenko *et al.*, 1986; Pearce and Purvis, 1986
1985G	N4451	II	Purvis *et al.*, 1987
1985H	N3359	II	Purvis *et al.*, 1987; Wheeler *et al.*, 1988
1985L	N5033	II	Filippenko and Sargent, 1986
1985P	N1433	II	Chalabaev and Christiani, 1987
1986A	N3367	Ia	Dekker *et al.*, 1988
1986E	N4302	II	G. Wegner, unpublished
1986G	N5128	Ia	Frogel *et al.*, 1987 (IR); Graham *et al.*, 1987 (IR); Phillips *et al.*, 1987
1986I	N4254	II	Wheeler *et al.*, 1987b; Filippenko, 1988a; Pennypacker *et al.*, 1989; R.B.C. Henry, unpublished
1986J	N891	II	Rupen *et al.*, 1987
1986O	N2227	Ia	Arsenault and D'Odorico, 1988
1986W	N1667	I	R.B.C. Henry, unpublished
1987A	LMC	II	Blanco *et al.*, 1987; Bouchet *et al.*, 1987 (IR); Cassatella *et al.*, 1987; Catchpole *et al.*, 1987; Danziger *et al.*, 1987; Fosbury *et al.*, 1987; Hanuschik and Dachs, 1987; Kirshner *et al.*, 1987 (UV); Larson *et al.*, 1987 (IR); Menzies *et al.*, 1987; Wamsteker *et al.*, 1987; Williams, 1987; Catchpole *et al.*, 1988; Danziger *et al.*, 1988; Elias *et al.*, 1988 (IR); Erickson *et al.*, 1988; Fransson *et al.*, 1989; Meikle *et al.*, 1988 (IR); Phillips *et al.*, 1988; Whitelock *et al.*, 1988
1987B	N5850	IIpec	Harkness and Wheeler, 1989; A.V. Filippenko, unpublished
1987C	Mrk90	II	A.V. Filippenko, unpublished
1987D	U7370	Ia	Schneider *et al.*, 1988
1987F	N4615	IIpec	Filippenko, 1989
1987K	N4651	IIb	Filippenko, 1988b
1987L	N2336	Ia	A.V. Filippenko, unpublished; Pearce *et al.*, 1988
1987M	N2715	Ib	Filippenko *et al.*, 1989
1987N	N7606	Ia	A.V. Filippenko, unpublished
1987O	U4060	Ia	A.V. Filippenko, unpublished
1987P	Anon	Ia	A.V. Filippenko, unpublished
1988A	N4579	II	A.V. Filippenko, unpublished

Table 2.1 (continued)

SN	Galaxy	Type	References
1988B	N3191	Ia	A.V. Filippenko, unpublished
1988C	U3933	Ia	A.V. Filippenko, unpublished
1988D	Anon	Ia	A.V. Filippenko, unpublished
1988F	U2988	Ia	A.V. Filippenko, unpublished
1988G	Anon	Ia	Reid et al., 1988
1988I	Anon	IIpec	Filippenko, 1989
1988L	N5850	Ib	A.V. Filippenko, unpublished
1988M	N4496B	II	Filippenko et al., 1988

Except for SN 1885A and 1987A, the reference list is intended to be complete through December, 1988. The intent is to update this list continually, so I would be grateful to hear of whatever omissions there may be. References are given to papers which display previously published spectra in previously unpublished formats, but not to mere repetitions. References to "Greenstein," "Zwicky," and "Lick," without a year, refer to my microphotometer tracings of Palomar photographic plates made available by Jesse Greenstein, some of Zwicky's plates loaned by Wallace Sargent, and the Lick Observatory plate collection provided by George Herbig.

Of the 133 supernovae listed in Table 2.1, 51 are Type Ia and 44 are Type II. The 11 classified as Type Ib include SN 1966J, not previously associated with Type Ib, on the basis of its apparent lack of the red Si II absorption feature (Chincarini and Perinotto, 1968). Another 24 are listed simply as Type I, because the available spectra do not establish whether they are Ia or Ib. Following Filippenko (1988b), SN 1987K is classified as a Type IIb, because it showed hydrogen lines near maximum light but developed a nebular-phase spectrum that lacked hydrogen lines and resembled Type Ib (thus disproving the notion that the presence or absence of hydrogen exhausts the possibilities). SNe 1962E and 1967H are listed as of unknown type in view of the low quality of the published spectra, and SN 1984N is simply labeled "pec" because its its strange spectrum lacks conspicuous hydrogen lines but does not closely resemble Type Ia or Ib.

2.4. Spectrum Formation

2.4.1. Photospheric Phase

During the first months after its explosion a supernova is optically thick, and the spectrum forms at and above the photosphere. Information on the outer layers of the ejected matter, relating to the nature of the progenitor star, is revealed gradually as the photosphere recedes with respect to the matter. During this phase the outstanding complication is that radiative transfer effects are important—the emergent spectrum is the result of a complex series of photon emissions, absorptions, and scatterings; the outstanding simplification is that matter and radiation are sufficiently closely coupled that an appeal to equilibrium approximations—

Planckian radiation field and local thermodynamic equilibrium (LTE) atomic level populations—is useful for first interpretations. In the 1980s, the escape-probability treatment of radiative transfer (Sobolev, 1960; Castor, 1970) and the equilibrium approximations began to be used to calculate synthetic spectra of supernovae in their photospheric phase (Branch, 1980, and subsequent papers). This section begins with a brief review of the assumptions and approximations upon which those exploratory calculations were based.

The velocity structure of a supernova is, to first order, very simple. Hydrodynamical effects die out shortly after the explosion, and each element of matter then coasts at constant velocity. By the time the supernova is observed, the radial distance that has been traveled by matter is just proportional to its velocity. From then on, velocity is proportional to radius and the matter distribution expands homologously with a radial density law that retains its shape and scales as the inverse cube of the time. Hydrodynamicists like to use mass as the radial coordinate but spectroscopists need to keep track of Doppler shifts and prefer to use velocity.

The spectrum consists of P Cygni lines, formed in the rapidly expanding outer layers, superimposed on a thermal continuous spectrum emitted from the photosphere. To a first approximation the photosphere can be regarded as a sharp layer that emits a blackbody continuum. The sharp-photosphere approximation can be used because the density gradient in the outer layers of the supernova matter is rather steep. For example, Colgate and McKee (1969) showed that the explosion of a polytrope of index 3 will produce a density law $\rho(v) \sim v^{-7}$ in the outer half of the ejected mass. The assumption that the continuous spectrum has a blackbody shape is, of course, just a convenient representation of a thermal continuum whose real shape depends on the wavelength dependence of the opacity and the extension of the atmosphere.

An individual line profile consists of an emission peak centered on the rest wavelength (in the frame of the supernova), accompanied by a blueshifted absorption. In the $v \sim r$ velocity field the surfaces of constant radial velocity with respect to the observer are just planes oriented perpendicular to the line of sight (Kirshner and Kwan, 1974), so there is a direct correspondence between each point on a line profile and one of the planes. For example, line photons emerging at the rest wavelength come from the plane that passes through the center of the supernova, and the blueshifted absorption occurs in the approaching matter located between the photosphere and the observer.

A line profile can be calculated when the radial dependence of the source function and the optical depth are specified. The source function depends on the ratio of the populations of the upper and lower levels of the transition. To a first approximation (for most lines) the source function is that of resonant scattering, in which line photons are only scattered, neither created nor destroyed, in the atmosphere. Then the source function is given simply by the product of the continuum intensity and the dilution factor for the radiation field above the photosphere. In the $v \sim r$ velocity field, a photon redshifts with respect to matter as it propagates, so a photon emitted in a line transition either is reabsorbed in the immediate vicinity or it redshifts out of resonance with the transition (hence the concept of escape probability); thus the line optical depth becomes a local quantity. A first approximation for the radial dependence of the line optical depth is that it is proportional to the adopted density law, $\rho(v)$, modified by whatever radial composition profile might be assumed. This approximation neglects any radial dependence of excitation and ionization. The neglect of excitation and ionization gradients falls between two simple extreme alternatives:

(1) that the level populations above the photosphere are fully thermalized; and
(2) that the level populations are determined entirely by the dilute radiation field.

In both cases the excitation decreases outwards, but the ionization behavior differs. In the first case, we could adopt a temperature distribution consistent with radiative equilibrium in a spherical atmosphere and calculate the LTE level populations. This would predict a radial decrease of ionization, and some lines having negligible optical depths at the photosphere could become important at higher levels in the atmosphere. In the radiative case the ionization would increase outwards because the decreasing flux of ionizing radiation would be more than compensated for by the decreasing recombination rate in the steep density gradient. In practice, the decreasing excitation and increasing ionization of the radiation-driven case, together with the steep density gradient, would produce decreasing optical depths for all lines of interest (in the absence of strong composition gradients); lines having negligible optical depths at the photosphere would be unimportant everywhere. The correct approach to the level populations is, of course, to solve the rate equations, but in exploratory work the simple approximation of $\tau \sim \rho$ with constant level populations is useful. Resonant scattering profiles calculated on the basis of these simple considerations (Branch, 1980) show that for lines having moderate optical depths at the photosphere ($1 < \tau < 10$) the blueshift of the absorption minimum corresponds closely to the velocity at the photosphere; the blueshift is smaller for weaker lines and larger for stronger ones.

The line blending that is such an observational nuisance in supernova spectra corresponds physically to multiple scattering in the atmosphere. A photon that escapes from one transition can redshift into resonance with another of longer wavelength, and scatter again before leaving the atmosphere. Mathematically, the line source functions become coupled. Olson's (1982) exact expressions for the coupled source functions of two lines are easily extended to an arbitrary number of interacting lines. A characteristic feature of calculated blends is that absorptions tend to dominate emissions—i.e., individual absorptions appear at their expected blueshifted locations, but emission peaks are not necessarily at the rest wavelengths of the individual lines. During the photospheric phase, therefore, it usually is best to base line identifications on absorption features rather than emissions.

Given the optical depths and the (coupled) source functions of the lines, synthetic spectra can be calculated. The relative optical depths of lines of a given ion can be fixed by assuming LTE excitation at the photosphere, and for each ion the optical depth scale can be controlled by an adjustable scale parameter, with the corresponding LTE element abundance worked out later in a separate calculation (this is the efficient approach for trial and error line identifications), or the optical depths can be fixed from the outset by assuming a composition and LTE.

A more sophisticated and much more computationally intensive approach to the calculation of photospheric spectra has been taken by Harkness (1985, 1986, 1987), who has calculated synthetic spectra by solving the comoving special-relativistic radiative transfer equations in spherical geometry, following numerical methods outlined by Mihalas, Kunasz, and Hummer (1976) and Mihalas (1980). A separate calculation transforms the comoving radiation field to the stationary observer's frame. Instead of assuming a sharp blackbody photosphere, opacity data is included and the emergent continuous spectrum is calculated. The assumptions of resonant scattering and LTE optical depths are retained, and level populations above the photosphere are taken to be fully thermalized.

If a supernova explodes inside a circumstellar shell created by the stellar wind of the progenitor star, the interaction between the rapidly expanding supernova matter and the slow circumstellar shell can produce special velocity laws and effects that cannot be described by resonant scattering. Fransson (1984) gave a detailed discussion of line formation under these circumstances. When the supernova shock breaks through the photosphere of the progenitor, a burst of hard ultraviolet radiation accelerates the circumstellar matter; because of the radial dilution of the radiation the inner parts of the shell are accelerated to higher velocities than the outer parts, and the initial velocity distribution in the shell is $v \sim r^{-2}$. This converging velocity distribution leads to a shock wave in the circumstellar gas, and then to high-energy radiation by free–free emission and Compton scattering. The high-energy radiation produces ions such as N V, Si IV, and C IV that have strong ultraviolet lines. Fransson used the escape-probability approximation to calculate profiles of such lines, taking collisional excitation into account, and predicted nearly symmetrical emission lines rather than the P Cygni lines characteristic of resonant scattering. For line formation in the circumstellar shell, which is more complex because the special case of $v \sim r$ does not apply, Fransson calculated line profiles by replacing the escape-probability approximation with the formalism of Rybicki and Hummer (1978). Spectrum effects arising from circumstellar interactions are discussed by Chevalier in this volume.

2.4.2. Nebular Phase

Hundreds of days after the explosion, the ejected matter begins to become optically thin to continuum photons and the transparent supernova enters its nebular phase, during which the spectrum forms primarily in the dense, deep layers of the ejected matter, which are important for nucleosynthesis. Now spectrum formation is radically different from the photospheric phase; radiative transfer effects are minimal—emitted photons escape directly, and the spectrum consists of emission lines whose profiles reflect the radial distribution of level populations (significant optical thickness in lines does produce some transfer effects, especially during the early nebular phase and at ultraviolet wavelengths). However, ionization fractions can be far out of LTE, and the problem of predicting the spectrum reduces to that of calculating the non-LTE level populations.

Meyerott (1978, 1980) considered the source of ionization and heating in Type I supernovae to be the γ-rays and positrons emitted during the radioactive decay of ^{56}Co to ^{56}Fe. He showed that the decay energy was likely to be converted to optical light with high efficiency, and concluded that the most prominent features in the nebular spectra of Type I are likely to be blends of [Fe II] and [Fe III] lines. Axelrod (1980a, b) used a numerical model of a homologously expanding shell of initially pure ^{56}Ni, along with realistic, self-consistent assumptions, to calculate synthetic spectra for Type I. He calculated the fraction of the decay energy that is deposited in the shell at each time, the ionization produced by fast electrons resulting from Compton scattering of the gammas, and the temperature of the thermal electrons— determined by a balance between heating via secondary electrons resulting from ionization, and cooling via collisional excitation of the forbidden-line radiation. Given the ionization level and the electron temperature, the non-LTE level populations of the first several ions of cobalt and iron and the emergent spectrum could be calculated. Fransson (1986) later extended this approach to the helium-rich and oxygen-rich cores of massive stars, in which the different composition produces a completely different emergent spectrum, but the heating and ionization is still controlled by cobalt decay.

2.5. Spectral Types Ia, Ib, and II

2.5.1. Type Ia

Interpretation of the optical spectra of Type Ia during the photospheric phase has concentrated heavily on the McDonald Observatory observations of SN 1981B in NGC 4536. The spectrum near maximum light was matched by synthetic spectra containing resonance scattering lines of Ca II, Si II, S II, Mg II, and O I (Branch *et al.*, 1982; Branch, 1985). Within 2 weeks of maximum light the spectrum developed blends of permitted Fe II lines which dominated the optical spectrum for at least 100 days (Branch *et al.*, 1983). After the basic nature of Type Ia spectra had been established, consideration began to be given to the spectra that would be predicted on the basis of particular explosion models for Type Ia (for the physics of deflagrations and detonations in accreting white dwarfs, see the chapter by Woosley in this volume). As shown in Figure 2.6, synthetic spectra computed for the composition and density structure of the carbon-deflagration model of Nomoto, Thielemann, and Yokoi (1984) were found to match SN 1981B spectra well, provided that the strongly stratified composition in the outer layers of the model was homogenized down to the level at which the velocity equals 8000 km s^{-1} (see Branch *et al.*, 1985; Branch, 1985). The spectrum synthesis technique of Harkness was used to confirm and extend the interpretation of the optical spectrum (Harkness, 1985, 1986) and to show that the ultraviolet spectrum can be explained in the same way, but with very heavy line blanketing (Wheeler and Harkness, 1986; Wheeler *et al.*, 1986; Harkness, 1987).

From broad-band infrared photometry, Elias *et al.* (1981) inferred the presence of a very broad absorption feature near 1.2 μ in Type Ia weeks after maximum light. Graham (1986) suggested that a blend of strong Si I lines might produce the feature, but spectra of SN 1986G in NGC 5128 (Frogel *et al.*, 1987) showed that the Si I lines were not present at detectable strength, and the nature of the broad absorption remains unknown. Narrower P Cygni features in the infrared, attributed by Frogel *et al.* to Na I and Mg II, are consistent with the interpretation of the optical

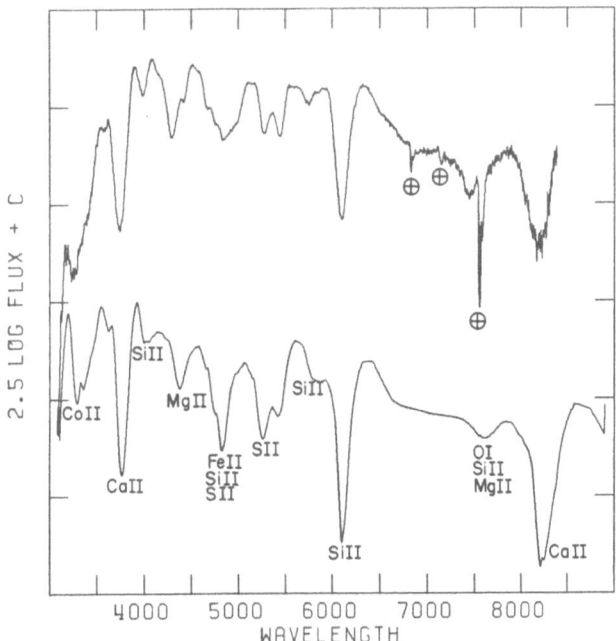

Figure 2.6. The maximum-light spectrum of the Type Ia SN 1981B (Branch *et al.*, 1983) is compared to a synthetic spectrum based on the carbon-deflagration model W7 of Nomoto, Thielemann, and Yokoi (1984). The composition in the outer layers of the model is assumed to be mixed. Terrestrial absorptions in the observed spectrum are indicated, and the vertical displacement between the two spectra is arbitrary. Reproduced from Branch *et al.* (1985), by permission.

spectrum, but the tentative identification of one infrared feature with He I is in conflict with the optical interpretation.

Axelrod's (1980a, b) synthetic spectra for the nebular phase, consisting mainly of forbidden emission lines from the first several ionization states of iron, as well as a [Co III] line near 6000 Å, gave a good account of the Kirshner and Oke (1975) spectra of SN 1972E. Figure 2.7 shows that one of Axelrod's synthetic spectra also resembles the spectrum of SN 1981B. Woosley, Axelrod, and Weaver (1984) concluded that the spectral match achieved by Axelrod for model parameters appropriate to a carbon deflagration would be lost if the supernova expanded as fast as in a detonation model.

It is not yet clear just when the supernova spectrum evolves from the photospheric to the nebular phase: the spectra appear to make a smooth transition from being dominated by permitted Fe II scattering features at 100 days to the nebular spectrum fit by Axelrod after a few hundred days. The nebular spectra appear to contain absorptions by Ca II, and the question of whether the Na I absorption also persists and blends with the [Co III] line has not yet been resolved.

The interpretation of the Type Ia spectra during both the photospheric and nebular phases provides support for the carbon deflagration model, which entails explosion at the Chandrasekhar mass, but it has become clear recently that Type

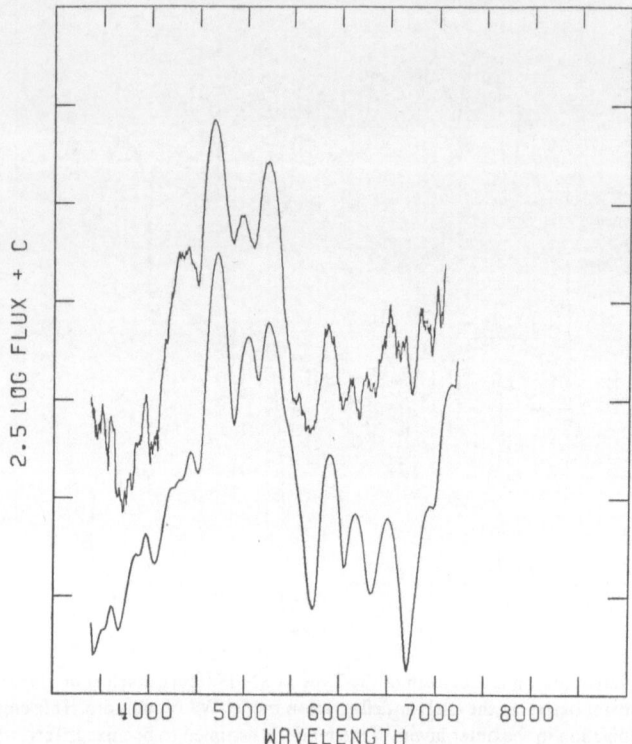

Figure 2.7. A spectrum of the Type Ia SN 1981B, obtained 250 days after maximum light by Beverly Wills at the McDonald Observatory, is compared to a synthetic spectrum computed by Axelrod (1980b). The synthetic spectrum is dominated by lines of [Fe II] and [Fe III], but the feature near 6000 Å is from [Co III]. Reproduced from Branch (1987b), by permission.

Ia supernovae are not strictly homogeneous. SN 1984A ejected an exceptional amount of high-velocity matter (Branch, 1987a), SNe 1986G and 1987D deviated from the norm at optical wavelengths (Phillips *et al.*, 1987; Schneider *et al.*, 1988), and SN 1986G also was unusual in the infrared (Frogel *et al.*, 1987). Branch, Drucker, and Jeffery (1988) present evidence of a significant range in expansion velocity among Type Ia. It has become clear that nature has a way to produce variations on the carbon deflagration theme (so do Graham, 1987 and Canal, Isern, and Lopez, 1988).

2.5.2. Type Ib

Strong lines of He I in the photospheric spectra of the Type Ib SNe 1983N and 1984L were identified by Harkness *et al.* (1987), who calculated synthetic spectra for compositions characteristic of the helium zone in a massive star and found that in order to reconcile the presence of He I lines with the red continuum, a large non-LTE overpopulation of the excited states of He I had to be invoked (Figure 2.8). Other spectral features were attributed to C II, O I, Ca II, and Fe II. The spectra of two other SNe Ib, 1983I and 1983V, were interpreted by Wheeler *et al.*

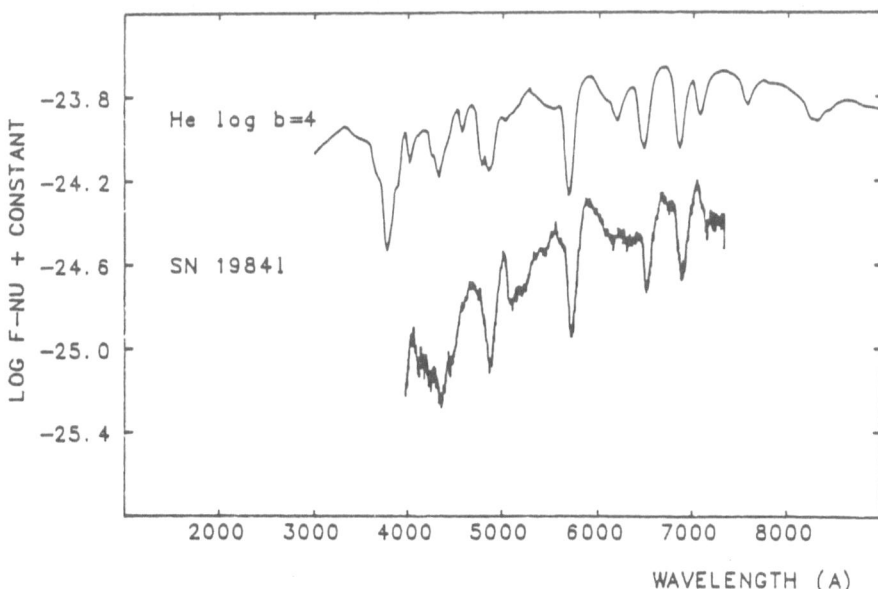

Figure 2.8. The spectrum of the Type Ib SN 1984L, about 20 days after maximum light, is compared to a synthetic spectrum in which the He I lines are strengthened by a factor of 10^4 with respect to their LTE strengths. The five deepest absorptions in the observed spectra correspond to He I $\lambda\lambda 4471$, 5015, 5876, 6678, and 7065, blueshifted by 10,000 km s^{-1}. Reproduced from Harkness *et al.* (1987), by permission.

(1987a) in the same way, but with more oxygen and less helium. A different point of view was taken by Branch and Nomoto (1986) and Branch (1988), who interpreted the early spectra of SNe 1984L and 1983V as blends of Fe II and He I lines and speculated that the outer layers might be an iron–helium mixture (nickel–helium, at the time of the explosion) arising from the off-center detonation of the helium zone in a white dwarf. Harkness *et al.* (1987) argue that an iron-rich composition would lead to excessively strong Fe II lines. In the helium-star scenario, it is not clear how to get the huge non-LTE excitation of He I, while in the detonation scenario it could come from the high-energy photons from nickel–cobalt decay.

In SNe 1983N and 1984L, the first signs of the nebular phase were detected only 20 days after maximum light, with the appearance of a line of [O I] 5577 (Harkness *et al.* 1987). By 8 months after maximum the spectra of SNe 1983N and 1985F (Figure 2.9) had developed strong broad emission lines of [O I] 6300, 6364, [Ca II] 7291, 7324, Na I 5890, 5896, and [Mg I] 4562 (Filippenko and Sargent, 1986; Gaskell *et al.*, 1986). Early estimates of the mass of oxygen needed to account for the nebular spectra were very high (Begelman and Sarazin, 1986; Filippenko and Sargent, 1986; Gaskell *et al.*, 1986), but large oxygen masses are not consistent with the relatively narrow light curve. A detailed discussion of the physics of the nebular spectrum has been given by Fransson and Chevalier (1989), who conclude that the observed line strengths can be explained in terms of the explosion (via core collapse) of the 8 $M_\odot$ core of a star originally of 25 $M_\odot$; such a core ejects about 2 $M_\odot$ of oxygen. Fransson

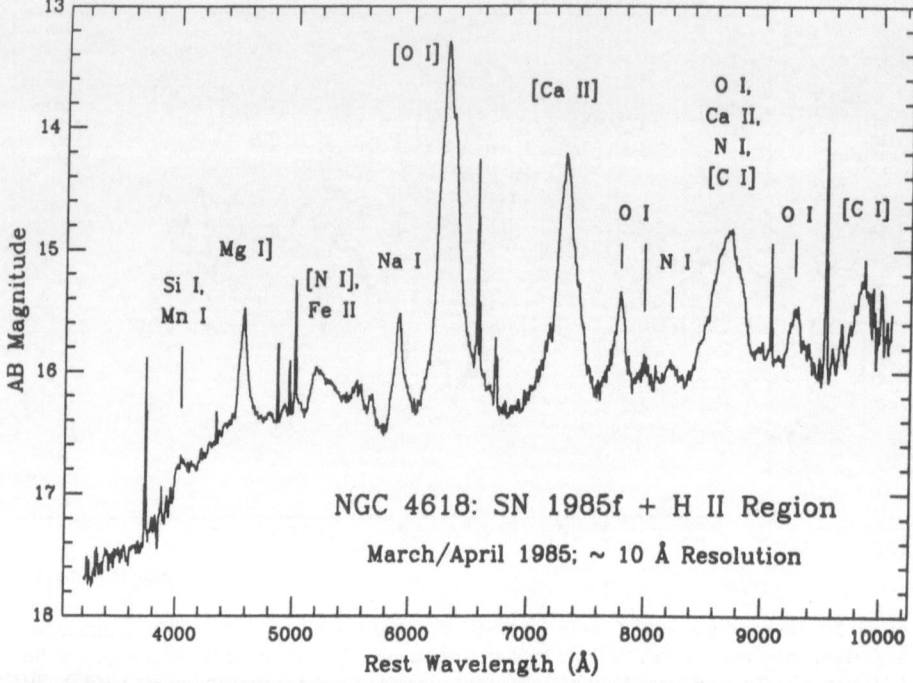

Figure 2.9. The spectrum of the Type Ib SN 1985F during the nebular phase. Reproduced from Filippenko and Sargent (1986), by permission.

and Chevalier also conclude that exploding white dwarfs cannot produce the absolute flux that is observed in the nebular phase of Type Ib.

An emission line in the infrared spectrum of SN 1983N, at 1.6 μ, was detected by Graham *et al.* (1986), and attributed to [Fe II]. The identification implied 0.3 $M_\odot$ of ejected iron. Oliva (1987), however, has suggested an alternative identification with lines of Si I.

The problem of identifying the progenitors and explosion mechanisms of Type Ib involves many observational constraints that are beyond the scope of this article, e.g., light-curve shape, association with H II regions, and radio emission (Porter and Filippenko, 1987; Branch, 1988; Ensman and Woosley, 1988; Panagia and Laidler, 1988), but it is clear that one of the critical needs is a quantitative analysis of the photospheric spectra, to complement the more reliable analysis of the nebular phase.

2.5.3. Type II
Type II supernovae obviously are explosions of stars that retain hydrogen-rich envelopes, and their optical spectra during the photospheric phase are qualitatively well understood. Initially the spectrum is practically continuous, with weak lines of hydrogen and neutral helium. After maximum light lines of Ca II, Na I, Fe II, Ti II, and Sc II appear in the form of resonant scattering profiles (Patchett and Branch, 1972; Branch *et al.*, 1981); hydrogen lines, however, develop profiles that have net

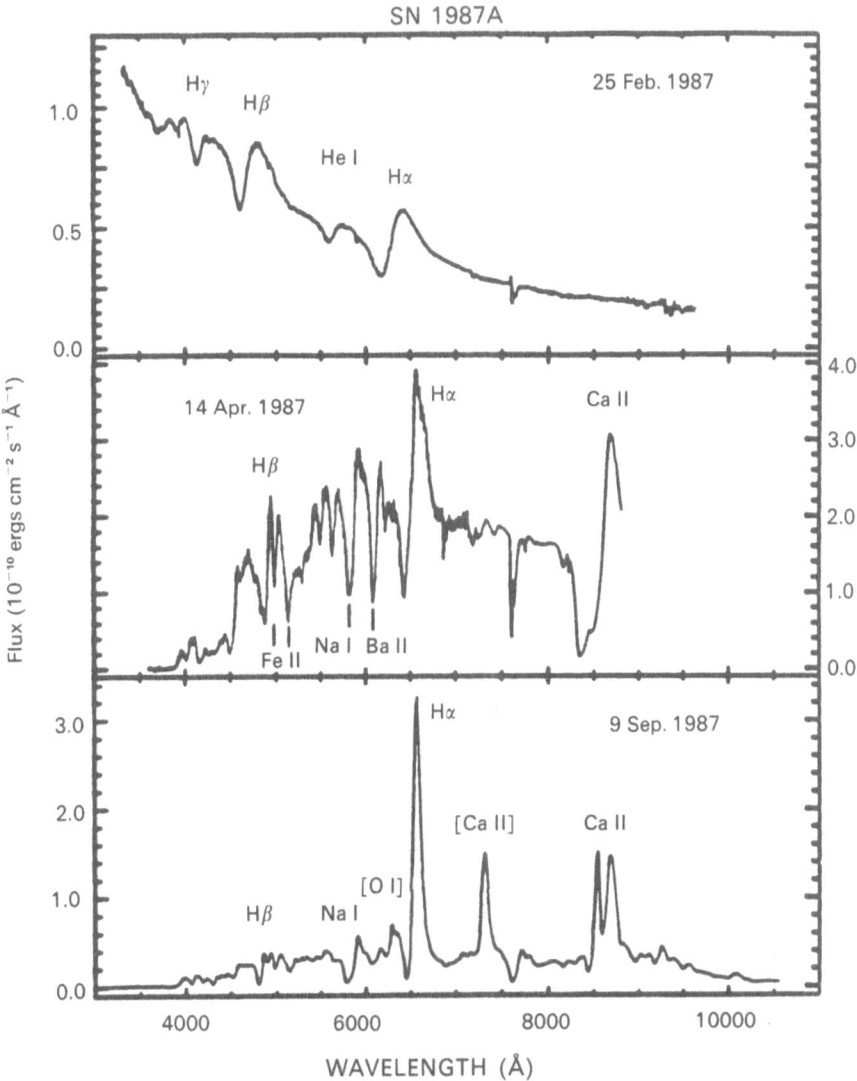

Figure 2.10. Spectra of the Type II SN 1987A in the Large Magellanic Cloud, obtained during the photospheric phase (top and middle panels) and during the transition to the nebular phase. Figure prepared by Mark M. Phillips, reproduced from McCray and Li (1988), by permission.

emission and are not well represented by resonant scattering. In SN 1979C the velocity at the photosphere was found to decrease from 10,000 to 8000 km s^{-1} during the first 6 weeks after maximum light. Panagia *et al.* (1980) initially interpreted features in the IUE ultraviolet spectra of SN 1979C as relatively high-excitation emission lines from ions, such as N V, Si IV, and C IV, formed in slow moving matter (< 4000 km s^{-1}) outside the photosphere, but after a recalibration of the IUE spectra Fransson *et al.* (1984) and Fransson (1984) concluded that the lines were broader, consistent with formation in the supernova ejecta.

Little interpretation of the nebular spectra of Type II supernovae was done prior to SN 1987A, although an important conclusion was reached by Uomoto and Kirshner (1986), who followed the nebular spectra of SN 1980K until 650 days and found that the flux in the Hα line decreased exponentially with a half-life closely corresponding to that of ^{56}Co. They inferred that about 0.1 $M_\odot$ of ^{56}Ni was ejected.

The optical spectra of SN 1987A in its photospheric phase (Figure 2.10) can be matched in detail by simple synthetic spectrum calculations based on resonance scattering (Jeffery and Branch, in preparation). The ultraviolet spectrum forms in the same way, but with very heavy line blanketing by many strong lines of Fe II, Co II, and Mg II (e.g., Lucy, 1987a, b, 1988), and the near infrared spectra show P Cygni profiles and line identifications that are generally consistent with the optical spectrum (Elias et al., 1988; Meikle et al., 1988). The nebular spectrum, both optical (Danziger and Guiffes, 1988) and infrared (Danziger and Bouchet, 1988; Rank et al., 1988) is developing more or less according to the expectations of Fransson and Chevalier (1987).

Several recent Type II supernovae have shown interesting individual peculiarities: for example, the premaximum spectra of SN 1983K in NGC 4699 contained features like those seen in Wolf–Rayet stars, rather than being continuous (Niemela, Ruiz, and Phillips, 1985), and spectra of SN 1984E in NGC 3169 showed narrow P Cygni features in the Balmer lines, apparently formed in a high mass-loss rate "superwind" from the progenitor, superimposed on the broader supernova features (Dopita et al., 1984).

2.6. Toward Quantitative Spectroscopy

In recent years the classic mysteries of line identifications and basic mechanisms of supernova spectrum formation have been resolved. The effort now will shift to deriving quantitative information about the composition of ejected matter as a function of velocity, which is needed to constrain the identification of the supernova progenitors and models of their explosion mechanisms.

Even before SN 1987A appeared, a number of attempts were made to extend and refine the basic picture of spectrum formation outlined in Section 2.4: e.g., Kolesov and Sobolev (1982) and Wagoner and his collaborators (Hershkowitz and Wagoner, 1987, and references therein) investigated the effects of electron scattering on the emergent continuum; Bychkov and Bychkova (1977), Feldt (1980), and Chugai (1982) solved the rate equations for the non-LTE hydrogen level populations; Sobolev and Chugai (1982) investigated the effects of electron scattering on line profiles; and Höflich, Wehrse, and Shaviv (1986) studied the effects of non-LTE level populations on the continuous spectrum.

The extensive spectral coverage of SN 1987A is now driving new developments in spectrum modeling. Synthetic spectra for the photospheric phase of SN 1987A have been calculated by Lucy (1987a, b, 1988) using a Monte Carlo spectrum synthesis code with approximate allowance for non-LTE effects, by Hauschildt et al. (1987) with an allowance for absorption as well as scattering in the lines, and by

Höflich (1987, 1988) with a full non-LTE solution for certain ions. Fransson and Chevalier (1988a) and Colgan and Hollenbach (1988) have made detailed predictions of the optical and infrared spectra in the nebular phase. Non-local thermodynamic equilibrium modeling of the spectra of Type Ia and Type Ib, with the effects of radioactivity γ-rays and down-scattered X-rays taken into account, also has begun (Pauldrach, Puls, Branch, Jeffery, and Kudritzki, in preparation). As this chapter is written none of the spectrum techniques have been refined and tested to the point of producing quantitatively reliable abundance estimates, but they soon will be.

As the spectrum modeling becomes more detailed, the limiting approximation eventually will become that of spherical symmetry. Spectropolarization observations of SN 1987A imply a significant shape asymmetry (Schwarz and Mundt, 1987; Cropper et al., 1988; Jeffery, 1988). Even worse, if the ejected matter clumps or breaks up into "fingers" (Müller and Arnett, 1982), future spectroscopic analysis will become complex, to put it mildly.

Acknowledgments

I would like to take this opportunity to salute all the observers whose hard work made Table 2.1 possible. I am especially grateful to Marshall McCall, Paul Rybski, Alan Uomoto, Craig Wheeler, and Beverley Wills for giving me the opportunity to work with them in the early 1980s on the analysis of their McDonald Observatory spectra of SN 1979C and SN 1981B, and to Alex Filippenko for providing numerous unpublished spectra of recent supernovae. My research on supernovae is supported by NSF grant AST 8620310.

References

Abdulwahab, M. and Morrison, P. 1978, Ap. J., 220, 1087.
Aller, L.H. and Ross, J. 1965, Pub. A.S.P., 77, 469.
Andrillat, Y. 1967, C.R. Acad. Sci. Paris, 265, 430.
Andrillat, Y. 1975, C.R. Acad. Sci. Paris, 280, 605.
Arp, H. 1961, Ap. J., 133, 883.
Arsenault, R. and D'Odoricio, S. 1988, Astr. Ap., 202, 55.
Assousa, G.E., Peterson, C.J., Rubin, V.C., and Ford, W.K., Jr. 1976, Pub. A.S.P., 88, 828.
Axelrod, T.A. 1980a, in Type I Supernovae, ed. J.C. Wheeler (Austin: University of Texas), p. 80.
Axelrod, T.A. 1980b, Thesis, University of California, Santa Cruz.
Baade, W. 1926, Astr. Nachr., 228, 359.
Baade, W. 1936, Pub. A.S.P., 48, 226.
Backhouse, T.W. 1888, M.N.R.A.S., 48, 108.
Barbieri, C., Bonoli, C., and Christiani, S. 1982, Astr. Ap., 114, 216.
Barbon, R., Ciatti, F., Iijima, T., and Rosino, L. 1989, Astr. Ap., 214, 131.
Barbon, R., Ciatti, F., and Rosino, L. 1973a, Astr. Ap., 29, 57.
Barbon, R., Ciatti, F., and Rosino, L. 1973b, Asiago Contr., No. 284..
Barbon, R., Ciatti, F., and Rosino, L. 1982a, Astr. Ap., 116, 35.
Barbon, R., Ciatti, F., Rosino, L., Ortolani, S., and Rafanelli, P. 1982b, Astr. Ap., 116, 43.
Barbon, R., Iijima, T., and Rosino, L. 1989, Astr. Ap., 220, 83.
Begelman, M.C., and Sarazin, C.L. 1986, Ap. J. (Letters), 302, L59.
Benvenuti, P., Sanz Fernandez de Cordoba, L., Wamsteker, W., Macchetto, F., Palumbo, G.G.C., and Panagia, N. 1982, An Atlas of UV Spectra of Supernovae (Paris: European Space Agency), ESA SP-1046.
Bertola, F. 1962, Asiago Contr., No. 128.
Bertola, F. 1963a, Asiago Contr., No. 135.

Bertola, F. 1963b, *Asiago Contr.*, No. 142.

Bertola, F. 1964, *Ann. d'Ap.*, **27**, 319.

Bertola, F. 1965, *Asiago Contr.*, No. 182.

Bertola, F. and Ciatti, F. 1971, *Asiago Contr.*, No. 253.

Bertola, F., Mammano, A., and Perinotto, M. 1965, *Asiago Contr.*, No. 174.

Blanco, V.M. and 11 others, 1987, *Ap. J.*, **320**, 589.

Bloch, M., Chalonge, D., and Dufay, J. 1964, *Ann. d'Ap.*, **27**, 315.

Bolton, C.T., Garrison, R.F., Salmon, D., and Geffken, N. 1974, *Pub. A.S.P.*, **86**, 439.

Borzov, G.G., Dibay, E.A., Esipov, V.F., and Pronik, V.I. 1969, *Sov. Astr.—AJ*, **13**, 423.

Bouchet, P., Stanga, R., Moneti, A., Le Bertre, T., Manfroid, J., Silvestro, G., and Slezak, E. 1987, in *ESO Workshop on Supernova 1987A*, ed. I.J. Danziger (Munich: European Southern Observatory), p. 159.

Branch, D. 1972, *Astr. Ap.*, **16**, 247.

Branch, D. 1980, in *Supernovae Spectra*, eds. R. Meyerott and G.H. Gillespie (New York: American Institute of Physics), p. 39.

Branch, D. 1984, in *Proceedings of the Eleventh Texas Symposium on Relativistic Astrophysics*, ed. D.S. Evans; *Ann. N.Y. Acad. Sci.*, **422**, 186.

Branch, D. 1985, in *Nucleosynthesis*, eds. W.D. Arnett and J.W. Truran (Chicago: University of Chicago), p. 261.

Branch, D. 1987a, *Ap. J. (Letters)*, **316**, L81.

Branch, D. 1987b, *Physica Scripta*, **35**, 787.

Branch, D. 1988, in *Proc. IAU Colloquim No. 108, Atmospheric Diagnostics of Stellar Evolution*, ed. K. Nomoto (Berlin: Springer-Verlag); Lecture Notes in Physics, Vol. 305, p. 281.

Branch, D., Buta, R., Falk, S.W., McCall, M.L., Sutherland, P.G., Uomoto, A., Wheeler, J.C., and Wills, B.J. 1982, *Ap. J. (Letters)*, **252**, L61.

Branch, D. and Doggett, J.B. 1985, *Astron. J.*, **90**, 2218.

Branch, D., Doggett, J.B., Nomoto, K., and Thielemann, F.-K. 1985, *Ap. J.*, **294**, 619.

Branch, D., Drucker, W., and Jeffery, D.J. 1988, *Ap. J. (Letters)*, **330**, L117.

Branch, D., Falk, S.W., McCall, M.L., Rybski, P., Uomoto, A.K., and Wills, B.J. 1981, *Ap. J.*, **224**, 780.

Branch, D. and Greenstein, J.L. 1971, *Ap. J.*, **167**, 89.

Branch, D., Lacy, C.H., McCall, M.L., Sutherland, P.G., Uomoto, A., Wheeler, J.C., and Wills, B.J. 1983, *Ap. J.*, **270**, 123.

Branch, D. and Nomoto, K. 1986, *Astr. Ap.*, **164**, L13.

Branch, D. and Patchett, B. 1973, *M.N.R.A.S.*, **161**, 71.

Branch, D. and Tull, R.G. 1979, *Astron. J.*, **84**, 1837.

Branch, D. and Venkatakrishna, K.L. 1986, *Ap. J. (Letters)*, **306**, L21.

Bychkov, V.S. and Bychkova, K.V. 1977, *Sov. Astr.—AJ*, **21**, 435.

Canal, R., Isern, J., and Lopez, R. 1988, *Ap. J. (Letters)*, **330**, L113.

Cannon, A.J. 1916, *Ann. Harvard Coll. Obs.*, **76**, 34.

Cassatella, A., Fransson, C., van Santwoort, J., Gry, C., Talavera, A., Wamsteker, W., and Panagia, N. 1987, *Astr. Ap.*, **177**, L29.

Castor, J.I. 1970, *M.N.R.A.S.*, **149**, 111.

Catchpole, R.M. and 19 others, 1987, *M.N.R.A.S.*, **229**, 15P.

Catchpole, R.M. and 15 others, 1988, *M.N.R.A.S.*, **231**, 75P.

Chalabaev, A.A. and Christiani, S. 1987, in *ESO workshop on SN 1987A*, ed. I.J. Danziger (Munich: European Southern Observatory), p. 655.

Chalonge, D. and Burnichon, M.L. 1968, *J. Observateurs*, **51**, 5.

Chincarini, G. and Margoni, R. 1964, *Asiago Contr.*, No. 149.

Chincarini, G. and Perinotto, M. 1968, *Asiago Contr.*, No. 205.

Chugai, N.N. 1982, *Sov. Astr.—Astron. J.*, **26**, 683.

Ciatti, F. 1973, *Astr. Ap.*, **22**, 465.

Ciatti, F. and Barbon, R. 1971, *Mem. Soc. Astr. Ital.*, **42**, 145.

Ciatti, F., Barbon, R., Cappellaro, E., and Rosino, L. 1988, *Astr. Ap.*, **202**, 15.

Ciatti, F. and Rosino, L. 1977, *Astr. Ap.*, **57**, 73.

Ciatti, F. and Rosino, L. 1978, *Astr. Ap. Suppl.*, **34**, 387.

Ciatti, F., Rosino, L., and Bertola, F. 1971, *Asiago Contr.*, No. 255.

Colgan, S.W.J. and Hollenbach, D.J. 1988, *Ap. J. (Letters)*, **329**, L25.

Colgate, S.A. and McKee, C. 1969, *Ap. J.*, **157**, 623.

Cropper, M., Bailey, J., McCowage, J., Cannon, R.D., Couch, W.J., Walsh, J.R., Straele, J.O., and Freeman, F. 1988, *M.N.R.A.S.*, **231**, 695.

Danziger, I.J. and Bouchet, P. 1988, preprint.

Danziger, I.J. Bouchet, P., Fosbury, R.A.E., Gouiffes, C., Lucy, L.B., Moorwood, A.F.M., Oliva, E., and Rufener, F. 1988, in *Supernova 1987A in the Large Magellanic Cloud*, eds. M. Kafatos and A.G. Michalitsianos (Cambridge: University of Cambridge), p. 37.

Danziger, I.J., Fosbury, R.A.E., Alloin, D., Cristiani, S., Dachs, J., Gouiffes, C., Jarvis, B., and Sahu, K.C. 1987, *Astr. Ap.*, **177**, L14.

Danziger, I.J. and Guiffes, C. 1988, preprint.

Dekker, M., D'Odorico, S., and Arsenault, R. 1988, *Astr. Ap.*, **189**, 353.

de Robertis, M.M. and Pinto P.A. 1985, *Ap. J. (Letters)*, **293**, L77.

de Vaucouleurs, G. and Corwin, H.G., Jr. 1985, *Ap. J.*, **295**, 287.

de Vaucouleurs, G., de Vaucouleurs, A., and Odewahn, J. 1981, *Pub. A.S.P.*, **93**, 181.

Dibai, E.A., Esipov, V.F., and Pronik, V.I. 1967, *Sov. Astr.—AJ*, **10**, 728.

Dopita, M.A., Evans, R. Cohen, M., and Schwartz, R.D. 1984, *Ap. J. (Letters)*, **287**, L69.

Dwek, E. and 11 others, 1983, *Ap. J.*, **274**, 168.

Elias, J.H., Frogel, J.A., Hackwell, J.A., and Persson, S.E. 1981, *Ap. J. (Letters)*, **251**, L13.

Elias, J.H., Gregory, B., Phillips, M.M., Williams, R.E., Graham, J.R., Meikle, W.P.S., Schwartz, R.D., and Wilking, B. 1988, *Ap. J. (Letters)*, **331**, L9.

Elias, J.H., Mathews, K., Neugebauer, G., and Persson, S.E. 1985, *Ap. J.*, **296**, 379.

Elliot, K.H., Blades, J.C., Zealey, W.J., Tritton, S. 1978, *Nature*, **275**, 198.

Ensman, L.M. and Woosley, S.E. 1988, *Ap. J.*, **333**, 754.

Erickson, E.F., Haas, M.R., Colgan, S.W.J., Lord, S.D., Burton, M.G., Wolf, J., Hollenbach, D.J., and Werner, M. 1988, *Ap. J.*, **330**, L39.

Fairall, A.P. 1975, *Mon. Not. Astr. Soc. South. Africa*, **34**, 94.

Feldt, A.N. 1980, Thesis, University of Oklahoma.

Filippenko, A.V. 1988a, in *Supernova 1987A in the Large Magellanic Cloud*, eds. M. Kafatos and A.G. Michalitsianos (Cambridge: University of Cambridge), p. 106.

Filippenko, A.V. 1988b, *Astron. J.*, **96**, 194.

Filippenko, A.V. 1989, *Astron. J.*, **97**, 726.

Filippenko, A.V., Porter, A.C., and Sargent, W.L.W. 1989, *Pub. A.S.P.* **100**, 1233.

Filippenko, A.V., Porter, A.C., Sargent, W.L.W., and Schneider, D.P. 1986, *Astron. J.*, **92**, 1341.

Filippenko, A.V. and Sargent W.L.W. 1985, *Nature*, **316**, 407.

Filippenko, A.V. and Sargent W.L.W. 1986, *Astron. J.*, **91**, 691.

Filippenko, A.V., Shields, J.C., and Sargent, W.L.W. 1988, *Pub. A.S.P.*, **100**, 1233.

Ford, W.K., Jr, and Rubin, V.C. 1968, *Pub. A.S.P.*, **80**, 466.

Fosbury, R.A.E., Danziger, I.J., Lucy, L.B., Gouiffes, C., and Cristiani, S. 1987, in *ESO Workshop on SN 1987A*, ed. I.J. Danziger (Munich: European Southern Observatory), p. 139.

Fransson, C. 1984, *Astr. Ap.*, **132**, 115.

Fransson, C. 1986, in *Highlights of Astronomy*, ed. J.P. Swings (Dordrecht: Reidel), p. 611.

Fransson, C. 1987, in *ESO Workshop on Supernova 1987A*, ed. I.J. Danziger (Munich: European Southern Observatory), p. 467.

Fransson, C., Benvenuti, P., Gordon, C., Hempe, K., Palumbo, G.G.C., Panagia, N., Reimers, D., and Wamsteker, W. 1984, *Astr. Ap.*, **132**, 1.

Fransson, C., Casatella, A., Gilmozzi, R., Kirshner, R.P., Panagia, N., Sonneborn, G., and Wamsteker, W. 1989, *Ap. J.*, **336**, 429.

Fransson, C. and Chevalier, R.A. 1987, *Ap. J. (Letters)*, **332**, L15.

Fransson, C. and Chevalier, R.A. 1989, *Ap. J.*, **343**, 323.

Frogel, J.A., Gregory, B., Kawara, K., Laney, D., Phillips, M.M., Terndrup, D., Vrba, F., and Whitford, A.E. 1987, *Ap. J. (Letters)*, **315**, L129.

Gaskell, C.M., Cappellaro, E., Dinerstein, H.L., Garnett, D., Harkness, R.P., and Wheeler, J.C. 1986, *Ap. J. (Letters)*, **306**, L77.

Gates, H.S., Zwicky, F., Bertola, F., Ciatti, F., and Rudnicki, F. 1967, *Astron. J.*, **72**, 912.

Gordon, C. 1972, *Astr. Ap.*, **20**, 79.

Graham, J.R. 1986, *M.N.R.A.S.*, **220**, 27P.

Graham, J.R. 1987, *Ap. J. (Letters)*, **318**, L47.

Graham, J.R., Meikle, W.P.S., Allen, D.A., Longmore, A.J., and Williams, P.M. 1986, *M.N.R.A.S.*, **218**, 93.

Graham, J.R., Smith, M.G., Longmore, A.J., and Williams, P.M. 1987, preprint.

Graham, W.R.M. and Duley, W.W. 1971, *Nature Phys. Sci.*, **232**, 43.

Greenstein, J.L. and Minkowski, R. 1973, *Ap. J.*, **182**, 225.

Greenstein, J.L. and Zwicky, F. 1962, *Pub. A.S.P.*, **74**, 35.

Hanuschik, R.W. and Dachs, J. 1987, in *ESO Workshop on Supernova 1987A*, ed. I.J. Danziger (Munich: European Southern Observatory), p. 153.

Harkness, R.P. 1985, in *Supernovae as Distance Indicators*, ed. N. Bartel (Berlin: Springer-Verlag); Lecture Notes in Physics, vol. 224, p. 183.

Harkness, R.P. 1986, in *Radiation Hydrodynamics in Stars and Compact Objects*, eds. D.H. Mihalas and K.-H. Winkler (Berlin: Springer-Verlag), p. 166.

Harkness, R.P. 1987, in *Relativistic Astrophysics*, ed. M.P. Ulmer (Singapore: World Scientific), p. 413.

Harkness, R.P. and Wheeler, J.C. 1989, this volume.

Harkness, R.P. and 10 others, 1987, *Ap. J.*, **317**, 355.

Harris, G.L.H., Hesser, J.E., Massey, P., Peterson, C.J., and Yamanaka, J.M. 1983, *Pub. A.S.P.*, **95**, 607.

Hauschildt, P., Spies, W., Wehrse, R., and Shaviv, G. 1987, in *ESO Workshop on Supernova 1987A*, ed. I.J. Danziger (Munich: European Southern Observatory), p. 433.

Hempe, K. 1986, *Astr. Ap.*, **158**, 329.

Henry, R.B.C. and Branch, D. 1987, *Pub. A.S.P.*, **99**, 112.

Hershkowitz, S. and Wagoner, R.V. 1987, *Ap. J.*, **322**, 967.

Hill, P.W. 1965, *M.N.R.A.S.*, **131**, 155.

Höflich, P. 1987, in *ESO Workshop on Supernova 1987A*, ed. I.J. Danziger (Munich: European Southern Observatory), p. 449.

Höflich, P. in *Atmospheric Diagnostics of Stellar Evolution*, ed. K. Nomoto (Berlin: Springer-Verlag); Lecture Notes in Physics, Vol. 305, p. 281.

Höflich, P., Wherse, R., and Shaviv, G. 1986, *Astr. Ap.*, **163**, 105.

Hubble, E. 1941, *Pub. A.S.P.*, **53**, 141.

Huffman, D.R. 1970, *Nature*, **225**, 833.

Humason, M.L. 1936, *Pub. A.S.P.*, **48**, 110.

Humason, M.L. and Minkowski, R. 1941, *Pub. A.S.P.*, **53**, 131.

Iye, M., Kodaira, K., Kikuchi, S., and Ohtani, H. 1975, *Pub. Astr. Soc. Japan*, **27**, 571.

Jeffery, D. J. 1987, *Nature*, **329**, 419.

Johnson, W.A. 1936, *Harvard. Bull.*, No. 902.

Kikuchi, S. 1971, *Pub. Astr. Soc. Japan*, **23**, 593.

Kirshner, R.P. 1985, in *Supernovae as Distance Indicators*, ed. N. Bartel (Berlin: Springer-Verlag); Lecture Notes in Physics, Vol. 224, p. 171.

Kirshner, R.P., Arp, H.C., and Dunlap, J.R. 1976, *Ap. J.*, **207**, 44.

Kirshner, R.P. and Kwan, J. 1974, *Ap. J.*, **193**, 27.

Kirshner, R.P. and Kwan, J. 1975, *Ap. J.*, **197**, 415.

Kirshner, R.P. and Oke, J.B. 1975, *Ap. J.*, **200**, 574.

Kirshner, R.P., Oke, J.B., Penston, M.V., and Searle, L. 1973a, *Ap. J.*, **185**, 303.

Kirshner, R.P., Sonneborn, G., Crenshaw, D.M., and Nassiopoulos, G.E. 1987, *Ap. J.*, **320**, 602.

Kirshner, R.P., Willner, S.P., Becklin, E.E., Neugebauer, G., and Oke, J.B. 1973b, *Ap. J.*, **180**, L97.

Kolesov, A.K. and Sobolev, V.V. 1982, *Sov. Astr.—AJ*, **26**, 255.

Larson, H.P., Drapatz, S., Mumma, M.J., and Weaver, H.A. 1987, in *ESO Workshop on SN 1987A*, ed. I.J. Danziger (Munich: European Southern Observatory), p. 147.

Lucy, L.B. 1987a, *Astr. Ap.*, **182**, L31.

Lucy, L.B. 1987b, in *ESO Workshop on SN 1987A*, ed. I.J. Danziger (Munich: European Southern Observatory), p. 417.

Lucy, L.B. 1988, in *Supernova 1987A in the Large Magellanic Cloud*, eds. M. Kafatos and A.G. Michalitsianos (Cambridge: University of Cambridge), p. 323.

Manning, P.G. 1970, *Nature*, **228**, 844.

Mayall, N.U. 1948, *Pub. A.S.P.*, **60**, 266.

McCall, M.L. Reid, N., Bessell, M.S., and Wickramasinghe, D. 1984, *M.N.R.A.S.*, **210**, 839.

McCarthy, M.F. 1974, in *Supernovae and Supernova Remnants*, ed. C.B. Cosmovici (Dordrecht: Reidel), p. 135.

McCarthy, M.F. and Araya, G. 1973, *Ric. Astr. Spec. Vaticana*, **8**, 439.

McCray, R. and Li, H.-W. 1988, in *Structure and Evolution of Galaxies*, ed. F.L. Zhi (Singapore, World Scientific), in press.

McLaughlin, D.B. 1937, *Publ. Obs. Univ. Mich.*, **6**, 107.

McLaughlin, D.B. 1963, *Pub. A.S.P.*, **75**, 133.

Meikle, W.P.S., Allen, D.A., Spyromilio, J., and Varani, G.-F. 1989, *M.N.R.A.S.*, **238**, 193.

Menzies, J.W. and 15 others, 1987, *M.N.R.A.S.*, **227**, 39P.

Meyerott, R.E. 1978, *Ap. J.*, **221**, 975.

Meyerott, R.E. 1980, *Ap. J.*, **239**, 257.

Mihalas, D. 1980, *Ap. J.*, **237**, 574.

Mihalas, D., Kunasz, P.B., and Hummer, D.G. 1976, *Ap. J.*, **206**, 515.

Minkowski, R. 1939, *Ap. J.*, **89**, 156.

Minkowski, R. 1940, *Pub. A.S.P.*, **52**, 206.

Minkowski, R. 1941, *Pub. A.S.P.*, **53**, 224.

Minkowski, R. 1963, *Pub. A.S.P.*, **75**, 505.

Morrison, P. and Sartori, L. 1969, *Ap. J.*, **158**, 541.

Müller, E. and Arnett, W.D. 1982, *Ap. J. (Letters)*, **261**, L109.

Mustel, E.R. 1971, *Sov. Astr.—AJ*, **15**, 1.

Niemela, V.S., Ruiz, M.T., and Phillips, M.M. 1985, *Ap. J.*, **289**, 52.

Nomoto, K., Thielemann, F.-K., and Yokoi, K. 1984, *Ap. J.*, **286**, 644.

Oke, J.B. and Searle, L. 1974, *Ann. Rev. Astr. Ap.*, **12**, 315.

Oliva, E. 1987, *Ap. J. (Letters)*, **321**, L45.

Olson, G.L. 1982, *Ap. J.*, **255**, 267.

Panagia, N. and Laidler, V.G. 1988, in *Supernova Shells and Their Birth Events*, ed. W. Kundt (Berlin: Springer Verlag); Lecture Notes in Physics, Vol. **316**, p. 187.

Panagia, N. and 26 others, 1980, *M.N.R.A.S.*, **192**, 861.

Panagia, N. and 29 others, 1985, preprint.

Patchett, B. and Branch, D. 1972, *M.N.R.A.S.*, **158**, 375.

Patchett, B. and Wood, R. 1976, *M.N.R.A.S.*, **175**, 595.

Payne-Gaposchkin, C. 1936a, *Ap. J.*, **83**, 173.

Payne-Gaposchkin, C. 1936b, *Ap. J.*, **83**, 243.

Payne-Gaposchkin, C. and Whipple, F.L. 1940, *Proc. Nat. Acad. Sci.*, **26**, 264.

Pearce, G., Patchett, B., Allington-Smith, J., and Parry, I. 1988, *Ap. Sp. Sci.*, **150**, 267.

Pearce, G. and Purvis, A. 1986, *Ap. Sp. Sci.*, **125**, 175.

Pennypacker, C.R. and 23 others, 1989, *Astron. J.*, **97**, 186.

Penston, M.V. and Blades, J.C. 1980, *M.N.R.A.S.*, **190**, 51P.

Phillips, M.M. 1988, in *Supernova 1987A in the Large Magellanic Cloud*, eds. M. Kafatos and A.G. Michalitsianos (Cambridge: University of Cambridge), p. 16.

Phillips, M.M., Heathcote, S.R., Hamuy, M., and Navarette, M. 1988, *Astron. J.*, **95**, 1087.

Phillips, M.M. and 27 others, 1987, *Pub. A.S.P.*, **99**, 592.

Popper, D.M. 1937, *Pub. A.S.P.*, **49**, 283.

Porter, A.C. and Filippenko, A.V. 1987, *Astron. J.*, **93**, 1372.

Prabhu, T.P. 1981, *Bull. Astr. Soc. India*, **9**, 60.

Prabhu, T.P. 1983, *Bull. Astr. Soc. India*, **11**, 240.

Pronik, V.I., Chuvaev, K.K., and Chugay, N.N. 1977, *Sov. Astr.—AJ*, **20**, 666.

Pskovskii, Yu. P. 1969, *Sov. Astr.—AJ*, **12**, 750.

Purvis, A., Pearce, G., and Reid, I.N. 1987, *Ap. Sp. Sci.*, **134**, 329.

Rank, D.M., Pinto, P.A., Woosley, S.E. Bregman, J.D., Witteborn, F.C., Axelrod, T.A., and Cohen, M. 1988, *Nature*, **331**, 505.

Reid, N., Mould, J., Wegner, G., Picard, A., and Coker, R. 1988, *Pub. A.S.P.*, **101**, 95.

Richter, T. and Sadler, E.M. 1983, *Astr. Ap.*, **128**, L3.

Rosino, L. 1963, *Coelum*, **31**, 52.

Rosino, L. and Bertola, F. 1961, *Asiago Contr.*, No. 116.

Rousseau, C., Prevot, L., and Bardin, C. 1963, *C.R. Acad. Sci. Paris*, **256**, 5284.

Rubin, V.C. and Ford, W.K., Jr. 1967, *Pub. A.S.P.*, **79**, 322.

Rudnicki, K. 1966, *Astr. Nachr.*, **289**, 247.

Rudnicki, K. and Zwicky, F. 1967, *Astron. J.*, **72**, 407.

Rupen, M.P., van Gorkom, J.H., Knapp, G.R., Gunn, J.E., and Schneider, D.P. 1987, *Astron. J.*, **94**, 61.

Rybicki, G.R. and Hummer, D.G. 1978, *Ap. J.*, **219**, 654.

Schild, H. 1985, *Astr. Ap.*, **142**, 401.

Schneider, D.P., Mould, J.R., Porter, A.C., Schmidt, M., Bothun, G.D., and Gunn, J.E. 1988, *Pub. A.S.P.*, **99**, 1167.

Schwarz, H.E. and Mundt, R. 1987, *Astr. Ap.*, **177**, L4.

Seddon, H. 1969, *Nature*, **222**, 757.

Sherman, O. T. 1888, *M.N.R.A.S.*, **47**, 14.

Sistero, R.F. and Castore de Sistero, M.E. 1972, *Ap. J.*, **176**, L123.

Smith, H.A. 1981, *Astron. J.*, **86**, 998.

Sobolev, A.M. and Chugai, N.N. 1982, *Sov. Astr. Lett.—AJ*, **8**, 327.

Sobolev, V.V. 1960, *Moving Envelopes of Stars* (Cambridge, MA: Harvard University Press).

Terlevich, R. and Melnick, J. 1988, *Nature*, **333**, 239.

Uomoto, A. and Kirshner, R.P. 1985, *Astr. Ap.*, **149**, L7.

Uomoto, A. and Kirshner, R.P. 1986, *Ap. J.*, **308**, 685.

Vorontsov-Velyaminov, B.A. and Savel'eva, M.V. 1961, *Sov. Astr.—AJ*, **5**, 416.

Wamsteker, W., Panagia, N., Barylak, M., Cassatella, A., Clavel, J., Gilmozzi, R., Gry, C., Lloyd, C., va Santvoort, J., and Talavera, A. 1987, *Astr. Ap.*, **177**, L21.

Wegner, G. and McMahan, R.K. 1987, *Astron. J.*, **93**, 287.

Wellmann, P. 1954, *Zeit. Ap.*, **35**, 205.

Wheeler, J.C. 1985, in *Supernovae as Distance Indicators*, ed. N. Bartel (Berlin: Springer-Verlag); Lecture Notes in Physics, Vol. 224, p. 34.

Wheeler, J.C. and Harkness, R.P. 1986, in *Galaxy Distances and Deviations from Universal Expansion*, eds. B.F. Madore and R.B. Tully (Dordrecht: Reidel), p. 45.

Wheeler, J.C., Harkness, R.P., and Barkat, Z. 1988, in *Proc. IAU Coll. No. 108, Atmospheric Diagnostics of Stellar Evolution*, ed. K. Nomoto (Berlin: Springer-Verlag); Lecture Notes in Physics, Vol. 305, p. 305.

Wheeler, J.C., Harkness, R.P., Barkat, Z., and Schwartz, D. 1986, *Pub. A.S.P.*, **98**, 1018.

Wheeler, J.C., Harkness, R.P., Barker, E.S., Cochran, A.L., and Wills, B.J. 1987a, *Ap. J. (Letters)*, **313**, L69.

Wheeler, J.C., Harkness, R.P., and Cappellaro, E., 1987b, in *Relativistic Astrophysics*, ed. M.P. Ulmer (Singapore: World Scientific), p. 402.

Wheeler, J.C. and Levreault, R. 1985, *Ap. J. (Letters)*, **294**, L17.

Whipple, F.L. and Payne-Gaposchkin, C. 1941, *Proc. Amer. Phil. Soc.*, **84**, 1.

Whitelock, P.A. and 20 others, 1988, *M.N.R.A.S.*, **234**, 5P.

Williams, R.E. 1987, *Ap. J. (Letters)*, **320**, L117.

Wood, R. and Andrews, P.J. 1974, *M.N.R.A.S.*, **167**, 13.

Woosley, S.E., Axelrod, T.S., and Weaver, T.A. 1984, in *Stellar Nucleosynthesis*, eds. C. Chiosi and A. Renzini (Dordrecht: Reidel), p. 263.

Wyckoff, S. and Wehinger, P.A. 1977, *Ap. Sp. Sci.*, **48**, 421.

Zwicky, F. 1940, *Rev. Mod. Phys.*, **12**, 66.

Zwicky, F. 1958, in *Handbuch der Physik* (Berlin: Springer-Verlag), Vol. LI, 774.

Zwicky, F. 1964a, *Ann. d'Ap.*, **27**, 300.

Zwicky, F. 1964b, *Ap. J.*, **139**, 514.

Zwicky, F. 1965, in *Stars and Stellar Systems*, Vol. VIII (Stellar Structure), eds. L.H. Aller and D.B. McLaughlin (Chicago: University of Chicago Press), p. 367.

Zwicky, F. and Barbon, R. 1967, *Astron. J.*, **72**, 1366.

Zwicky, F. and Karpowicz, M. 1964, *Astron. J.*, **69**, 759.

Zwicky, F. and Karpowicz, M. 1965, *Astron. J.*, **70**, 564.

3. Supernova Light Curves

ROBERT P. KIRSHNER

3.1. Introduction

The supernova phenomenon was discovered because stellar explosions are prodigious sources of optical light. Although we know today that the energy budget for the core collapse of a massive star allocates 99% of the energy released to neutrinos and 1% to kinetic energy, the 10^{-4} part that becomes light makes supernovae the most luminous stars in the universe, and provides essential clues to the nature of supernova explosions. Even the pocket change of millionaires is significant: a typical supernova emits a substantial fraction of the luminosity of a typical galaxy, and the brightest have been detected a third of the way across the observable universe. This review summarizes recent work that uses the supernova light curve, the plot of luminosity versus time, to help derive a picture of supernova explosions and to explore the dimensions of the universe.

The light output of supernovae provides valuable clues to the underlying stellar events. Near maximum light, diffusion through the exploded star's expanding debris reflects the size, mass, and composition of the star, as well as the energy source. Far past maximum light, the decline of a supernova gives strong clues to the chemical composition of the progenitor's interior, evidence concerning the prompt synthesis of radioactive elements in the supernova explosion, and hints or limits to the emergence of a pulsar.

Despite the fact that supernovae are unscheduled and brief events, while observatories cast their schedules months ahead, the effort required to observe them is worthwhile not only to understand supernova explosions, but also to establish the cosmological distance scale. Though Cepheid variable stars are more predictable, supernovae are 10^6 times brighter! In this review, progress toward the cosmological use of supernovae will be described. Historical supernovae at known distances in our own galaxy provide some calibration of the distances to well-observed modern cases. Recent work on the distances to Tycho's supernova remnant (SNR) and to SN 1006 through study of their remnants will be reviewed. Observations of SN Ia in galaxies of well-established distances, and a deep understanding of the origin of SN Ia events may allow the light curves of those objects to be used as a powerful tool in erecting the cosmological scale. The basis for this belief is briefly examined here.

An independent path to extragalactic distances comes from the Expanding Photosphere Method (EPM) for Type II supernovae, which allows the distance to a SN II to be determined without reference to other rungs in the extragalactic distance ladder. Improvements in that technique based on atmosphere models for SN 1987A are summarized here.

Two general comments on supernova light curves are in order. First, the discussion of supernova light curves for supernovae of various types assumes that each supernova has been properly classified by the use of spectra. Supernova types are spectroscopic types (Zwicky, 1968; Oke and Searle, 1974; Harkness and Wheeler, in this volume). Assigning supernova types by their light curves or from circumstantial evidence such as association with H II regions or absolute magnitude leads to circular arguments concerning the homogeneity of light curves, confuses SN Ib with SN II, and renders the cosmological use of supernova magnitudes a treacherous matter. Supernova types are spectroscopic types.

Second, we often have to choose between analyzing large sets of fragmentary or heterogeneous data, and the smaller sets of well-observed cases. Each type of data has its uses: Large and qualitative differences among supernovae appear even in the heterogeneous data, while small and quantitative differences can be seen only in the best observations. For the present, each set of data needs to be used with an understanding of its limitations. In the future, we can hope to obtain a large, accurate, and well-calibrated set of observations so that we do not have to choose between the extensive and the reliable.

3.2. Light Curves of Type IA Supernovae

The earliest supernovae detected by Zwicky (1938) in his initial supernova search using the Palomar 18-inch Schmidt formed a remarkably homogeneous class. It was not until SN 1940B that the first example of a different type was clearly seen by Minkowski (1941), requiring the first refinement of supernova classification into the original type (SN I) and the new type (SN II). Recent work has shown that the SN I include two distinct types: the original SN Ia, and the newly described SN Ib (Uomoto and Kirshner, 1985; Harkness *et al.*, 1987; Porter and Filippenko, 1987). The SN Ia prototypes, SN 1937C in IC 4182, SN 1972E in NGC 5253, and SN 1981B in NGC 4536 have similar spectra that show no strong lines of hydrogen, and light curves which are reported by Minkowski (1964), Aardeberg and de Groot (1973), and by Buta and Turner (1983). More extensive compilations include the composite light curve for 38 SN Ia constructed by Barbon, Ciatti, and Rosino (1973), and the work of Barbon, Ciatti, and Rosino (1979) and of Cadonau, Sandage, and Tammann (1985). Useful analyses of the light curves have been carried out by Leibundgut (1988) and by Doggett and Branch (1985). Figure 3.1 illustrates these results for SN Ia.

Some limitations of these data need to be noted. First, for most supernovae, discovery comes after maximum light, so the light curves have been adjusted in time to minimize the dispersion. This is reliable if there are good observations in the rapidly descending phase that extends for the first 20 days past maximum, during which the blue light output decreases by nearly 2 mag. A few SN Ia have been measured well before maximum light, with the earliest data extending about 15 days earlier and over 3 mag below the maximum (Pskovskii, 1977). The data in Figure 3.1

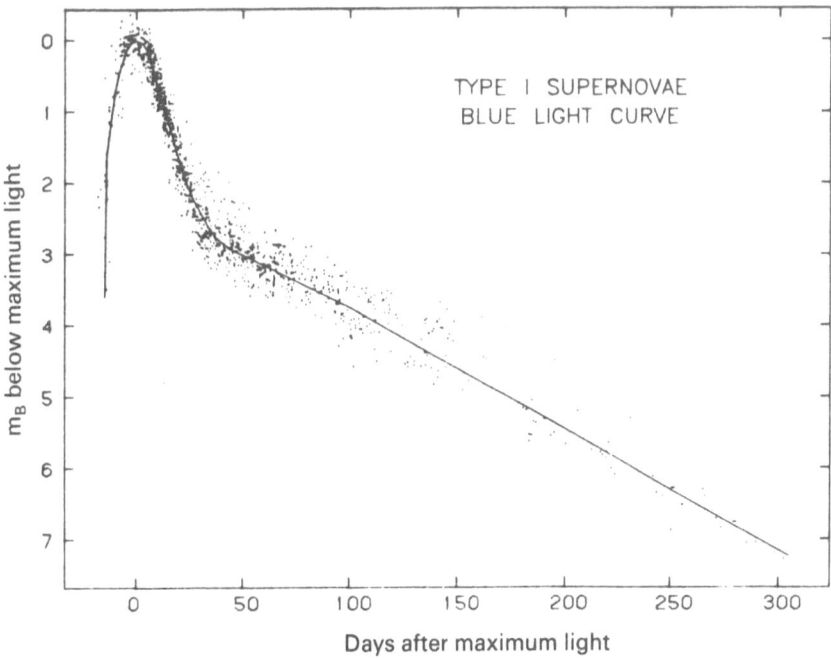

Figure 3.1. Composite blue light curve for SN I, from Dogget and Branch (1985), based on the compilation of Barbon, Ciatti, and Rosino (1973).

emphasize photographic measurements and are confined to the B bandpass. Photo-electric data have intrinsically higher accuracy, and a recent study by Younger and van den Bergh (1985) of photoelectric data only provides an assessment of rates of decline and of color evolution for supernovae. They conclude that the mean rate of decline for SN I at age 20 days is 0.065 ± 0.007 mag day^{-1}, and that the color evolution shows an initial reddening of B–V at a rate of 0.04–0.05 mag day^{-1}. Comparison to spectrophotometry (Kirshner *et al.*, 1973) shows that the initial reddening corresponds to cooling in an underlying continuum (Branch and Patchett, 1974) and that the B and V bands carry most of the energy emitted by SN Ia during the first month. Observations in the infrared (Elias *et al.*, 1981) establish that SN Ia have characteristic light curves when observed at those wavelengths.

While photographic data suffer from the intrinsic nonlinearity of photographic emulsions, the photoelectric data are not without their difficulties since the entrance aperture of the photometer unavoidably includes background light from the parent galaxy of the supernova. While the contribution of this diffuse light is small compared to the supernova at maximum, its relative contribution grows as the object fades. For example, the galaxy light through a 10 arc second aperture on a galaxy of surface brightness 20 mag arcsec^{-2} is roughly 15 mag, and can make a significant difference to the magnitude and color measured for a supernova unless the "sky" measurements are made at regions of comparable galaxy background.

Luckily, the solution to this problem is in hand since CCD detectors provide linear images which contain the data needed for accurate supernova magnitudes and for precise galaxy subtraction, although care is needed to assure that the CCD observations are transformed to the standard systems. As described by Filippenko *et al.* (1986) and by Schlegel and Kirshner (1989), an image of the parent galaxy taken after the supernova has faded below the detection threshold can be used to subtract the galaxy background from the frames with the supernova present. Other stars in the image, and the galaxy itself provide useful flux calibration, so that observations which are not obtained under optimum sky conditions still can provide useful flux data. The only drawback to this scheme is that it entails waiting a year or more before the reference image can be obtained.

There have been several attempts to examine the variation among individual SN Ia light curves more closely. For example, Rust (1974) sought to find evidence for time dilation due to the redshift of supernovae on the assumption that SN I were intrinsically identical. Since the effect is only of order the redshift, which for most cases is less than 0.03, only very good data, or data at higher redshift would reveal this. More empirically, Barbon, Ciatti, and Rosino (1973) divided their sample of SN I light curves into two classes, the fast and the slow, based on the rate of decline after maximum. Pskovskii (1977) has made even finer distinctions based on the rate of decline. In this area of small quantitative differences, large but heterogenous samples may prove deceptive. Since supernova data is likely to have a dispersion due to reddening, unknown time of maximum, and galaxy background in addition to any intrinsic dispersion, the correlations among rate of decline and other properties should be viewed with caution until they are firmly established through extensive sets of high quality observations. On the other hand, there can be little question that at least one well-observed SN Ia, SN 1986G, declined much more rapidly than could be accounted for by any combination of errors (Phillips *et al.*, 1987).

Careful examination of the best data sets for SN Ia reveals a small "bump" in the light curve at about 40 days which is masked in Figure 3.1 by the dispersion among the low-accuracy measurements. The bump, of amplitude 0.1 mag, and duration 10 days, appears at a time when the spectrophotometry shows the spectrum is shifting from lines superposed on a continuum to the superposition of many blended lines of [Fe II] and [Fe III] (Kirshner and Oke, 1975). The explanation for this small bump may lie with a change in the continuum opacity or it may be more closely connected to the way that photometric bands respond to a changing spectrum.

At times past 50 days, the most conspicuous feature of SN Ia light curves is their linear decline. In this long decline, the spectrum changes very little (Kirshner and Oke, 1975). While some variations in the decline rate were noted by Baade (quoted in Zwicky, 1968), the modern evidence seems to favor a decline at a rate of 0.01516 ± 0.0024 mag day^{-1} based on data between 200 and 400 days (Barbon *et al.* (1984). The longest SN Ia light curve is that of SN 1972E reported by Kirshner and Oke (1975) based on spectrophotometry for 660 days during which the supernova declined from 8 to 21 in flux at the V band wavelength. The exponential decline of 0.013 mag day^{-1} is consistent with the observations obtained at the B band for other objects. A bolometric flux constructed from spectrophotometry from 3200 to 10,000 Å is reported by Kirshner (1980) for the same time interval.

The long exponential decline seen in the B band for many supernovae is also seen in the bolometric data for SN 1972E. This suggests that radioactivity may play an important role in the energetics of SN Ia. Realistic suggestions have centered on ^{56}Ni, which decays to ^{56}Co, which then decays to stable ^{56}Fe with an e-folding time of 111 days. The difference between the energy supply and the decay rate as observed (0.015 mag day^{-1} corresponds to an exponential time of 77 days) is attributed to the mechanism of energy deposition from the radioactive decay in the form of γ-rays and positrons (Colgate and McKee, 1969).

Models for the late-time behavior of SN Ia by Axelrod (1980) show that the declining temperature in the expanding debris should eventually lead to a situation where most of the energy escapes from the supernova shell through fine structure lines of neutral species. This "infrared catastrophe" (really an infrared bonanza) is predicted to take place at about the time of the last observation of SN 1972E. It will be possible to test this prediction by sustained observation of a bright SN Ia with HST. A supernova which is at apparent magnitude 12 at maximum will reach 26 mag in 2 years, and observations beyond that level are feasible (though not trivial) with HST.

If SN Ia are sufficiently good standard candles, then the detection and measurement of supernovae at high redshift may be a useful way to determine the global geometry in the universe. The difference in apparent magnitude between $\Omega = 1$ and $\Omega = 0$ at $z = 0.3$ is about 0.13 mag. Since the dispersion in absolute magnitudes for SN Ia may be just a little larger, even a few measurements of very distant SN Ia could help constrain this elusive parameter of cosmology. Since the apparent magnitude is expected to be about 22 mag, this is a challenging technical problem, but a very promising beginning has been made by surveying distant clusters by Hansen *et al.* (1989). They have successfully detected a supernova in a cluster at $z = 0.28$, and further detections are likely. This work should provide direct confirmation of time dilation, test the calculations of K-corrections (Leibundgut, 1988), and provide an estimate of the supernova rate at earlier epochs. Spectroscopic data are required to show that the objects are in fact the same as the local SN Ia. While these data are not easy to obtain, they are within the limits of well-equipped 4-meter class telescopes.

3.3. The Light Curves of SN Ib

The recently recognized class of Type Ib supernovae has been adequately described spectroscopically (Wheeler and Levreault, 1985; Filippenko and Sargent, 1985, 1986; Uomoto and Kirshner, 1985; Gaskell *et al.*, 1986). They resemble SN Ia at maximum, since they have no hydrogen lines, but the strongest line in the SN Ia spectrum, a deep absorption at about 6150 Å is missing in these supernovae. This spectroscopic peculiarity was noted for SN 1962L in NGC 1073 (Bertola, 1964) and SN 1964L in NGC 3938 (Bertola, Mammano, and Perinotto, 1965) which were designated SN I peculiar. But recent work on SN 1982R in NGC 1187 (Graham, 1986), SN 1983I in NGC 4051 (Elias *et al.*, 1985; Tsvetkov, 1985), SN 1983N in NGC 5236 (Panagia *et al.*, 1989), SN 1983V in NGC 1365 (Wheeler *et al.*, 1987), SN 1984L in NGC 991 (Schlegel and Kirshner, 1989), and SN 1985F in NGC 4618 (Filippenko and Sargent, 1985, 1986; Filippenko *et al.*, 1986) makes it clear that the

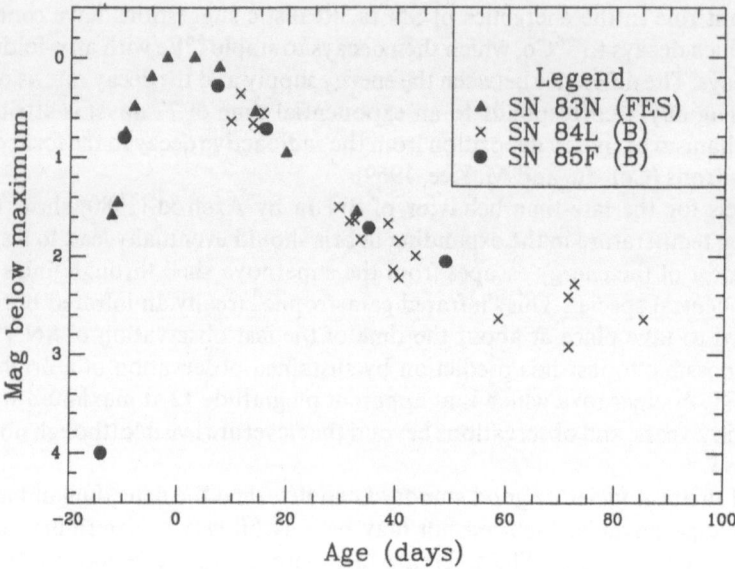

Figure 3.2. The early light curves for SN Ib.

differences between SN Ia and SN Ib are not confined to a single spectroscopic detail.

So far, SN Ib have been found only in spiral galaxies near regions of recent star formation (Porter and Filippenko, 1987; Panagia and Laidler, 1988; van den Bergh, 1988). In contrast, the SN Ia are found in all types of galaxies, including ellipticals with no obvious sign of recent star formation. Unlike SN Ia, the SN Ib show radio emission (Sramek, Panagia, and Weiler, 1984; Panagia, Sramek, and Weiler 1986) which may have its origin in a shock running through circumstellar matter (Chevalier, 1984). Infrared photometry (Elias *et al.*, 1985) shows that SN Ib can be distinguished from SN Ia by their light curves at 1.2 μ. In addition, the decline after maximum in the infrared at J, H, and K bands is slower in SN Ib than in SN Ia. Here we examine optical photometry of SN Ib, emphasizing the late-time results of Filippenko *et al.* (1986) for SN 1985F and Schlegel and Kirshner (1989) for 1984L.

Early light curves for SN 1983N (Panagia *et al.*, 1989), SN 1984L (Buta, 1984; Tsvetkov, 1987), and SN 1985F (Tsvetkov, 1986) are shown in Figure 3.2. The shape of the maximum appears to be well defined, with enough points before maximum for SN 1985F and especially SN 1983N to establish the time of maximum for SN 1984L from comparison with those two, as shown. The rate of decline observed in the first 40 days is approximately 0.057 mag day^{-1}, which is similar to the rate reported above for SN Ia. However, the subsequent behavior of SN 1984L and SN 1985F differs markedly from that of SN Ia. Figure 3.3 shows a composite (and somewhat schematic) light curve for SN Ib. Although the SN Ib show good evidence for an exponential decline, the rate of decline is much less for the SN Ib than for SN Ia. Where SN Ia typically decline by 0.015 mag day^{-1}, the observed rate here for SN Ib is about 0.010 mag day^{-1}. This is remarkably close to the value of 0.0095

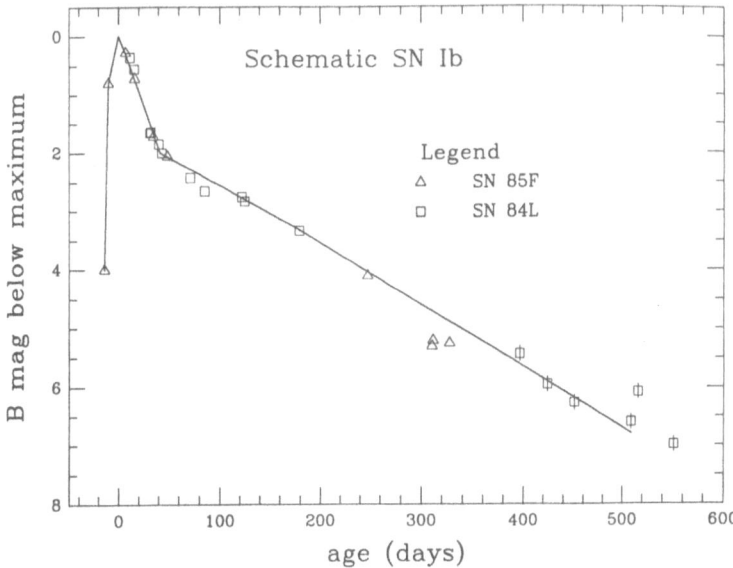

Figure 3.3. A composite light curve for the late phases of SN Ib based on SN 1985F and SN 1984L.

mag day^{-1} observed for the Type II SN 1987A (Catchpole *et al.*, 1988) where there is no question that the deposition of energy from ^{56}Co dominates the energetics at late times. Put more simply, it is very close to the energy input rate from ^{56}Co which has an e-folding time of 111 days that corresponds to 0.0098 mag day^{-1}. It seems plausible to suppose, along with Weaver and Woosley (1980), that radioactive decay from the Ni–Co–Fe chain powers the late-time evolution of all supernovae, including the SN Ib which were not recognized in 1980. The required mass of radioactive cobalt is about 0.1 $M_\odot$.

The light curves for SN Ib have been modeled by Ensman and Woosley (1988). They find that the mass of the star, at the time of explosion, must be in the range of 4–8 $M_\odot$ in order to reproduce the shape of the light curve. This could be the core of a 20 $M_\odot$ star which has undergone extensive mass loss, as suggested by the match between late-time spectra and models by Fransson (1988) and by Axelrod (1988). An alternative view, that SN Ib might have their origins in white dwarfs has been proposed (Branch and Nomoto, 1986; Branch, 1988; Nomoto, Shigeyama, and Hashimoto, 1988).

3.4. The Cosmological Uses of SN Ia

The utility of SN I lies in the possibility that they are good standard candles for the Hubble diagram. This could lead to estimates of the Hubble constant if the absolute magnitude of SN I can be accurately established, and estimates of Ω if the Hubble diagram for supernovae can be constructed out to sufficiently large redshift so that the cosmological effect is at least as large as the variance among SN I.

Separating SN Ib from SN Ia helps with this process, since SN Ib are fainter at maximum light by about 1.5 mag (Uomoto and Kirshner, 1986). Confining attention to the SN Ia, Cadonau, Sandage, and Tammann (1985) and Leibundgut (1988) have examined the evidence concerning the intrinsic variation in absolute magnitude of SN Ia. They find the scatter about the mean in the Hubble relation in a suitably selected sample is < 0.3 mag, based principally on SN Ia in the Coma cluster.

The absolute magnitude is connected to the Hubble constant by

$$M_B = -18.19 + 5 \log(H/100).$$

Since there is consensus on a detailed model for SN Ia based on the deflagration of a carbon–oxygen white dwarf (Woosley and Weaver, 1986), there is some reason to believe that SN Ia are events with a well-defined energy source, and that the small dispersion in absolute magnitudes reflects this underlying physical mechanism. This view has been taken one step further by Arnett, Branch, and Wheeler (1985) who argue that the mass of radioactive cobalt produced by the destruction of a white dwarf can be calculated, and further that models for the emission from SN Ia (Arnett, 1982) allow the observable blue magnitude to be inferred. Their value of $M_B = -19.5$ requires, if taken at face value, a Hubble constant of 58 km s^{-1} Mpc^{-1}, and one in the range 40–70 km s^{-1} Mpc^{-1}. It is interesting to note that if this argument is taken seriously, it conflicts with some seriously proposed values for H_0 (Aaronson et al., 1980; de Vaucouleurs, 1983).

Another way to establish the absolute magnitude of SN Ia is to calibrate the distance to some events by other means. In the case of historical remnants widely attributed to SN I, this means determining the distance to the remnants of SN 185, SN 1006, SN 1572, and SN 1604. Of course, no spectra were available for these events, so the classification must be tentative, since it is based on light curves (Clark and Stephenson, 1977). As Doggett and Branch (1985) have shown, the separation of SN I from SN II-L on the basis of their visual light curves is not easy. There has been some progress on the distance to SN 1006 and to SN 1572. Recent studies (Kirshner, Winkler, and Chevalier, 1987) of the $H\alpha$ emission from these remnants and the proper motion of the filaments (Long, Blair, and van den Bergh, 1988) provide distances of 1.4–2.1 kpc for SN 1006 and 2.0–2.8 kpc for SN 1572. A systematic comparison of data for SN I by Strom (1988) suggests that for three of these four $M_v = -19.7 \pm 0.5$, while SN 1572 has $M_v = -17.6 \pm 0.5$. Strom suggests that Tycho's event may have been an SN Ib, although that seems inconsistent with the picture for SN Ib as the result of a massive star with mass loss sketched above. Strom's value for M corresponds to H_0 near 50 km s^{-1} Mpc^{-1}. In any case, the construction of an accurate distance scale for historical supernova remains a valuable way to help calibrate the absolute magnitude of supernovae. Although we cannot expect the photometry of the eleventh century to improve, we can hope to sharpen our estimates of the distance.

3.5. Light Curves for SN II

Barbon, Ciatti, and Rosino (1979) have demonstrated that the light curves of SN II can be usefully divided into two types: the SN II-P (for Plateau) and the Type

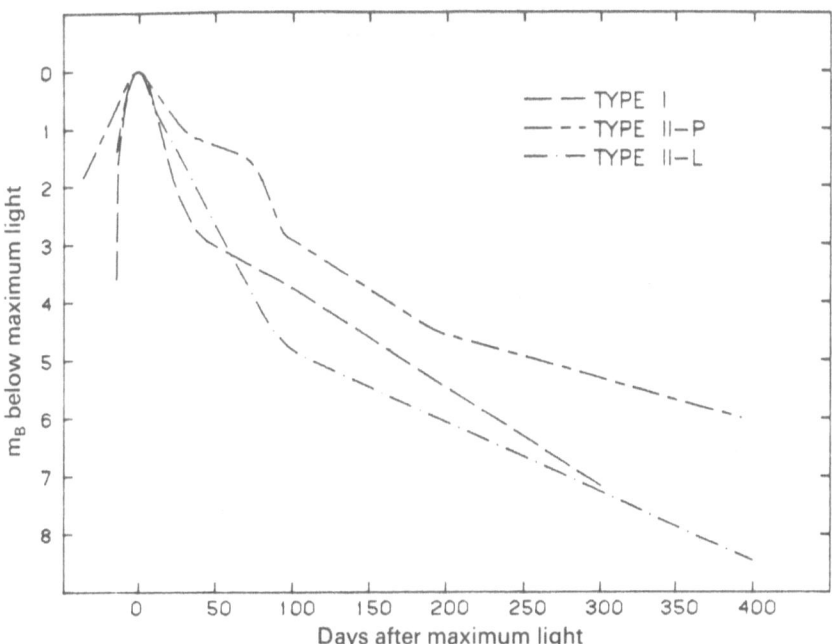

Figure 3.4. The light curves of Doggett and Branch (1985) for SN II-L and SN II-P compared to SN I.

II-L (for Linear). The difference between these two types is that the plateau type declines much more slowly after maximum, with a long plateau from age 25 days to about 80 days after maximum light a little more than 1 mag. below the maximum. In contrast, the blue light from SN II-L declines linearly from age 14 days to age 80 days at the rapid rate of 0.05 mag day^{-1}, so that at age 80 days, a SN II-L is typically 4 mag fainter than at maximum, while a SN II-P is less than 2 mag below its peak. This large and qualitative difference is conspicuous even in heterogeneous samples and the utility of these classes seems clear. A more elaborate classification by Pskovskii (1978) has been proposed, but the evidence for its reality suffers from the modest quality of many of the light curves.

As shown in Figure 3.4, Doggett and Branch (1985) have compared the schematic light curves for SN I, SN II-P, and SN II-L. The resemblance between SN I and SN II-L is strong, and demonstrates why the classification of supernovae should be by their spectra and not by their light curves. Interestingly, the spectroscopic differences, if any, between SN II-P and SN II-L remain to be defined. A new spectroscopic class of SN II delineated by Filippenko (1989) has a spectrum which resembles a Seyfert galaxy, and a slower decline in its light curve than SN II-P.

Light curves at late times are available for SN 1979C and SN 1980K in the compilation of Younger and van den Bergh (1985). While Doggett and Branch have examined the slopes of the B light curves at late times, these are much less likely to give an idea of the supernova energetics than in the case of SN I because the spectrum at late times is different. While the SN I have many blended lines in the B and V bands that carry much of the flux, the spectra of SN II contain strong hydrogen lines, and at times past 100 days the hydrogen lines carry much of the energy. Uomoto and Kirshner (1986) examined the flux in the $H\alpha$ line for SN 1980K, an SN II-L in NGC 6946. They found that the decay was nearly exactly exponential up to a year past maximum light. Interestingly enough, the decay rate of the exponential matched within the errors the radioactive decay rate for ^{56}Co. Uomoto and Kirshner suggested that the energy source for the late-time emission from SN II might come from about 0.1 $M_\odot$ of radioactive cobalt synthesized in the explosion, and mixed to some extent with the hydrogen debris.

The study of SN 1987A is of such superior quality and detail to all previous work on SN II that it is not possible to do justice to it in the available space. For the general context, readers could consult Arnett *et al.* (1989) and the many references listed there. Outstanding photometric investigations have been carried out in Chile at CTIO (Blanco *et al.*, 1987; Hamuy *et al.*, 1988) and in South Africa at SAAO (Catchpole *et al.*, 1987; Menzies *et al.*, 1987; Catchpole *et al.*, 1988; Whitelock *et al.*, 1988; Catchpole *et al.*, 1989). Figure 3.5 shows the light curves for SN 1987A as observed by the SAAO group.

The interpretation of data for SN 1987A is helped by the identification of the

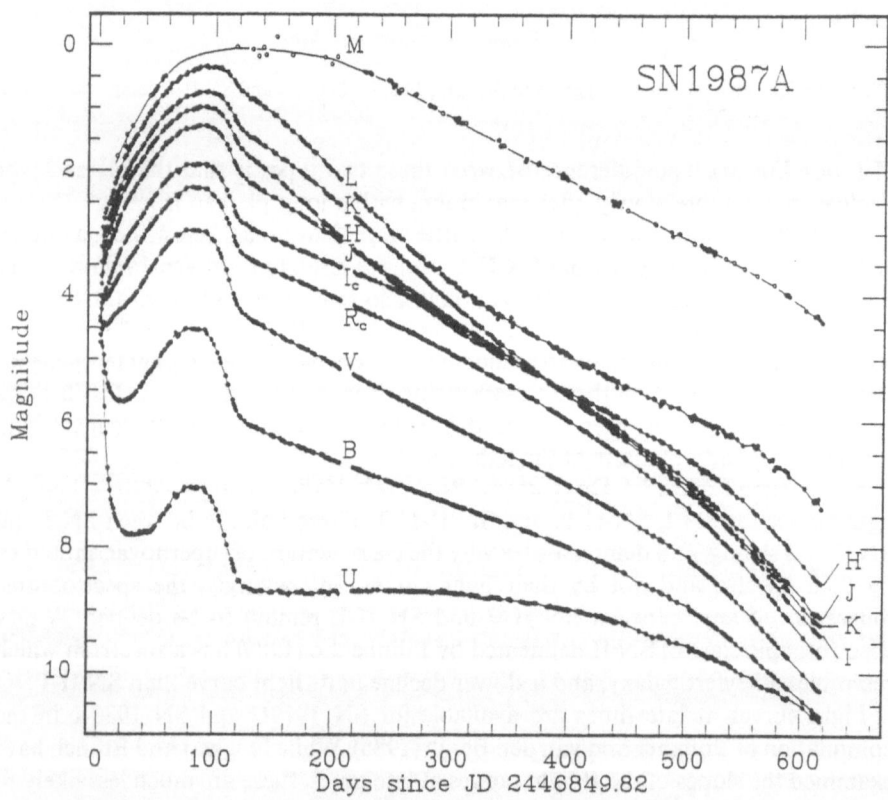

Figure 3.5. The light curve for SN 1987A from the ground-based ultraviolet to the infrared as observed by the South African collaboration (Catchpole *et al.*, 1989).

progenitor as the B3 supergiant SK $-69\ 202$, by knowledge of the moment of core collapse (February 23.316 1987), by a known distance to the LMC ($M - m = 18.5$) and a well-determined reddening to the supernova of $A_v = 0.6$. The light curve for SN 1987A is very far from typical in the first 100 days, which is generally attributed to the fact that SN 1987A was an explosion in a blue supergiant rather than in a red supergiant, the picture which accounts for the observations in most SN II. However, the behavior after day 100 resembles that of SN II-L, and the spectrum resembles those of SN 1980K obtained by Uomoto and Kirshner (1986).

In the case of SN 1987A, an integration from U band to M band provides a bolometric flux. Although the fact that the spectrum consists of emission lines causes some technical difficulties (see Menzies, 1988), the comparison to models shows that

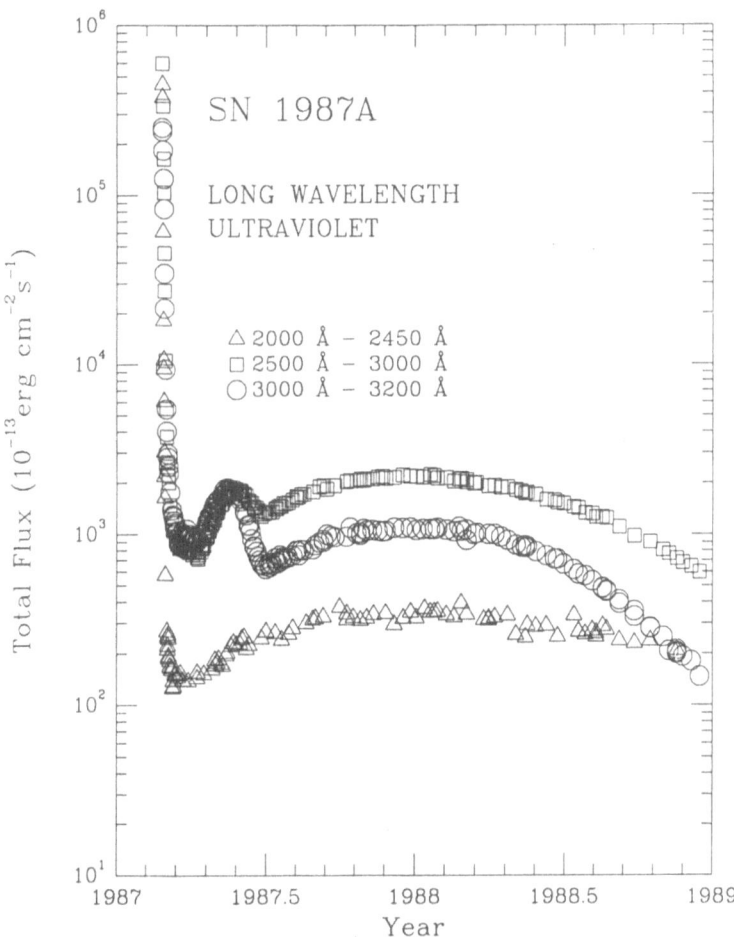

Figure 3.6. The satellite ultraviolet light curve for SN 1987A. The inital plunge as the photosphere cools is followed by a local maximum at the time of optical maximum. The long exponential decline seen in the optical and the bolometric flux is not seen in these IUE observations. This is presumably due to a gradual, but significant decrease in the ultraviolet opacity.

the long exponential decline, at least from day 100 to day 300, fits very well to the ^{56}Co decay slope and requires approximately 0.078 $M_\odot$ of radioactive cobalt. After day 300, the supernova declined more rapidly than the cobalt decay rate, presumably as a result of increasing transparency to γ-rays. The hypothesis that ^{56}Co is responsible for the energy of SN 1987A at late times received a decisive confirmation from the observations of the 837 and 1240 keV γ-ray lines of ^{56}Co.

SN 1987A was the first supernova where an extensive ultraviolet light curve has been obtained (for a summary of other data, see Blair and Panagia, 1987). Figure 3.6 shows data from the International Ultraviolet Explorer (IUE) in the wavelength range 2000–3000 Å (Kirshner et al, 1987). The very rapid initial decline is the result of cooling of the supernova atmosphere following the shock breakout at the surface. Subsequent fluxes reflect the very large line blanketing in the ultraviolet due to the effects of many overlapping resonance lines of common elements in the supernova atmosphere. The slow increase in the ultraviolet flux relative to the optical and near infrared where most of the flux emerges may have its origin in a slow decrease in this opacity.

A subtle change in the rate of decline for the optical flux began in late-1988, and continues at the time of writing. As shown in Figure 3.7, this small excess of flux above the extrapolated light curve amounts to roughly 8×10^{37} ergs s^{-1}. This could be the contribution of a pulsar or accretion on a neutron star. The report of pulses from a 1986 Hz pulsar in SN 1987A by Middleditch et al. (1989) makes the calorimetry of the pulsar by means of the optical light curve especially intriguing.

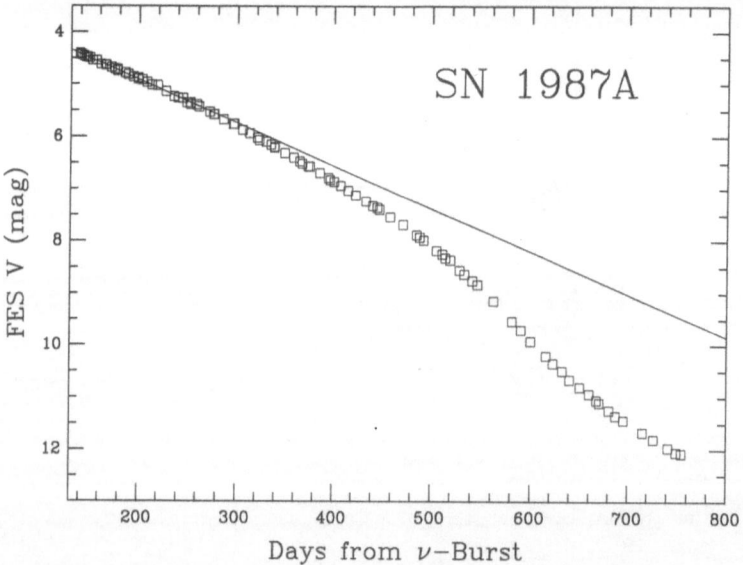

Figure 3.7. The flattening of the light curve for SN 1987A as observed with the Fine Error Sensor (FES) on the IUE satellite. The solid line is the ^{56}Co decay rate, normalized to fit the earlier observations. About 0.07 $M_\odot$ of radioactive cobalt is required. The energy in the optical light curve required to produce the observed flattening is roughly 7×10^{37} erg s^{-1}, and could be due to accretion on a neutron star, or the electromagnetic contribution of a pulsar, or a variation in absorption.

3.6. Cosmology and SN II

The absolute magnitudes of SN II cover a very wide range, overlapping SN I at the bright end, but more generally 1.5 mag fainter. Combined with their association with H II regions in spiral galaxies, where uncertainties due to reddening compound the problem, SN II seem poor candidates for cosmologically useful distance indicators. However, because they have a composition in their atmospheres which is not very different from stars that are well understood, there is some hope of using the observed properties of each SN II as a "custom yardstick" rather than a "standard candle." The basic idea of the Expanding Photosphere Method (EPM) is simple (Kirshner and Kwan, 1974, 1975). This method is sometimes referred to as the Baade–Wesselink method. However, neither Baade nor Wesselink ever used it on supernovae, and their method for Cepheids relies on the fact that periodic variables, unlike supernovae, both expand and contract. We observe the flux and the energy distribution for an SN II. We assume that the supernova emits a flux which is well represented by a dilute blackbody

$$ f = \frac{r^2}{D^2} A\pi B(T), $$

where r is the supernova's radius, D is the distance to the supernova, A is some correction factor to be determined from models, and $B(T)$ is the Planck function. Since the expansion velocity of material at the photosphere can be inferred from the spectra, given a suitable model atmosphere and a theory of line formation, the angular rate of expansion can be compared to the linear expansion rate at the distance D. This method has been reasonably successful in estimating some extragalactic distances (Branch et al., 1981, 1985, 1987; Kirshner, 1985), but the accurate evaluation of the dilution factor, A, requires good model atmospheres for SN II (Wagoner, 1989).

In the case of SN 1987A, detailed models have been calculated by a number of authors (Eastman, 1989; Höflich, 1989; Wagoner, 1989). The models of Eastman and Kirshner (1989) are based on a computer code that simultaneously solves the non-LTE equations of statistical equilibrium and the time-independent equation of radiation transport in an expanding atmosphere. The model spectra resemble those observed, as shown in Figure 3.8, and some parameters of the supernova atmosphere can be determined. Figure 3.9 shows that a dilute blackbody fit provides a good representation of one of the detailed models: in practice the color temperature could be accurately determined from BVRI photometry. Figure 3.10 shows the results for the distance to the LMC obtained at each of five times during the first 10 days after core collapse. It is very encouraging that the derived distance is constant, while the color temperature changes from 13,000 K to 6000 K, the age of the supernova changed by a factor of 5, and A varied by nearly a factor of 4. The mean distance derived is 49 ± 6 kpc, in good accord with the recent results of Walker and Mack (1988), who obtained an LMC distance of 48.8 kpc from RR Lyrae stars, and with Walker (1987) who measured the distance with Cepheids to be 49.4 kpc. This result differs significantly from the determination of Branch (1987) who found 55 ± 5 kpc. The 10% difference arises entirely from Branch's neglect of the dilution factor A. The

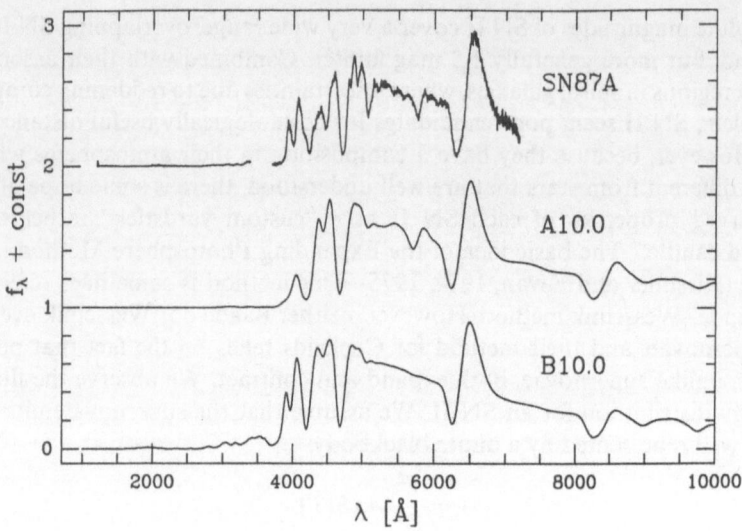

Figure 3.8. Two models of Eastman and Kirshner (1989) compared to the spectrum of SN 1987A at age 10 days. Model B has a lower density in the atmosphere than model A.

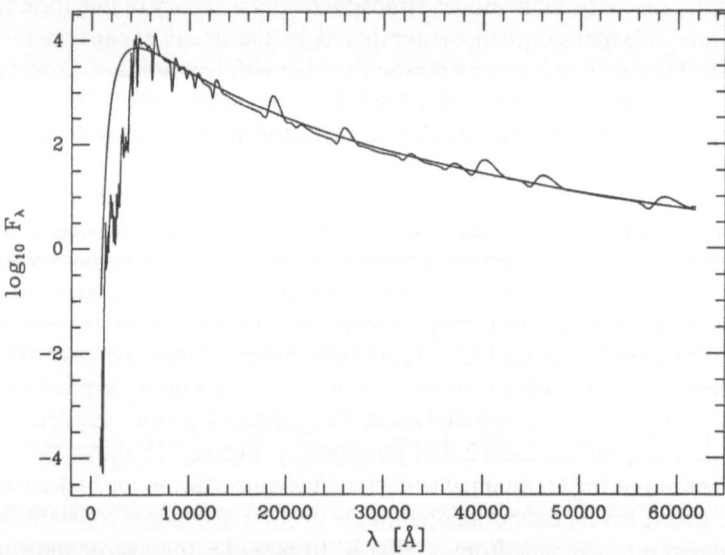

Figure 3.9. A blackbody fit to a model atmosphere for SN 1987A. The distance determination uses the fact that the detailed spectrum is well approximated by a dilute blackbody.

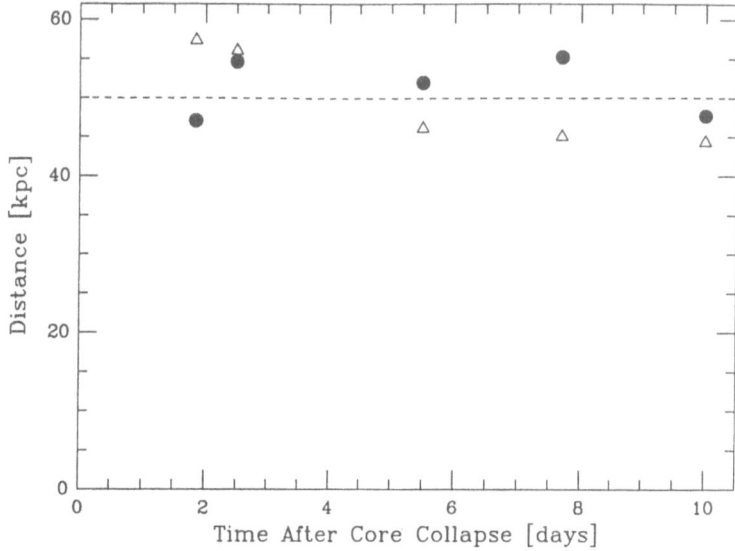

Figure 3.10. The results of repeated distance determinations for SN 1987A, using photometry, and the models of Eastman and Kirshner (1989). The filled circles and triangles refer to different photometric bands. The fact that the resulting distance is the same during epochs when the physical conditions change dramatically is a good test of the method. The derived distance to the LMC is 49 ± 6 kpc.

performance of the EPM in this test case is good, and raises the prospect that it may prove equally accurate when applied to more distant cases. If that hope is realized, then "custom yardstick" distances to SN II may move distances derived from the light curves of SN II to the center of the discussion of the extragalactic scale. The measurement of some SN I in the same galaxies or the same clusters as well-measured SN II would provide an absolute calibration for those standard candles which could provide accurate distances to $z = 0.3$, and perhaps a global value for Ω from supernovae alone.

Acknowledgments
R.P.K.'s research on supernovae is supported by the National Science Foundation through grant AST-8516537, and by NASA through grants NAG5-645 and NAG 5-841. Collaboration with Eric Schlegel and with Ron Eastman was essential to the work reported here.

References
Aardeberg, A. and de Groot, M. 1973, *Astron. Astrophys.*, **28**, 295.
Aaronson, M. *et al.* 1980, *Ap. J.*, **239**, 12.
Arnett, W.D. 1982, *Ap. J.*, **253**, 785.
Arnett, W.D., Bahcall, J.N., Kirshner, R.P., and Woosley, S.E. .989, *Ann. Rev. Astron. Astrophys.*, **27**, 629.
Arnett, W.D., Branch, D., and Wheeler, J.C. 1985, *Nature*, **314**, 337.
Axelrod, T.S. 1980, Ph.D. thesis, University of California, Santa Cruz.
Axelrod, T.S. 1988, in *IAU Colloquium 108*, ed. K. Nomoto (Berlin: Springer-Verlag).
Barbon, R., Ciatti, F., and Rosino, L. 1973, *Astron. Astrophys.*, **25**, 241.
Barbon. R., Ciatti, F., and Rosino, L. 1979, *Astron. Astrophys.*, **72**, 287.

Barbon, R. et al. 1984, Astron. Astrophys. Suppl., 58, 735.

Bertola, F. 1964, Ann. Astrophys., 27, 319.

Bertola, F., Mammano, A., and Perinotto, M. 1965, Contrib. Asiago Obs., 174, 51.

Blair, W.P. and Panagia, N. 1987 in Exploring the Universe with the IUE, ed. Y. Kondo (Dordrecht: Reidel).

Blanco, V.M. et al. 1987, Ap. J., 320, 589.

Branch, D. 1985, in Supernovae as Distance Indicators, ed. N. Bartel (Berlin: Springer-Verlag).

Branch, D. 1987, Ap. J. (Letters), 320, L23.

Branch, D. et al. 1981, Ap. J., 244, 780.

Branch, D. 1988, in IAU Colloquium 108, ed. K. Nomoto (Berlin: Springer-Verlag), p. 281.

Branch, D. and Nomoto, K. 1986, Astr. Ap., 164, L13.

Branch D. and Patchett B. 1973, M.N.R.A.S., 161, 71.

Buta, R. J. and Turner, A. 1983, Pub. A.S.P., 95, 72.

Buta, R. 1984, IAU Circulars, 3981, 3983.

Cadonau, R, Sandage, A.R., and Tammann, G.A. 1985, in Supernovae as Distance Indicators, ed. N. Bartel (Berlin: Springer-Verlag), p. 151.

Catchpole, R. et al. 1987, M.N.R.A.S. 229, 15P.

Catchpole, R. et al. 1988, M.N.R.A.S. 231, 75P.

Catchpole, R. et al. 1989, M.N.R.A.S. 237, 55.

Chevalier, R. 1984, Ap. J. (Letters), 285, L63.

Clark, D.H. and Stephenson, F.R. 1977, The Historical Supernovae (New York: Pergamon).

Colgate, S. and McKee, C. 1969, Ap. J., 157, 623.

de Vaucouleurs, G. 1983, Ap. J., 268, 468.

Doggett, J.B. and Branch, D. 1985, Astron. J., 90, 2303.

Eastman, R.G. 1989, in Proceedings of the Berkeley Workshop on Particle Astrophysics, ed. E. Norman (Singapore: World Scientific).

Eastman, R.G. and Kirshner, R.P. 1989, Ap. J., 347, 777.

Elias, J.H., Frogel, J.A., Hackwell, J.A., Persson, S.E. 1981, Ap. J., 251, L13.

Elias, J.H., Matthews, K. Neugebauer, G., and Persson, S.E. 1985, Ap. J., 296, 379.

Ensman, L. and Woosley, S.E. 1988, Ap. J., 333, 754.

Filippenko, A.V. 1989, Astron. J., 97, 726.

Filippenko, A.V. and Sargent, W.L.W. 1985, Nature, 316, 407.

Filippenko, A.V. and Sargent, W.L.W. 1986, Astron. J., 91, 691.

Filippenko, A.V., Porter, A.C., Sargent, W.L.W., and Schneider, D.P. 1986, Astron. J., 92, 1341.

Fransson, C. 1988, in IAU Colloquium 108, ed. K. Nomoto (Berlin: Springer-Verlag), p. 385.

Gaskell, C.M. et al. 1986, Ap. J. (Letters), 306, L77.

Graham, J.R. 1986, M.N.R.A.S. 220, 27P.

Hamuy, M., Suntzeff, N.B., Gonzales, R., and Martin, G. 1988, Astron. J. 95, 63.

Hansen, L., Jorgensen, H.E., Norgaard-Nielsen, H.U., Ellis, R.S., and Couch, W.J. 1989, Astron. Astrophys., in press.

Harkness, R. et al. 1987, Ap. J., 317, 355.

Höflich, P. 1989, in Proceedings of the Berkeley Workshop on Particle Astrophysics, ed. C. Pennypacker, in press.

Kirshner, R.P. 1980, in Supernovae Spectra eds. R. Meyerott and G.H. Gillespie (New York: American Institute of Physics).

Kirshner, R.P. 1985, in Supernovae as Distance Indicators, ed. N. Bartel (Berlin: Springer-Verlag), p. 171.

Kirshner, R.P. and Kwan, J. 1974, Ap. J., 193, 27.

Kirshner, R.P. and Kwan, J. 1975, Ap. J., 197, 415.

Kirshner, R.P. and Oke, J.B. 1975, Ap. J., 200, 574.

Kirshner, R.P., Oke, J.B., Penston, M. and Searle, L. 1973, Ap. J., 185, 303.

Kirshner, R.P., Sonneborn, G., Crenshaw, D.M., and Nassiopoulos, G.E. 1987, Ap. J., 320, 602.

Kirshner, R.P., Winkler, P.F., and Chevalier, R.A. 1987, Ap. J. (Letters), 315, L135.

Leibundgut, B. 1988, Ph.D. inaugural dissertation, University of Basel.

Long, K.S., Blair, W.P., and van den Bergh, S. 1988. Ap. J., 333, 749.

Menzies, J.W. et al. 1987, M.N.R.A.S., 227, 39P.

Menzies, J.W. 1988, *M.N.R.A.S.*, in press.

Middleditch, *et al.* 1989, *IAU Circ.*, No. 4735.

Minkowski, R. 1941, *Pub. A.S.P.*, **53**, 224.

Minkowski, R. 1964, *Ann. Rev. Astron. Astrophys.*, **2**, 247.

Nomoto, K., Shigeyama, T., and Hashimoto, M. 1988 in *IAU Colloquium 108*, ed. K. Nomoto (Berlin: Springer-Verlag), p. 319.

Oke, J.B. and Searle, L. 1974, *Ann. Rev. Astron. Astrophys.*, **12**, 315.

Panagia, N. and Laidler, V. 1988, in *Supernova Shells and Their Birth Events*, ed. W. Kundt, Lecture Notes in Physics, Vol. 316 (Berlin: Springer-Verlag), p. 187.

Panagia, N., Sramek, R.A., and Weiler, K.W. 1986, *Ap. J. (Letters)*, **300**, L55.

Panagia, N. *et al.* 1989, private communication.

Phillips, M.M. *et al.* 1987, *Pub. A.S.P.*, **99**, 592.

Porter, A.C. and Filippenko, A.V. 1987, *Astron. J.*, **93**, 1372.

Pskovskii, Y.P. 1977, *Soviet Astr.—AJ* **26**, 675.

Pskovskii, Y.P. 1978, *Soviet Astr.—AJ* **22**, 201.

Rust, B.W. 1974, Ph.D. thesis, University of Illinois.

Schlegel, E.M. and Kirshner, R.P. 1989, *Astron. J.*, **98**, 577.

Sramek, R.A., Panagia, N., and Weiler, K.W. 1984, *Ap. J. (Letters)*, **285**, L59.

Strom R.G. 1988, *M.N.R.A.S.*, **230**, 331.

Tsvetkov, D.Yu. 1985, *Astron. Zh.*, **62**, 365.

Tsvetkov, D.Yu. 1986, *Astron. Zh. Pis'ma*, **12**, 784.

Tsvetkov, D.Yu. 1987, *Astron. Zh. Pis'ma*, **13**, 894.

Uomoto, A.K. and Kirshner, R.P. 1985, *Astron. Astrophys.*, **149**, L7.

Uomoto, A.K. and Kirshner, R.P. 1986, *Ap. J.*, **308**, 685.

van den Bergh, S. 1988, *Ap. J.*, **327**, 156.

Wagoner, R.V. 1989, in *Proceedings of the Berkeley Workshop on Particle Astrophysics*, ed. C. Pennypacker, in press.

Walker, A.R. 1987, *M.N.R.A.S.*, **225**, 627.

Walker, A.R. and Mack, P. 1988, *Astron. J.*, **96**, 1362.

Weaver, T.A. and Woosley, S.E. 1980, in *Ninth Texas Symposium*, eds. J. Ehler, J. Perry, and M. Walker (New York: New York Academy of Sciences).

Wheeler, J.C., Harkness, R.P., Barker, E.S., Cochran, A.L., and Wills, D. 1987, *Ap. J. (Letters)*, **313**, L69.

Wheeler, J.C. and Levreault, R. 1985, *Ap. J. (Letters)*, **294**, L17.

Whitelock, P.A. *et al.* 1988, *M.N.R.A.S.*, **234**, 5P.

Woosley, S.E. and Weaver, T.A. 1986, *Ann. Rev. Astron. Astrophys.*, **24**, 205.

Younger, P.F. and van den Bergh, S. 1985, *Astron. Astrophys. Suppl.*, **61**, 365.

Zwicky, F. 1938, *Phys. Rev.*, **53**, 1019.

Zwicky, F. 1968, in *Stars and Stellar Systems*, Vol. VIII (Chicago: University of Chicago Press).

4. Radio Supernovae

RICHARD A. SRAMEK and KURT W. WEILER

4.1. Introduction

Considering the large amount of energy released by a supernova explosion we would expect that the effects of the explosion would be seen over a wide range in the electromagnetic spectrum. However, until recently the only detected radio emission associated with supernovae has been from pulsars, the spinning condensed remains of the stellar explosions, and from supernova remnants (SNR). The SNRs arise either from the interaction of the supernova ejecta with the interstellar medium or they are driven by the rotational energy lost by the central pulsar. The SNRs probably take a couple of hundred years for the level of radio emission to build up. Pulsars may possibly be radio emitters at an earlier age, but it is unlikely that their radio radiation can escape through the ejected supernova envelope during the first few decades.

A question that has often been asked is whether there is radio radiation coincident with the very luminous optical flash of the explosion itself, or perhaps arising in the first several years as a direct result of the explosion. This question was answered with the radio detection of SN 1970G (Gottesman et al., 1972) less than a year after the observed optical explosion. Radio emission from this supernova was seen as a weak increase in the radio flux density from the giant H II region NGC 5455 in which the supernova was situated. Nearly a decade passed before another supernova was detected in the radio portion of the spectrum. In 1980 the bright supernova SN 1979C was detected (Weiler et al., 1981) and for the first time the supernova was well isolated from the surrounding extended radio emission and a detailed radio light curve could be constructed. SN 1979C remains the best studied radio supernova.

With the exception of SN 1970G, the pre-1980 searches for radio emission from optical supernovae produced no detections (de Bruyn, 1973; Brown and Marscher, 1978; Ulmer et al., 1980). In retrospect, we now know that the radio telescopes of that time had only marginally sufficient sensitivity to detect the brightest radio supernovae. A bright supernova exploding at the distance of the Virgo cluster might produce a radio flux density of a few milli-Janskys, which is about the r.m.s. noise level of the early searches (a Jansky is the unit of flux density used in radio astronomy; one Jansky equals 10^{-26} W m^{-2} Hz^{-1}).

Progress in the study of radio supernovae reflects the history of instrumental improvement that radio astronomy saw in the 1970s. Improved cooled receivers became common, wide bandpass correlators were developed, and two large imaging arrays, the Westerbork Synthesis Radio Telescope (WSRT) and the NRAO Very

Large Array (VLA), came into operation (Hogbom and Brouw, 1974; Napier, Thompson, and Ekers, 1983). For the first time there were instruments that had both the sensitivity and the resolution to detect the brighter supernovae easily. The sensitivity was such that sub-milli-Jansky detection levels could be reached with just a few minutes observation. This allowed observations of a new supernova to be slipped easily into an observing schedule with minimum disruption. Responding to supernova discoveries became routine. The improved resolution of these imaging arrays also allowed positional agreement between an optically discovered supernova and a radio source to be established on the arcsecond angular scale, while mitigating the effects of the large scale radio structure of the host galaxy.

4.2. Radio Emission from SN 1970G

Radio emission was first detected from the supernova SN 1970G with both the NRAO Greenbank Interferometer (Gottesman *et al.*, 1972) and the Westerbork Synthesis Radio Telescope (Goss *et al.*, 1973). This was a Type II supernova in the nearby spiral galaxy M101, which reached a peak brightness of 11.7 mag.

The measurement of the supernova radio flux density was complicated by the fact that the supernova occurred near the giant H II region NGC 5455. The radio flux from the H II region had to be subtracted before a clear detection of the supernova could be made. However, even with this limitation, a rough radio light curve was produced for this object (Allen *et al.*, 1976) and a few general characteristics of supernova radio emission were given. The emission was nonthermal, appeared about a year after the explosion, lasted for a few years, and at its peak the radio luminosity at 6 cm was about 10^{+26} ergs s^{-1} Hz^{-1}. This is an order of magnitude greater than that of the brightest Galactic SNR Cassiopeia A. Although subsequent radio supernovae showed a wide range of characteristics, this summary developed in 1976 gives the basic picture.

4.3. Further Radio Detections of Type II Supernovae

The next radio detection of a supernova was the Type II object SN 1979C in M100 (NGC 4321). Observations made at the VLA a few weeks after the optical discovery in April 1979 showed no radio emission at the location of the supernova. However, when reobserved a year later, a bright unresolved source dominated the radio image (Weiler *et al.*, 1981). This radio image of M100 made at a 6 cm wavelength is shown in Figure 4.1. The nucleus of the galaxy is seen at the center, but the brightest feature is the radio supernova at the lower left.

In a nearby normal galaxy, most of the general disk radio emission is resolved with sub-arcsecond resolution. Outside of a possible compact core in the nucleus, the only unresolved features are a few compact H II regions and some old SNRs, and these seldom have the brightness of the more powerful radio supernovae. It is therefore not too difficult to distinguish radio supernovae against the general galactic background. This process is aided by the positional accuracy of the radio

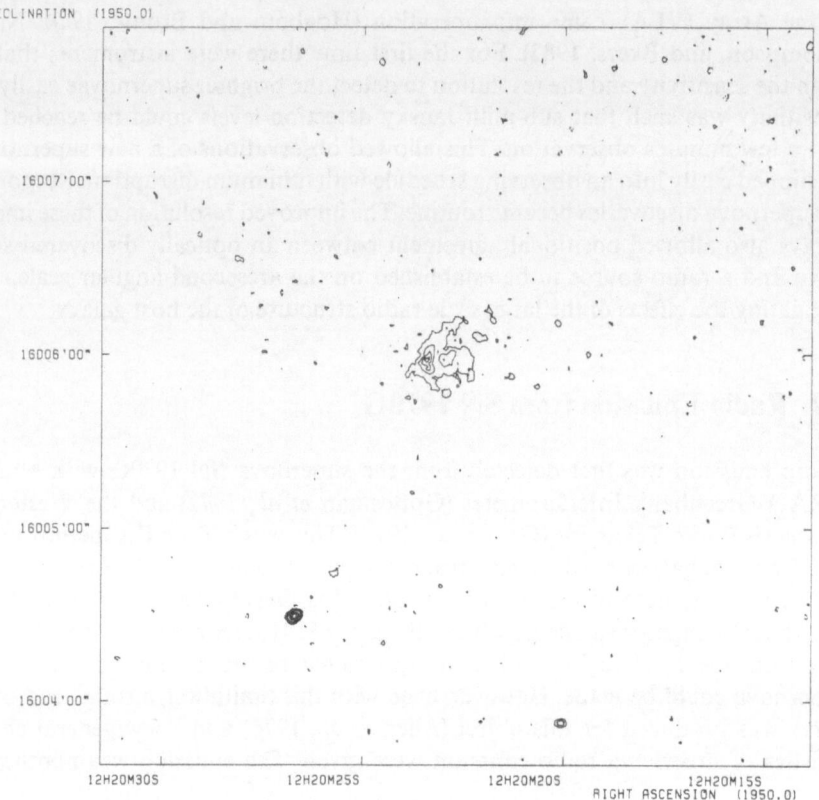

Figure 4.1. A VLA image of M100 at 6 cm wavelength taken in April 1980 showing SN 1979C, the first well-studied radio supernova. The peak brightness of the galactic nucleus seen at the center of the image is about 1 mJy beam^{-1}. The brightness of the supernova SN 1979C at the lower left is 5.3 mJy beam^{-1}. (From Weiler *et al.*, 1981.)

measurements which are typically a few tenths of an arcsecond. When optical measurements are available with comparable accuracy, the identification of a radio feature with an optically detected supernova is readily made. Detecting radio supernovae with a modern high resolution array telescope is primarily a matter of obtaining enough sensitivity.

The radio data for SN 1979C is more complete than for any other supernova. After the initial detection, VLA observations were made on an average of once a month for 3 years, usually at 20 cm and 6 cm and occasionally at 2 cm wavelength. Since 1984 observations have been made four times a year. With such extensive observations, detailed radio light curves can be obtained, and since two or more frequencies were observed, the evolution of the radio spectrum can be studied. Figure 4.2 shows the radio light curves of SN 1979C at 20 cm and 6 cm wavelengths, and Figure 4.3 shows the resulting spectral index changes over the same period. These radio light and spectral index curves are thought to be typical for Type II supernovae.

The scenario vaguely implied by the observations of SN 1970G, are clearly stated

Figure 4.2. The radio light curves of SN 1979C from VLA data at two wavelengths. The rapid rise at progressively lower frequencies is thought to be due to free–free thermal absorption in the circumstellar shell produced by the stellar wind of the supernova progenitor. After the shell becomes optically thin, the luminosity decays like a power law in time. (From Weiler *et al.*, 1986.)

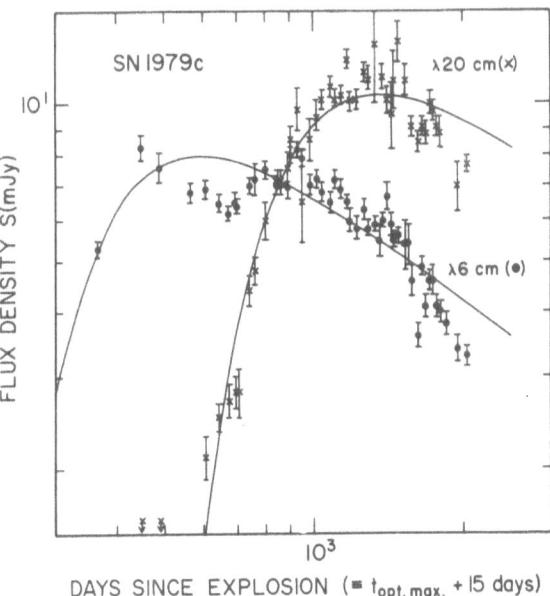

Figure 4.3. The spectral index between 6 and 20 cm of SN 1979C plotted as a function of time. The radio spectrum steepens as the shell becomes transparent to centimeter radio radiation. (From Weiler *et al.*, 1986.)

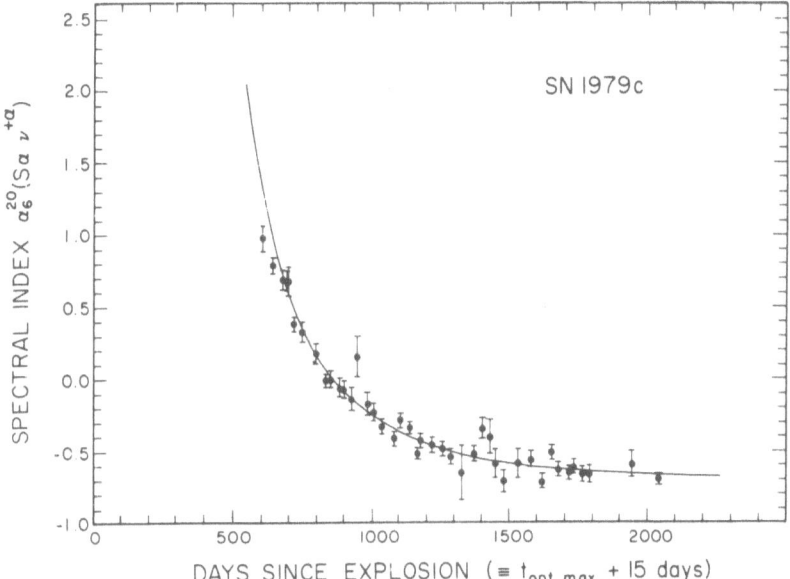

by the SN 1979C observations. The radio emission comes significantly after the optical outburst; there is a very fast rise of flux density when the source does turn on; the emission turns on first at short wavelengths and later at progressively longer wavelengths; after the rapid turn on there is a period of slow decay; during this period of decay the emitting region is optically thin and the radio spectral index is constant and reasonably typical of a nonthermal synchrotron source. The most complete radio data published so far for SN 1979C is in Weiler *et al.* (1986), and more recent measurements show a continuing slow decline in flux density.

The radio flux density of SN 1979C was sufficiently high that very long baseline interferometry (VLBI) was done. With this technique, a radio interferometer is created using independent radio telescopes several hundred or even thousands of kilometers apart. The signals are recorded on magnetic tape and played back and multiplied together at a later time to produce interference fringes. In this way an interferometer with a fringe spacing or resolution of a few milli-arcseconds can be constructed.

Using VLBI, Bartel *et al.* (1985) have not only measured the diameter of SN 1979C, but successive observations over a few years showed the supernova ejecta to be expanding. The diameter at 6 cm wavelength increased from 1.05 milli-arcseconds in December 1982 to 1.43 milli-arcseconds in December 1983. The observed radio angular expansion rate combined with the physical expansion velocity inferred from optical spectral line velocity measurements of the supernova envelope give a direct measurement of the distance to NGC 4321. This in turn provides an independent determination of the Hubble constant and the extragalactic distance scale (Bartel *et al.*, 1985).

The next supernova to be detected in the radio was again a Type II supernova, SN 1980K, in NGC 6946. Radio monitoring began soon after its optical discovery in October 1980, and a detection was made at 6 cm only 35 days after the optical maximum (Sramek, Weiler, and van der Hulst, 1982). SN 1980K was less luminous than SN 1979C and turned on sooner, but in general it showed the same characteristics as SN 1979C (Weiler *et al.*, 1986).

4.4. Radio Detection of Type I Supernovae

SN 1970G, SN 1979C, and SN 1980K were all luminous Type II supernovae. Several attempts were made at the VLA to detect Type I supernovae with no success until the explosion of the Type Ib supernova SN 1983N in NGC 5236 (M83). This supernova was detected at the VLA 11 days before optical maximum and evolved extremely quickly (Sramek, Panagia, and Weiler, 1984). While the radio emission from Type II supernovae could be studied for several years after their detection, SN 1983N fell below the detection limit at the VLA in about a year. Figure 4.4 shows the radio light curves for SN 1983N at 6 cm and 20 cm wavelengths.

Not only did SN 1983N evolve very quickly in the radio, its spectrum during the declining phase was significantly steeper than that of the Type II supernovae. The

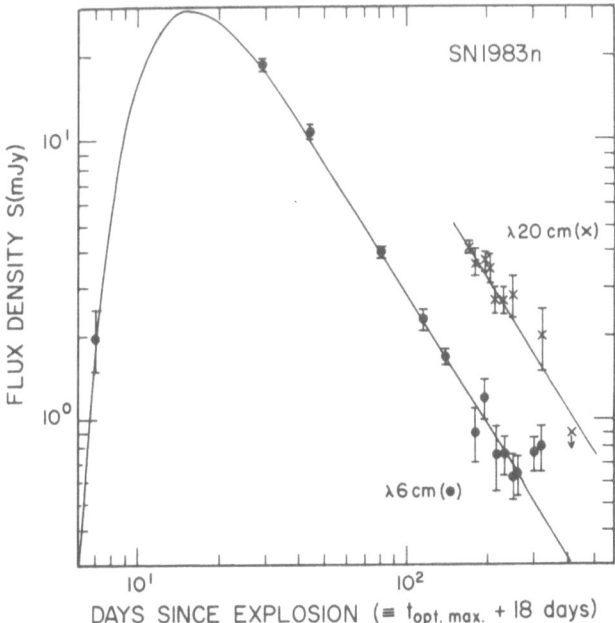

Figure 4.4. The radio light curve of the Type Ib supernova SN 1983N. The age of the supernova is measured from the estimated date of explosion taken to be 18 days before the maximum optical luminosity. The radio emission was detected 11 days before optical maximum and showed a very rapid rise and fall compared to the Type II radio supernovae. (From Weiler *et al.*, 1986.)

spectral index of SN 1983N was -1.03 while that of the Type II supernovae approached about -0.7.

In October 1984 a second Type I supernova was detected in the radio, SN 1984L in NGC 991 (Panagia, Sramek, and Weiler, 1986). This radio source was weaker than SN 1983N and less data was taken, but it showed the same rapid decline in radio flux density and the same steep spectral index. Apart from the rapid spectral evolution and the steep spectral index, the Type I supernovae share the general radio characteristics of the Type II supernovae. This suggests that the radio emission mechanism may be the same for the two types, with differences just being of degree or environment.

The two Type I supernovae detected by their radio emission, SN 1983N and SN 1984L, are both members of the subluminous and spectroscopically peculiar subclass of this type, usually identified as Type Ib. The nature of the stellar progenitors of this subclass is still uncertain, but the presence of radio emission offers another distinguishing characteristic.

To date there have been no radio detections of the more prevalent Type Ia supernovae, with an upper limit of about 10^{+26} ergs s^{-1} Hz^{-1} being set on their radio luminosity (see Table 4.1). This limit is comparable to the luminosity of the detected Type II supernovae. A high sensitivity radio image of a nearby Type Ia supernova would be very useful to establish their radio properties better.

Table 4.1. Properties of radio supernovae (adapted from Weiler et al. 1986).

SN name	SN type	Observed optical		Observed radio maximum at 6 cm					
		Distance (Mpc)	maximum (m)	Age (yr)	Flux density (mJy)	Spectral luminosity (ergs s^{-1} Hz^{-1})	Ratio to Cas A	Optically thin radio index α_6^{20} (S $\propto \nu^{+\alpha}$)	Rate of flux density decline β (S $\propto t^{\beta}$)
A. Detections									
SN 1950B	II?	~7.0	≲14.5	30.0	0.5[a]	~3 × 10^{25}	~5.0	−0.4[a]	—
SN 1957D	II?	~7.0	≲15.0	23.0	1.9[a]	~1 × 10^{26}	~15.0	−0.25[a]	—
SN 1961V	V	12.0	12.5	23.0	0.11[d]	~8 × 10^{24}	1.0	−0.4	—
SN 1970G	II	7.2	11.7	1.4	~2.5	~1 × 10^{26}	~15.0	−0.7 ± 0.1[b]	—
SN 1979C	II-L	17.0	11.6	1.2	8.3	~2 × 10^{27}	~250.0	−0.72 ± 0.05	−0.71 ± 0.08
SN 1980K	II-L	7.0	11.6	0.4	2.6	~1 × 10^{26}	~15.0	−0.50 ± 0.06	−0.65 ± 0.10
SN 1981K	II?	6.6	<16.0	0.5	~2.0[c]	~1 × 10^{26}	~15.0	−0.91 ± 0.07	−0.73 ± 0.06
SN 1983N	Ib	~7.0	11.8	0.08	18.5	~1 × 10^{27}	~15.0	−1.03 ± 0.06	−1.59 ± 0.08
SN 1984L	Ib	~24.0	13.9	0.14	0.7	~4 × 10^{26}	~125.0	−1.0 ± 0.2	−1.5 ± 0.3
SN 1986J	V	10.0	—	3.0	124.0[e]	~8 × 10^{27}	~60.0	−0.7	—
SN 1987A	II	0.05	3.0	0.01	150.0[f]	~4 × 10^{23}	0.05	−1.0	—
B. New upper limits (3·σ)									
SN 1980N	Ia	33.0	12.5	0.2	<0.6	<7 × 10^{26}	<100.0	—	—
SN 1981B	Ia	33.0	12.0	0.05	<0.3	<3 × 10^{26}	<50.0	—	—
SN 1983G	Ia	22.0	13.0	1.6	<0.2	<2 × 10^{26}	<35.0	—	—
SN 1984A	Ia	22.0	<13.1	0.2	<0.2	<1 × 10^{26}	<15.0	—	—

[a] Cowan and Branch (1985).
[b] Allen et al. (1976).
[c] van der Hulst et al. (1983).
[d] Cowan et al. (1988).
[e] Rupen et al. (1987).
[f] Turtle et al. (1987).

4.5. Supernovae Discovered by Their Radio Emission

Until recently, supernovae have been discovered exclusively by their optical emission. There are now two supernovae that have been discovered by their radio emission, SN 1981K in NGC 4258 (van der Hulst *et al.*, 1983; Weiler *et al.*, 1986) and SN 1986J in NGC 891 (Rupen *et al.*, 1987).

SN 1981K was first detected in a VLA radio study of NGC 4258 in January 1982. This source was not there in previous images made in the late 1970s using the WSRT. Radio monitoring at the VLA and the WSRT produced the light curve shown in Figure 4.5 which shows this source to have a nonthermal spectrum and to be decreasing in flux density. A search of archival optical photographic plates revealed a transient object which appeared near the beginning of August 1981 and was located within 2 arcseconds of the radio source. Although the scant optical evidence does not allow a determination of the type of this supernova, the slow radio decline is characteristic of Type II supernovae. This object, designated SN 1981K in retrospect, was missed in the optical search programs because it was only 16th and 17th magnitude when the plates were taken.

The other supernova discovered by its radio emission was SN 1986J in NGC 891. A VLA radio image made by Rupen *et al.* (1987) at 20 cm in August 1986 shows a dominant point source 65 arcseconds south-west of the galactic nucleus (Figure 4.6). Earlier images made at the WSRT showed no such source. Based on available radio

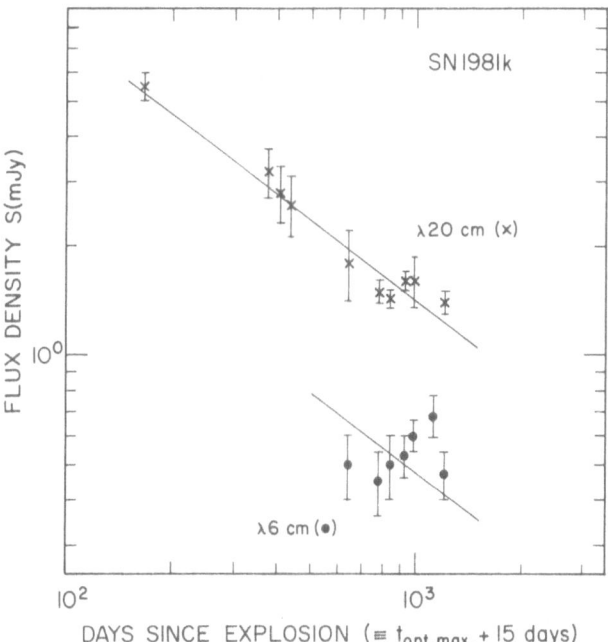

Figure 4.5. The radio light curve of the first supernova discovered by its radio emission, SN 1981K. The long and slow radio decay is indicative of a Type II supernova. (From Weiler *et al.*, 1986.)

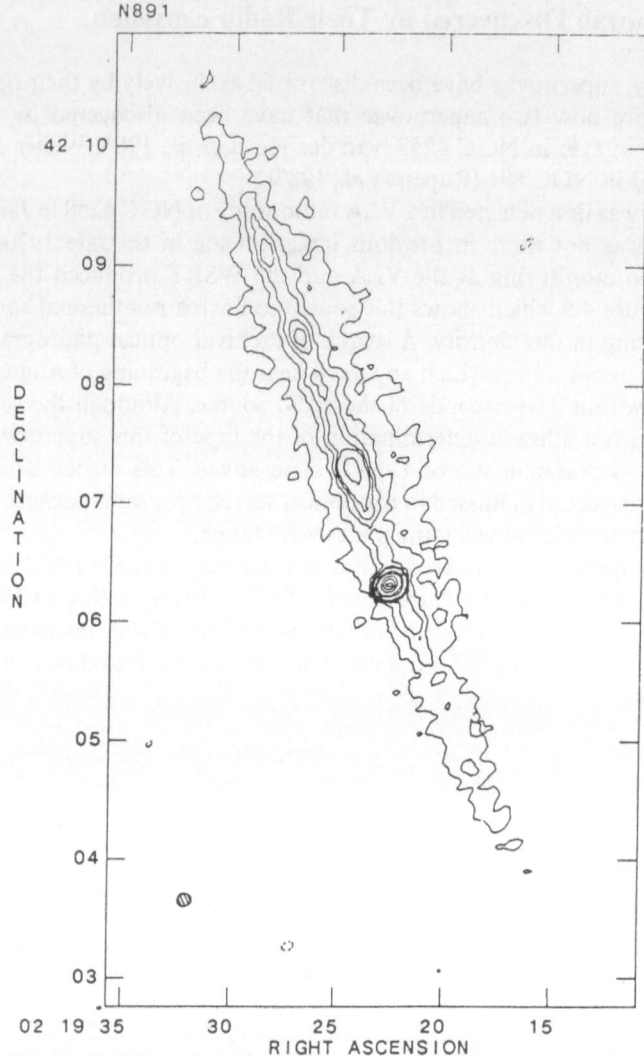

Figure 4.6. A 21-cm wavelength VLA image of NGC 891 showing the bright radio supernova SN 1986J. The disk of emission is symmetric about the galactic nucleus which has a brightness of 10 mJy beam^{-1}. The supernova, to the southwest of the nucleus, has a brightness of 64 mJy beam^{-1}. (From Rupen *et al.*, 1987.)

and optical data, the supernova explosion probably occurred sometime in 1982 or 1983. SN 1986J remains the most luminous of the radio supernovae. In absolute luminosity it is about one thousand times the luminosity of Cassiopeia A, and over four times the peak radio luminosity of SN 1979C, the next brightest radio supernova. The flux density of SN 1986J reached 128 mJy at 6 cm compared to the 10–20 mJy maximum for other radio supernovae. The VLBI was done on SN 1986J giving a diameter of 1.4 milli-arcseconds (Bartel, Rupen, and Shapiro, 1987).

The optical counterpart of the radio supernova was found in September 1986 when photometric and spectroscopic measurements were made. Its luminosity in r band (6500 Å) was 19.5 mag. Re-examination of earlier observations made in January 1984 showed it with an r-band magnitude of 18.4. An unusual feature of the spectrum was the narrow 1000 km s^{-1} emission line widths. Most supernovae show line widths of at least several thousand kilometers per second. Although the spectrum is similar to a Type II supernova, the narrow line widths are like those seen in SN 1961V, the prototype of the Type V supernova. SN 1961V is also a radio emitter (see below). The high optical luminosity at an age of 4 years suggests that a pulsar may be the source of energy for the optical emission (Chevalier, 1987).

4.6. Intermediate Age Radio Supernovae

Observations of SN 1970G and SN 1979C tell us that the centimeter radio emission from Type II supernovae can last for at least several years. A hundred or more years after the explosion, as the supernova ejecta interact with the interstellar medium, the radio emission is again significant in the form of an SNR. But what is the luminosity of supernovae with an intermediate age of 10–100 years? There is at least a partial answer with the detection of radio emission from SN 1957D and probably from SN 1950B (Cowan and Branch, 1982, 1985) and from SN 1961V (Branch and Cowan, 1985).

M83 (NGC 5236) is a galaxy with a high rate of star formation and a high number of cataloged supernovae. Five known supernovae ranging from SN 1923A to SN 1983N are located within this galaxy. In 1981, and again in 1983 and 1984, Cowan and Branch made deep VLA radio images of this galaxy and found unresolved nonthermal radio sources located within a few arcseconds of the published positions of three of the supernovae. SN 1983N was already known to be a radio emitter (see above), but SN 1957D and SN 1950B were new detections and were much older than any previous radio supernovae.

The optical position of SN 1957D is known with arcsecond accuracy (Pennington and Dufour, 1983) and its identification with the radio source is considered certain. The accuracy of the optical position of SN 1950B is only about 10 arcseconds and the radio identification is less certain but still likely. These supernovae are sufficiently aged that no change in flux density would be expected over several years. However, the positional coincidence and nonthermal spectrum of the radiation make them good candidates for supernovae. The 6 cm luminosity of SN 1950B is about five times that of Cas A and SN 1957D is about fifteen times Cas A. The optical classifications are unknown, but the radio properties resemble Type II supernovae.

An unresolved radio source coincident to better than 1 arcsecond with the position of SN 1961V was detected at the VLA at 20 cm (Branch and Cowan, 1985; Klemola, 1986). This source is very weak, 0.19 mJy, which gives an absolute luminosity about equal to Cas A. Later observations at 6 cm (Cowan, Henry, and Branch, 1988) show the radio spectrum to be nonthermal, with a spectral index of −0.4. Since the supernova is embedded in an H II region, some portion of the radio emission may be thermal. However, a contribution from the H II region probably does not alter the radio spectrum appreciably (Cowan, Henry, and Branch, 1988).

SN 1961V is considered to be the prototype of the Type V supernovae, characterized by optical spectra like a Type II object, but with comparatively narrow lines corresponding to an expansion velocity of 2000 km s^{-1} and a very slowly declining light curve. The presupernova star for SN 1961V was seen for nearly 40 years before the explosion and the supernova itself was studied optically for about 8 years after the explosion. The enormous luminosity of the progenitor star and the extended high luminosity light curve suggest that the progenitor star was very massive, perhaps as much as a thousand solar masses. The optical longevity and low expansion velocity of the Type V supernova is also seen in SN 1986J which is currently the most luminous radio supernova.

It is not known if sources like SN 1961V and SN 1957D are declining radio supernovae or very young SNRs that are brightening as they sweep up interstellar matter (Gull, 1973). Observations over the next few decades should reveal if these sources are decaying in flux.

4.7. Very Early Radio Emission from SN 1987A

The supernova SN 1987A in the Large Magellanic Cloud was the first supernova in four centuries visible to the naked eye. It was discovered within a day of the explosion and offered a unique chance to study the early supernova processes. SN 1987A was a Type II supernova, but was unusual optically and in the radio. Optical emission lines showed a very high velocity, with Doppler widths of more than 20,000 km s^{-1}; at optical maximum the supernova was underluminous by 2 mag compared to other Type II supernovae.

Radio emission was detected with the first observations made at 0.843 GHz and 1.4 GHz, 2 days after explosion (Turtle *et al.*, 1987). Observations 1 day later at 2.3 GHz also detected the supernova but by then the flux density was declining. Although the peak flux density was over 100 mJy, SN 1987A was underluminous by a factor of 1000 compared to a bright radio supernova like SN 1979C. Also, except for the 0.843 GHz radiation, which was detectable for several weeks, the radio emission fell below the detection limits within a few days. SN 1987A was so underluminous and passed away so quickly that it would have gone undetected in any more distant galaxy.

The explanation of the radio quiet nature of SN 1987A generally lies with the progenitor having been a B3 supergiant which would not be expected to have a large stellar wind. Thus there is no extensive circumstellar shell necessary for generating powerful radio emission (Storey and Manchester, 1987; Chevalier and Fransson, 1987; Manchester, 1987).

4.8. Summary of Radio Supernovae

A summary of the properties of the known radio supernovae is given in Table 4.1, a modified version of an earlier table in Weiler *et al.* (1986). Of particular interest

Figure 4.7. The 6 cm radio luminosity versus age of supernovae and young SNRs. The Type I supernovae are shown as open squares and the Type II are shown as filled squares. The shell type SNRs are shown as open circles, and the filled center SNRs are shown as filled circles. Where radio light curves have been measured for the supernovae, the best fit curve and maximum measured point at 6 cm are shown. Upper limits for the Type Ia supernovae are shown, along with an upper limit for the light curve for SN 1981B. (From Weiler *et al.*, 1986.)

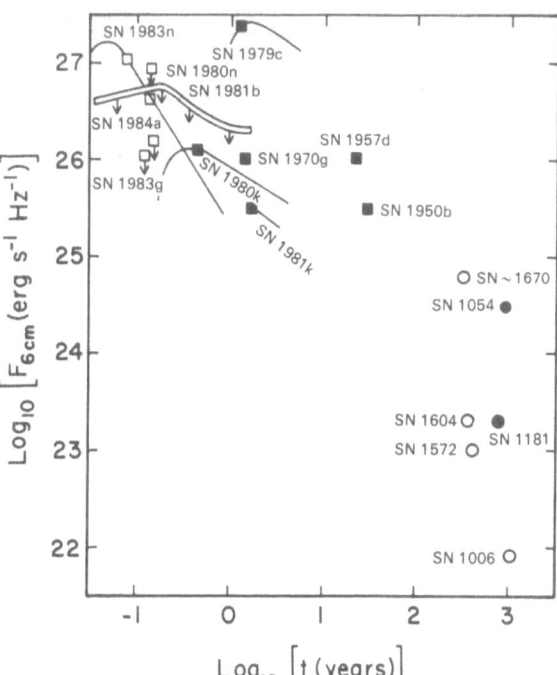

is the wide range of luminosities seen in the radio supernovae and the correlation of steep spectral index with steep temporal index (fast decay rate) as predicted in the models discussed below. SN 1950B, SN 1957D, and SN 1981K have rather poor optical data but, based on their radio spectral indices and decay rates, are likely to have been Type II supernovae. As the table shows, there have still been no detections of a supernova of Type Ia.

The relative luminosities of the different types of radio supernovae and the younger SNRs can be seen in Figure 4.7. The detected radio supernovae are far more luminous than the young SNRs. It is apparent that continued decay of the luminosity of the Type Ib supernovae SN 1983N could not directly evolve into the SNRs. It is possible that the Type II objects could do so. What is missing is an observed evolutionary track of radio supernovae at age 10–100 years.

4.9. Models for the Supernova Radio Emission

Given the steep radio spectra and the very high brightness temperatures seen in radio supernovae, the emission process is almost certainly nonthermal. For example, both the direct measurements using VLBI and a source model assuming a 10^{+4} km s^{-1} expanding shell give an angular diameter for SN 1979C of about 1 milliarcsecond. To produce the measured flux density then requires a source brightness temperature of about 10^{+9} K (Weiler *et al.*, 1981; Bartel *et al.*, 1985). This is far higher than typical astrophysical thermal radio sources, which usually have a brightness temperature of about 10,000 K. Therefore, models for supernova radio emission need to explain the source of energetic particles and the magnetic field that produce the nonthermal synchrotron radiation, and they need to explain the time evolution of the ratio luminosity.

Two classes of models have been developed thus far to explain the radio observations; those that utilize a central pulsar formed by the supernova to power the radio emission (Pacini and Salvati, 1981), and those that utilize shock acceleration in the region exterior to the supernova photosphere (Chevalier, 1982, 1984; also Chapter 5 in this volume). Both classes of models produce a power-law radio spectrum and a power-law decay of luminosity with time. In fact, in a limiting case, the pulsar and the shock acceleration models predict identical radio light curves. A detailed comparison of these two models is given in Weiler *et al.* (1986). For most of the radio supernovae it is likely that the delayed appearance of the radio emission is caused by free–free absorption within the circumstellar envelope arising from the stellar wind of the progenitor star.

The pulsar driven model utilizes the rotational energy loss of a rapidly spinning and slowing pulsar left over from the supernova explosion. This model can fit the observed radio light curves but there are a few problems. The pulsar and thus the magnetic field and energetic particles generated by the pulsar are located within the several solar masses of material ejected by the supernova explosion. This ejecta will be opaque to radio waves for several decades after the explosion. Only if the ejecta clump into filaments at a very early age can the radiation escape. However, this would not explain why VLBI observations of SN 1979C and SN 1986J show extended radio sources.

The second class of models utilize shock acceleration in the region where the supernova ejecta interact with a pre-existing circumstellar shell of material that was created by the stellar wind in the latter stages of the star's life. The region between the shock propagating into the circumstellar material and the reverse shock propagating back into the supernova envelope is Rayleigh–Taylor unstable, and the instability can drive turbulence which results in magnetic field enhancement and particle acceleration (Chevalier, 1984).

Since the circumstellar shell is ionized and optically thick at radio frequencies, the synchrotron radiation generated in the shock region does not escape until the shock wave passes far enough through the shell that the optical depth is unity. Then there is a very rapid radio turn-on first at high frequencies and then at progressively lower frequencies. Thus, there is a very natural process that ties together the particle acceleration and the early absorption of the radio radiation. From models of supernova explosions the density gradient of the supernova envelope is a power law which in turn predicts a radio light curve with a power-law decay with time. This also agrees well with observations (Weiler *et al.*, 1986). Both the intensity and the time scale of the radio emission are dependent on the characteristics of the circumstellar shell.

One side result of the radio observations of supernovae is an estimate of the mass loss rates of the progenitor stars. Assuming that free–free thermal absorption by surrounding matter is the cause of the delayed radio turn-on, we can calculate the amount of ionized matter along the line of sight to the supernovae. If we further assume that this matter is produced from the stellar wind of either the supernova progenitor or from a binary companion of the supernova progenitor, we can estimate the mass loss rate (Weiler *et al.*, 1986). The estimated rates are quite high for both Type II ($5 \times 10^{-5}\ M_\odot\ \mathrm{yr}^{-1}$ for SN 1979C) and Type Ib ($2 \times 10^{-6}\ M_\odot\ \mathrm{yr}^{-1}$ for SN 1983N). For the very fast time scales of SN 1987A, Storey and Manchester (1987) find a low mass loss rate of less than $10^{-7}\ M_\odot\ \mathrm{yr}^{-1}$, suggesting that the progenitor did not go through a recent red giant phase.

A refinement of the model of Chevalier has been given by Lundqvist and Fransson (1988) where, instead of assuming a fully ionized and constant temperature circumstellar shell, they calculate the time-dependent temperature and ionization structure of the shell. This leads to a revision of the mass loss rates of earlier models by a factor of 2 or 3. Under some circumstances it leads to a cooling of the circumstellar material near the shock front as the shell turns optically thin. This produces a luminosity variation like that seen in SN 1979C at 6 cm between days 500 and 700 (Figure 4.2).

4.10. Conclusion

The study of radio supernovae has developed almost in its entirety since the detection of SN 1979C in 1980. The detailed characteristics of the radio light curves and spectral evolution are known and can be related to the properties of the progenitor stars. However, better statistics for radio supernovae still need to be etablished. As in any early exploration the most luminous objects receive the greatest scrutiny, but they may not be representative. With a more complete understanding of the range of radio luminosities, and the corresponding light curves, the radio studies combined with studies at other wave bands will play an important role in understanding the final stages of stellar evolution.

Acknowledgments
The National Radio Astronomy Observatory in operated by Associated Universities, Inc. under a cooperative agreement with the National Science Foundation.

References
Allen, R.J., Goss, W.M., Ekers, R.D., and de Bruyn, A.G. 1976, *Astron. Astrophys.*, **48**, 253–261.
Bartel, N., Rogers, A.E.E., Shapiro, I.I., Gorenstein, M.V., Gwinn, C.R., Marcaide, J.M., and Weiler, K.W. 1985, *Nature*, **318**, 25–30.
Bartel, N., Rupen, M.P., and Shapiro, I.I. 1987, *IAU Circ.*, No. 4292.
Branch, D. and Cowan, J.J. 1985, *Ap. J. (Letters)*, **297**, L33–L36.
Brown, R.L. and Marscher, A.P. 1978, *Ap. J.* **220**, 467–473.
Chevalier, R.A. 1982, *Ap. J.*, **259**, 302–310.
Chevalier, R.A. 1984, *Ann. N.Y. Acad. Sci.*, **422**, 215–232.
Chevalier, R.A. 1987, *Nature*, **329**, 611–612.
Chevalier, R.A. and Fransson, C. 1987, *Nature*, **328**, 44–45.
Cowan, J.J. and Branch, D. 1982, *Ap. J.*, **258**, 31–34.
Cowan, J.J. and Branch, D. 1985, *Ap. J.*, **293**, 400–406.
Cowan, J.J., Henry, R.B.C., and Branch, D. 1988, *Ap. J.*, **329**, 116–121.
de Bruyn, A.G. 1973, *Astron. Astrophys.*, **26**, 105–112.
Goss, W.M., Allen, R.J., Ekers, R.D., and de Bruyn, A.G. 1973, *Nature Phys. Sci.*, **243**, 42–44.

Gottesman, S.T., Broderick, J.J., Brown, R.L., Balack, B., and Palmer, P. 1972, *Ap. J.*, **174**, 383–388.
Gull, S.F. 1973, *M.N.R.A.S.*, **161**, 47–69.
Hogbom, J.A. and Brouw, W.N. 1974, *Astron. Astrophys.*, **33**, 289–301.
van der Hulst, J.M., Hummel, E., Davies, R.D., Pedlar, A., and van Albada, G.D. 1983, *Nature*, **306**, 566–568.
Klemola, A.R. 1986, *Pub. A.S.P.*, **98**, 464–466.
Lundqvist, P. and Fransson, C. 1988, *Astron. Astrophys.*, **192**, 221–233.
Manchester, R.N. 1987, in *Proceedings of the ESO Workshop on SN 1987A*, ed. I.J. Danziger (Garching: European Southern Observatory), pp. 177–185.
Napier, P.J., Thompson, A.R., and Ekers, R.D. 1983, *Proc. IEEE*, **71**, 1295–1322.
Pacini, F. and Salvati, M. 1981, *Ap. J. (Letters)*, **245**, L107–L108.
Panagia, N., Sramek, R.A., and Weiler, K.W. 1986, *Ap. J. (Letters)*, **300**, L55–L58.
Pennington, R.L. and Dufour, R.J. 1983, *Ap. J.*, **270**, L7–L11.
Rupen, M.P., van Gorkom, J.H., Knapp, G.R., Gunn, J.E., and Schneider, D.P. 1987, *Astron. J.*, **94**, 61–70.
Sramek, R.A., Panagia, N., and Weiler, K.W. 1984, *Ap. J. (Letters)*, **285**, L59–L62.
Sramek, R.A., Weiler, K.W., and van der Hulst, J.M. 1982, in *IAU Symposium 97, Extragalactic Radio Sources*, eds. D.S. Heeschen and C.M. Wade (Dordrecht: Reidel), p. 391.
Storey, M.C. and Manchester, R.N. 1987, *Nature*, **329**, 421–423.
Turtle, A.J., Campbell-Wilson, D., Bunton, J.D., Jauncey, D.L., Kesteven, M.J., Manchester, R.N., Norris, R.P., Storey, M.C., and Reynolds, J.E. 1987, *Nature*, **327**, 38–40.
Ulmer, M.P., Crane, P.C., Brown, R.L., and van der Hulst, J.M. 1980, *Nature*, **285**, 151–152.
Weiler, K.W., van der Hulst, J.M., Sramek, R.A., and Panagia, N. 1981, *Ap. J. (Letters)*, **243**, L151–L156.
Weiler, K.W., Sramek, R.A., Panagia, N., van der Hulst, J.M., and Salvati, M. 1986, *Ap. J.*, **301**, 790–812.

5. Interaction of Supernovae with Circumstellar Matter

ROGER A. CHEVALIER

5.1. Introduction

Stars of essentially all masses are known to lose mass during their evolution. The mass loss characteristics at any time are determined by the evolutionary status of the star. The region around a star that is modified by these processes is referred to as the circumstellar medium. In addition to mass loss from a single star, mass transfer and mass loss in a binary system may play a role in determining the circumstellar environment.

When a star explodes as a supernova, it initially interacts with its circumstellar medium. This interaction can either be through the hydrodynamic interaction of the expanding supernova gas, or through radiation emitted by the supernova. Multiwavelength observations have been particularly important for elucidating these phenomena. Since these observational techniques have become available only fairly recently, strong evidence for circumstellar interaction was first found for the supernovae SN 1979C and SN 1980K. Previous reviews of circumstellar interaction are Chevalier (1984a) and Fransson (1986b).

The plan for this chapter is as follows. Section 5.2 describes the expected circum-stellar environments around supernovae. The degree to which these can be specified depends on our knowledge of the supernova progenitors. The hydrodynamic inter-action of the supernova with the ambient medium is treated in Section 5.3. The outer density profile of the supernova plays a role in this problem. The radiative interaction is discussed in Section 5.4. In addition to the supernova light evolution, light travel time effects play a role in the interaction. Future prospects are discussed in Section 5.5.

5.2. Circumstellar Environments

The supernova type for which we have the best knowledge of the progenitor stars are the Type II events. With the occurrence of SN 1987A, it has become clear that there are at least two classes of progenitor types. The great majority of observed SN II (Type II supernovae) are thought to have red supergiant progenitor stars. This is in accord with model calculations for the evolution of massive stars. The outstanding exception is SN 1987A in the Large Magellanic Cloud (LMC), which is thought to have an observed B3 I progenitor star (West *et al.*, 1987; Walborn *et al.*, 1987). Whether the progenitor star is a red or blue supergiant at the time of the explosion is of great importance for the immediate circumstellar environment.

The evolution of massive stars is not well understood, particularly when the effects of mass loss are included. Wolf–Rayet stars are thought to be massive stars that have lost their hydrogen envelopes during their evolution. The evolution leading to Wolf–Rayet stars is still controversial (Chiosi and Maeder, 1986), but mass loss during a red supergiant phase is quite plausible (Maeder, 1981). Typical lifetimes of Wolf–Rayet stars are $(3–8) \times 10^5$ yr.

If Type Ib supernovae have Wolf–Rayet star progenitors (Harkness and Wheeler, this volume) and Type II supernovae have progenitors with hydrogen envelopes, it is plausible that most Type Ib events come from more massive stars than do SN II and the relative rates of these supernovae give information on the range of progenitor masses. Van den Bergh, McClure, and Evans (1987) estimate that the SN Ib rate is 0.36 times the SN II rate (see also Branch, 1986). For an assumed initial mass function like that of Kennicutt (1984), van den Bergh (1988) derives a lower mass limit of about $8–12$ $M_\odot$ for SN II and a lower limit of about $20–30$ $M_\odot$ for SN Ib. These numbers are for a Hubble constant in the range $50–75$ km s^{-1} Mpc^{-1}. The mass limit for the SN Ib is in rough agreement with expectations for Wolf–Rayet stars (Chiosi and Maeder, 1986), although it is somewhat low. The implication is that early B stars generally become SN II, while O stars generally become SN Ib (van den Bergh, 1988).

The SN II and SN Ib have massive star progenitors which are known to lose mass during their evolution and to emit ionizing radiation. For the main sequence evolution, the mass 15 $M_\odot$ is a critical stellar mass above which the circumstellar effects are large and below which they decrease rapidly. This is because the rate of emission of ionizing photons drops dramatically for stars later than type B0 (e.g., Panagia, 1973). For the massive stars, values of $r_s n_H^{2/3}$, where r_s is the radius of the Stromgren sphere and n_H is the density, are in the tens of pc cm^{-2}, while for the later-type stars the values drop to a few pc cm^{-2} (see Table 5.1). For O-type main sequence stars, strong winds are clearly present while for the B stars, effects of a wind are usually unobservable (Abbott, 1982). An exception is the Be stars, which are observed to have winds with velocities of $600–1100$ km s^{-1} and mass loss rates of $10^{-11}–3 \times 10^{-9}$ $M_\odot$ yr^{-1} (Snow, 1981). The wind luminosity is much lower than that of the O-type stars. Table 5.1 shows the radii of the bubbles that can be created by the winds assuming that adiabatic theory applies (Weaver et al., 1977). McKee, Van Buren, and Lazareff (1984) have examined the propagation of a wind bubble in a cloudy medium. They find that during the main sequence lifetime of an O4–B0 star, a region of radius

$$R_h = 53 n_m^{-0.3} \text{ pc,}$$

where n_m is the average density, is made homogeneous by the photoionizing radiation. A wind bubble expands adiabatically out to this radius, but suffers radiative losses at larger radius so that R_h may characterize the final bubble radius.

The mass range for SN II discussed above covers both O4–B0 stars with large bubbles, and the early B stars which are likely to have smaller regions affected by winds. The nature of the interstellar medium is likely to be important for the early B stars. In our scheme for SN Ib, they are expected to be in large bubbles created by the main sequence wind and photoionizing radiation. Some Wolf–Rayet stars

Table 5.1. Main sequence evolution.

Mass $(M_\odot)$	Spectral type	Lifetime $(10^6$ yr$)$	L_{wind}[a] (ergs s^{-1})	$r_{\text{wind}} n_H^{1/5}$ (pc cm$^{-3/5}$)	$r_s n_H^{2/3}$ [b] (pc cm^{-2})
100	O4 V	3.4	2.1×10^{37}	100	138
60	O6 V	4.2	5.3×10^{36}	86	82
20	O9 V	10.3	1.1×10^{35}	68	40
15	B0.5 V	11.1	2.4×10^{34}	53	10
10	B2 V	21.1	—		2.9

[a] Based on Abbott (1982).
[b] Based on Panagia (1973).

may be the result of the binary evolution of lower mass stars, in which case an extended cavity may not be present.

During the next evolutionary phase for a massive star, the red giant or supergiant phase, the star has a slow wind with velocity 5–50 km s^{-1} and a mass loss rate of 10^{-7}–10^{-4} $M_\odot$ yr^{-1} (Zuckerman, 1980). The total duration of the red giant phase is about 10% of the main sequence lifetime. The rate of mass loss may evolve during this phase; the highest rates of mass loss have been observed in OH/IR stars at the tip of the red giant branch with $M < -6$. The duration of the OH/IR phase may be about 5×10^5 yr and the wind velocity is about 15 km s^{-1} (e.g., Herman, 1985). However, most OH/IR stars may have initial masses in the range 2–5 $M_\odot$, which is below the range of interest for supernovae.

Most Type II supernovae are expected to explode at this point, but extreme mass loss can complicate the evolution. Loops can occur in the HR diagram (Chiosi and Maeder, 1986); while the star is relatively blue, a faster, lower density wind is expected. Loss of the hydrogen envelope leads to a blue Wolf–Rayet star; these stars have typical mass loss rates of 10^{-5}–10^{-4} $M_\odot$ yr^{-1} and wind velocities of 1000–2000 km s^{-1} (Chiosi and Maeder, 1986). This is the expected immediate environment of a Type Ib supernova.

The case of SN 1987A shows that massive stars with their hydrogen envelope can also explode as blue stars. The reason for this is still controversial. Suggestions include the low metallicity in the LMC (Arnett, 1987), low metallicity plus restricted convection (Woosley, 1988), mass loss (Maeder, 1987), and mixing throughout the stellar envelope (Saio, Nomoto, and Kato, 1988). Extreme mass loss does not seem to be the explanation because models for the light curve show that there was a substantial hydrogen envelope at the time of the explosion (Woosley, 1988). In evolutionary models with only low metallicity, the star remains blue throughout its life. However, the presence of red supergiants in the LMC with relative numbers comparable to those in our galaxy (Humphreys and Davidson, 1979) suggests that an earlier red supergiant phase is likely. The models of Woosley and of Saio *et al.* do go through a red supergiant phase. For an initial mass of 20 $M_\odot$, the time from the red phase to the explosion is about 10^4 yr.

The interaction of the fast wind from the blue star with the red supergiant wind creates a shocked, cool shell of the dense wind and a hot shell of shocked fast wind (Chevalier and Imamura, 1983; McCray, 1983). Ring nebulae have been observed around Wolf–Rayet stars. These typically have radii of 3–10 pc, velocities of 30–100 km s^{-1}, and densities of 300–1000 cm^{-3} (Chu, Treffers, and Kwitter, 1983). Some of the nebulae show evidence for abundance enhancements of nitrogen and helium (Kwitter, 1981, 1984). Such enhancements are expected in 15 and 25 $M_\odot$ red supergiants if stellar winds have removed at least 9 $M_\odot$ of the star (Lamb, 1978). Many of the wind-blown nebulae are asymmetrically distributed about the Wolf–Rayet star (Chu, Treffers, and Kwitter, 1983). A model in which this is due to stellar motion is attractive, but is not tenable if the fast wind only interacts with the red supergiant wind because both winds have the same space motion. Interaction with the interstellar medium is needed (e.g., Bandiera, 1987). Asymmetries in the winds may also be a factor.

On the assumption that the progenitor of SN 1987A had a red supergiant phase, a similar swept up shell is expected in this case (Chevalier, 1987b, 1988). In general, the shocked dense gas is radiative, whereas the shocked fast wind is nonradiative. Equations expressing conservation of mass and momentum in the dense shell and conservation of energy for the shocked fast wind lead to a cubic equation for the velocity, V_s, of the shell

$$\dot{M}_r V_s \left(\frac{V_s^2}{V_r} - 2V_s + V_r \right) = \frac{1}{2 + \beta} \dot{M}_b V_b^2,$$

where $\dot{M}$ is the mass loss rate for the red (r) and blue (b) phases, V_r and V_b are the wind velocities during these phases, and β is the fraction of the spherical volume of the shell occupied by the shocked fast wind and is likely to be close to 1. The values of V_s for parameters appropriate to SN 1987A can vary over a wide range: 30–150 km s^{-1}. A velocity of 30 km s^{-1} and an age of 10^4 yr corresponds to a shell radius of 10^{18} cm. The shell mass is

$$M_s = 0.5 \left(\frac{V_s}{V_r} - 1 \right) \left(\frac{\dot{M}_r}{5 \times 10^{-5} \ M_\odot \ \mathrm{yr}^{-1}} \right) \left(\frac{t_s}{10^4 \ \mathrm{yr}} \right) M_\odot,$$

where t_s is the shell age. The density in the shell depends on the low temperature that is attained behind the radiative shock; hydrogen densities of order 10^4 cm^{-3} or more are possible. Closer to the supernova, the density is simply determined by $\rho = \dot{M}_b / 4\pi r^2 V_b$, where $\dot{M}_b \sim 3 \times 10^{-6} \ M_\odot \ \mathrm{yr}^{-1}$ and $V_b \sim 550$ km s^{-1} (Chevalier and Fransson, 1987).

Our understanding of the progenitors of SN Ia is not sufficiently well developed to be able to predict their circumstellar environment. Iben and Tutukov (1984) have considered a number of paths that potentially lead to an SN Ia explosion through the evolution of binary stellar systems. Many of the paths involve mass loss from the binary at some point in the evolution, but the evolution timescale is sufficiently long and the progenitor velocities are likely to be sufficiently high that the exploding star may expand into the ambient interstellar medium. An exception is if the progenitor is a white dwarf accreting mass from the wind of a companion star.

5.3. Hydrodynamic Interaction

The nature of the outer parts of the exploded star plays a crucial role in the circumstellar interaction. The result of the supernova explosion is a radial flow in free expansion. The velocity field is given by $v = r/t$ where t is the age of the explosion. The density can be expressed as $\rho = Bt^{-3}f(v)$ where B is a constant and $f(v)$ is a function that depends on the initial hydrodynamic evolution. The pressure in the expanding gas is negligible because of adiabatic expansion.

Numerical simulations of supernova explosions typically show that the outer part of the density profile can be approximated by a power law in radius (e.g., Jones, Smith, and Straka, 1981). A recent example is Arnett's (1988) model for SN 1987A which shows an outer profile of the form $\rho \propto r^{-n}$, where n is about 9. Some insight into the production of power-law profiles can be gained from self-similar flow theory. Sakurai (1960) found self-similar solutions for the propagation of a shock wave in a medium with $\rho = Ax^{\beta}$ where A and β are constants and x is the distance from the boundary of the star. The solutions are planar, but they should apply to the thin layers at the surface of a star. The solution can be continued into the regime where the gas expands into the vacuum. In the limit $t \to \infty$, the density profile asymptotically approaches a homologous free expansion. It is of the power-law form $\rho \propto (-x)^{-(1+\lambda+\beta)/\lambda}$, where λ is an eigenvalue from the self-similar solution. For the adiabatic index $\gamma = \frac{4}{3}$ and $\beta = \infty$, which is the limit of an initially exponential medium, we have $\rho \propto (-x)^{-6.67}$. Raizer (1964) had already noted that the planar expansion of a shock wave in an exponential medium asymptotically approaches a power-law form. This solution can be expected to describe the propagation of a strong shock wave through the exponential atmosphere of a star if the density scale height is much less than the stellar radius and radiative losses are negligible. These assumptions should especially hold for the explosion of a relatively compact star like the progenitor of SN 1987A. The initially planar free expansion develops into a spherically symmetric expansion. Conservation of mass shows that this steepens the power law by two powers of radius to $\rho \propto r^{-8.7}$. This result generally applies to the expansion of an initially exponential atmosphere. The outer envelope structure of the SN 1987A progenitor described by Woosley (1988) can be approximated by $\beta = 5$; this case lead to a final density distribution $\rho \propto r^{-9.6}$, in good agreement with numerical results. Smaller values of β lead to steeper density profiles.

After a few doubling times of the initial radius (typically a few days), the shock wave in the circumstellar matter heats gas to a high temperature and the cool supernova matter freely expands. The exact time of formation of the shock wave depends on radiative preacceleration (Epstein, 1981) and has not yet been determined accurately. The outer shock wave cannot be freely expanding, so the supernova matter interacts with the hot gas and a second, inner (or reverse) shock front develops. By this time, the effects of radiative acceleration should be small and the red supergiant wind velocity is much less than the shock velocity so that the circumstellar medium can be regarded as stationary.

The expanding supernova gas can be described by the density profile $\rho_s = Bt^{-3}(r/t)^{-n}$, while the circumstellar medium is described by $\rho_c = Ar^{-2}$, where $A = \dot{M}/4\pi v_w$, if it is due to a steady stellar wind. Here $\dot{M}$ is the mass loss rate from the progenitor star and v_w is the wind velocity. The interaction between two media depends on two dimensional constants, A and B, and can be described by a self-similar flow, i.e., one in which the profiles of the physical variables remain constant in time. The radius of the contact discontinuity between the two shocked

Table 5.2. Properties of self-similar solutions.

n	R_1/R_c	R_2/R_c	α	ρ_2/ρ_1	p_2/p_1	u_2/u_1	M_2/M_1
6	1.377	0.958	0.62	3.9	0.21	1.006	0.44
7	1.299	0.970	0.27	7.8	0.27	1.058	0.82
9	1.250	0.981	0.096	19	0.33	1.090	1.6
12	1.226	0.987	0.038	46	0.37	1.104	2.7
14	1.218	0.990	0.025	70	0.38	1.108	3.4

flows is (Chevalier, 1982a; Nadyozhin, 1985)

$$R_c = \left(\alpha\frac{B}{A}\right)^{1/(n-2)} t^{(n-3)/(n-2)},$$

where α is a dimensionless constant. The solutions apply for $n > 5$; if $n < 5$, the flow is expected to approach that of a blast wave from an instantaneous explosion, for which the outer shock front radius increases as $t^{2/3}$. Table 5.2 lists some of the parameters for self-similar solutions in which the flow is energy-conserving and is adiabatic in the postshock region. In the table, the subscript 1 refers to the outer shock wave and 2 to the inner shock wave, except for the mass ratio which is the mass of shocked supernova gas divided by the mass of shocked circumstellar gas. It can be seen that larger values of n lead to larger density contrasts between the inner and outer regions. At the discontinuity, the density goes to infinity and the temperature goes to zero in the solutions; an actual situation will only approach this situation asymptotically in time. Numerical computations are in general agreement with the self-similar solution (Jones and Smith, 1983; Band and Liang, 1988).

The result of the interaction is a shell structure. While the shells have a finite width in the energy-conserving case, they become very thin if radiative processes are important. In the limit of a very thin shell ($R_1 \simeq R_c \simeq R_2$), the self-similar solution can be simply derived from conservation of mass and momentum (Chevalier 1982b) and is

$$\alpha = \frac{2}{(n-4)(n-3)}, \qquad \frac{M_2}{M_1} = \frac{n-4}{2},$$

$$\frac{\rho_2}{\rho_1} = \frac{(n-4)(n-3)}{2}, \qquad \frac{p_2}{p_1} = \frac{(n-4)}{2(n-3)}, \qquad \text{and} \qquad \frac{T_2}{T_1} = \frac{1}{(n-3)^2}.$$

In many cases, these relations provide an adequate description of the energy-conserving case, and will be used to estimate some of the shell properties.

The major cooling mechanisms for the hot gas are Compton cooling and radiative cooling (Chevalier, 1981, 1982b; Fransson, 1982, 1984; Lundqvist and Fransson, 1988). The outer shock velocity, v_s, is expected to be 10^4 km s^{-1} or greater; the postshock temperature is

$$T_1 = 1.39 \times 10^9 \left(\frac{v_s}{10^4 \text{ km s}^{-1}}\right)^2 \text{ K}.$$

The density ahead of the outer shock is

$$\rho_0 = 3.4 \times 10^{-18} \left(\frac{\dot{M}}{5 \times 10^{-5} \, M_\odot \, \text{yr}^{-1}} \right) \left(\frac{v_w}{10 \, \text{km s}^{-1}} \right)^{-1} \left(\frac{t}{100 \, \text{day}} \right)^{-2}$$

$$\times \left(\frac{v_s}{10^4 \, \text{km s}^{-1}} \right)^{-2} \text{gm cm}^{-3}.$$

During the first month, Compton cooling of both the inner and outer shells can be important. The ratio of Compton cooling time to flow time is

$$\frac{t_c}{t_f} = \frac{0.33 \beta_1 T_9^{1/2}}{T_{\text{eff}4}^4 R_{15} W \delta},$$

where β_1 is to account for multiple scattering and is of order unity, $T_9 = T_1/10^9$ K, $R_{15} = R_1/10^{15}$ cm, $T_{\text{eff}4}$ is the effective temperature of the photosphere in units of 10^4 K, $\delta = 1 + 0.43 T_9 + 0.032 T_9^2$ is a relativistic correction, and W is the dilution factor which varies from $\frac{1}{2}$ close to the photospheric radius, R_p, to $\frac{1}{4}(R_p/R_1)^2$ far from the photosphere. In the limit that $R_1 \gg R_p$, $t_c < t_f$ up to an age

$$t = 10 \frac{m \delta}{\beta_1} \left(\frac{L_{\text{sn}}}{10^{43} \, \text{ergs s}^{-1}} \right) \left(\frac{v_s}{10^4 \, \text{km s}^{-1}} \right)^{-2} \text{days},$$

where $m = (n - 3)/(n - 2)$ and L_{sn} is the photospheric luminosity.

Radiative cooling is not generally important for the shocked circumstellar gas because of its relatively high temperature and low density. However, it is important for the shocked supernova gas and the cooling time for the gas in the immediate post-reverse shock region is

$$t_{\text{cool}} = \frac{7 \times 10^9}{(n - 4)(n - 3)^3 m^2} \left(\frac{\dot{M}}{5 \times 10^{-5} \, M_\odot \, \text{yr}^{-1}} \right)^{-1} \left(\frac{v_w}{10 \, \text{km s}^{-1}} \right) \left(\frac{t}{100 \, \text{days}} \right)^2$$

$$\times \left(\frac{v_s}{10^4 \, \text{km s}^{-1}} \right)^4 \left(\frac{\Lambda}{2 \times 10^{-23} \, \text{ergs cm}^3 \, \text{s}^{-1}} \right)^{-1} \text{s},$$

where Λ is the cooling function. For free–free emission, $\Lambda \simeq 10^{-23}(T/10^8 \, \text{K})^{1/2}$; at a temperature of 10^7 K and lower, cooling by line emission dominates. For $n = 10$, the cooling time is less than the flow time up to an age

$$t = 2 \times 10^2 \left(\frac{\dot{M}}{5 \times 10^{-5} \, M_\odot \, \text{yr}^{-1}} \right) \left(\frac{v_w}{10 \, \text{km s}^{-1}} \right)^{-1} \left(\frac{v_s}{10^4 \, \text{km s}^{-1}} \right)^{-4}$$

$$\times \left(\frac{\Lambda}{2 \times 10^{-23} \, \text{ergs cm}^3 \, \text{s}^{-1}} \right) \text{days}.$$

Supernova remnants in the interstellar medium are thought to enter a radiative phase late in their life; for circumstellar interaction, the radiative phase occurs early.

We have been assuming that one temperature can describe both the electrons and ions. The electron temperature is of particular importance for the cooling processes. Coulomb collisions are able to bring about electron–ion equipartition in the shocked supernova gas for a typical case during the first few years or more of evolution. For the shocked circumstellar matter, Coulomb collision can only maintain equipartition for about the first 10 days, after which unequal electron and ion temperature are possible. However, as in galactic supernova remnants, collisionless heating due to plasma instabilities could lead to equipartition even at late times (e.g., McKee and Hollenbach, 1980).

If unimpeded by a magnetic field, thermal heat conduction is expected to have a significant effect on the structure of the shocked regions (Chevalier, 1982a). The main effect is the conduction of heat from the outer shocked circumstellar gas to the cooler shocked supernova gas. In the limits of either saturated heat conduction or an isothermal shocked layer, no new dimensional parameters are introduced so that self-similar flow might be expected. However, such solutions do not appear to exist (Band, 1988). Bedogni and D'Ercole (1988) have numerically computed the flow resulting from the interaction of a supernova envelope with constant density interstellar gas and found that when heat conduction was included, the flow did not settle down to a smooth flow. The pattern was that the inner shock wave moved in toward the center, while a new shock wave formed close to the contact discontinuity. It appears that if heat conduction is effective, it causes a complex time-dependent flow that has not yet been fully elucidated.

Another aspect of the flow that has not yet been fully studied is the presence of nonradial instabilities. General features of the solutions are that the inner parts of the shell are denser than the region close to the outer shock front and that the interaction shell decelerates. These features imply that the flows are subject to the Rayleigh–Taylor instability. The expectation is that the shell becomes irregular, but the final nonlinear outcome of the instability is unknown.

During the phases when cooling processes are effective, the shock wave energy is lost to radiation. The power liberated at the outer shock front is

$$\dot{E}_1 = \frac{\dot{M} v_s^3}{2 v_w} = 1.6 \times 10^{42} \left(\frac{\dot{M}}{5 \times 10^{-5}\ M_\odot\ \mathrm{yr}^{-1}} \right) \left(\frac{v_w}{10\ \mathrm{km\ s}^{-1}} \right)^{-1}$$

$$\times \left(\frac{v_s}{10^9\ \mathrm{km\ s}^{-1}} \right)^3 \mathrm{ergs\ s}^{-1},$$

and that liberated at the inner shock front is

$$\dot{E}_2 = \frac{n-4}{2(n-3)^2} \dot{E}_1.$$

For $n = 10$, $\dot{E}_2$ is a factor of 16 smaller than $\dot{E}_1$. If Compton cooling by photospheric photons is the relevant cooling mechanism, this energy emerges as ultraviolet radiation and if radiative cooling is the cooling mechanism, it emerges as X-ray radiation.

X-ray radiation provides the most direct observation of the hot interaction region. Because of its lower temperature and higher density, the shocked supernova gas is the dominant X-ray source. There are three main evolutionary phases of the X-ray light curve. In the first phase, there is substantial absorption by the surrounding gas. In the energy range 0.5–10 keV, the photoelectric effect is the dominant absorption process and the opacity is $\kappa^{-1} = 9.4 \times 10^{-3}\ (E/keV)^{8/3}$ gm cm^{-2} for a low ionization gas. With this opacity, optical depth unity to gas outside the outer shock front is reached at an age

$$t = 310m \left(\frac{\dot{M}}{5 \times 10^{-5} \, M_\odot \, \text{yr}^{-1}} \right) \left(\frac{v_w}{10 \, \text{km s}^{-1}} \right)^{-1} \left(\frac{v_s}{10^4 \, \text{km s}^{-1}} \right)^{-1} \left(\frac{E}{\text{keV}} \right)^{-8/3} \text{days,}$$

where $m = (n-3)/(n-2)$. If elements such as oxygen are fully ionized in the preshock region, the absorption is reduced. If a dense layer is formed in the interaction region by cooling processes, the absorption is increased. In the second phase, radiative cooling is still important and the X-ray luminosity is given by $\dot{E}_2$. The X-ray luminosity evolves slowly. The luminosity decreases more rapidly during the third, adiabatic phase. Free–free radiation results in luminosity $\propto t^{-1}$. A steeper decline of X-ray luminosity occurs if the shock wave breaks out of the red supergiant wind. Itoh and Masai (1989) note that the rapid expansion cooling can lead to an apparent overionization of the X-ray emitting gas.

Einstein Observatory observations of SN 1980K provided the first X-ray detection of an extragalactic supernova (Canizares, Kriss, and Feigelson, 1982) and the emission can be interpreted as thermal radiation from the shocked supernova gas (Chevalier, 1982b). At a distance of 10 Mpc, the X-ray luminosity about 40 days after the discovery of the supernova was 2×10^{39} ergs s^{-1}. Over the next 50 days, the flux declined by a factor of 2 or more, indicating that the shell was in the energy-conserving phase. The luminosity and the evolutionary phase are consistent with $\dot{M}$ of $(1-3) \times 10^{-5} \, M_\odot$ yr^{-1} for $v_w = 10$ km s^{-1} if $v_s \approx 1.3 \times 10^4$ km s^{-1} (from the line width measured by Barbon, Ciatti, and Rossino, 1982). Inverse Compton scattering of photospheric photons with relativistic electrons is another possible source of the X-ray emission (Beall, 1979; Canizares, Kriss, and Feigelson, 1982), but Chevalier (1982b) estimates that thermal emission dominates the inverse Compton mechanism.

Thermal X-ray emission from the nearby supernova SN 1987A was not detected soon after the explosion (Makino, 1987). This can be attributed to the low density that is expected in the blue supergiant wind (Chevalier and Fransson, 1987). As discussed in the next section, there is good evidence for a dense shell remnant of a red supergiant phase at a distance of about 5×10^{17} cm from the progenitor star. A substantial brightening of thermal X-ray emission is expected when the outer shock front reaches the dense gas in 10–20 years.

While X-ray emission from the interaction region can be calculated in detail, the production of radio synchrotron radiation from the region, which is also likely, is less well understood. Yet, it appears that nonthermal radio emission is the most common means by which the interaction region is observed (Weiler et al., 1986; Sramek and Weiler, this volume). The magnetic field in the interaction region may be built up by random motions generated by the Rayleigh–Taylor instability or it may be the compressed circumstellar field if the progenitor star had a strong magnetic field (Fedorenko, 1984). The relativistic electrons may also be generated by the random motions or may be accelerated in the shock waves. If the efficiency of the generation of magnetic fields and relativistic electrons is comparable to that known to occur in the galactic supernova remnants, the observed luminosity of radio supernovae can be reproduced (Chevalier, 1982b).

The observations show that there is a low-frequency turnover to the radio spectra at early times. In the circumstellar model, the turnover is due to free–free absorption in the circumstellar medium external to the interaction shell. This mechanism can explain the sharp turn on of the radio emission at one frequency. The basic properties of the radio light curves are described by a simple model in which it is assumed that the ratios of magnetic energy density and relativistic electron energy density to thermal energy density remain constant with time in the interaction shell. The radio flux at a particular frequency then varies as

$$F_v = \text{const.}\left(\frac{\lambda}{20 \text{ cm}}\right)^{(\gamma-1)/2}\left(\frac{t}{t_{20}}\right)^{-(\gamma+5-6m)/2} \exp\left[-\left(\frac{\lambda}{20 \text{ cm}}\right)^2\left(\frac{t}{t_{20}}\right)^{-3m}\right],$$

where λ is the wavelength of observation, γ is the power-law index of the electron energy spectrum, t_{20} is the time at which the circumstellar medium exterior to the emitting shell has a free–free absorption optical depth of unity at 20 cm, and $m = (n-3)/(n-2)$. In this model γ and the temperature of the absorbing medium are assumed to be constant in time. While this model does reproduce the basic features of observed radio light curves (Chevalier, 1984a, b; Weiler *et al.*, 1986), it is probably oversimplified. Values of γ for different radio supernovae are observed to range from 2 to 3 (Weiler *et al.*, 1986) so that some evolution of γ might be expected. Also, the preshock gas does undergo complex temperature and ionization evolution (Lundqvist and Fransson, 1988). At early times, Compton heating results in a gas temperature of about 10^5 K and at late times, recombination of the gas is important. Lundqvist and Fransson (1988) found that a feature near the maximum of the 20 cm light curve for SN 1979C can be reproduced when these effects are included.

The time it takes for the interaction region to move out to a point where the gas is optically thin gives a measure of the amount of circumstellar matter. There are presently five radio supernovae with fairly extensive data, including the rising part of the radio light curve (Sramek and Weiler, this volume). Table 5.3 lists the supernova, the supernova type, the time of optical depth unity at 20 cm, and an estimate of circumstellar density (factor of 2 or 3 accuracy). Of the five supernovae, only SN 1979C and SN 1980K show clear evidence for wind interaction outside of radio wavelengths. For SN 1983N and SN 1987A, this is attributable to the low circumstellar density and for SN 1986J to the late discovery. The results show that the winds around SN 1979C, 1980K, and 1986J are consistent with the dense slow winds expected around red supergiant stars. The density around the Type Ib event SN 1983N is considerably lower, but is roughly consistent with the value expected around a Wolf–Rayet star. A wind velocity of 1000 km s^{-1} and $\dot{M} = 10^{-4} M_\odot$ yr^{-1} leads to a value of $\dot{M}/v_w$ that is a factor of 5 below the estimated value. SN 1987A had an even earlier turn-on and was a faint radio supernova, but the estimated value of $\dot{M}/v_w$ is roughly consistent with the density expected around a B3 I star like the Sk-69 202 progenitor star (Chevalier and Fransson, 1987). The observational estimate is again a factor of a few larger than the expected value. If there is clumping in the circumstellar wind, the observational estimates are reduced.

An exciting development is the possibility of resolving radio supernovae with very long baseline interferometry (VLBI) techniques. The expansion of SN 1979C has

Table 5.3. Radio supernovae.

Supernova	Type	t_{20} (days)	$\dot{M}/v_w$ $(M_\odot \, yr^{-1})/(km \, s^{-1})$	Ref.
1979C	II	950	1×10^{-5}	a
1980K	II	190	3×10^{-6}	a
1983N	Ib	30	5×10^{-7}	b
1986J	II	1600	2×10^{-5}	c
1987A	II	2	1×10^{-8}	d

[a] Lundqvist and Fransson (1988).
[b] Chevalier (1984b); Sramek, Panagia, and Weiler (1984).
[c] Chevalier (1987a).
[d] Chevalier and Fransson (1987).

been measured (Bartel *et al.*, 1985) and Bartel (1988) has estimated that if the expansion follows $R \propto t^m$, then $m = 1.03 \pm 0.15$. The radius and the expansion law are consistent with circumstellar interaction. SN 1986J, which is currently the brightest radio supernova, has also been resolved by VLBI observations and there is evidence for asymmetry in the radio brightness distribution (Bartel, Rupen, and Shapiro, 1989). Measurements of the expansion of the radio source should allow the age of the supernova to be estimated. A prediction of the circumstellar inter-action model is that radio supernovae should have a shell structure. The develop-ment of observing techniques and the high probability of a nearby (~ 10 Mpc) Type II supernova in the next 10 years should allow this prediction to be tested.

The presence of relativistic electrons in the interaction region suggests that relativistic protons are also present. Theories of shock acceleration predict high efficiency for proton acceleration (Ellison and Eichler, 1985). Relativistic protons interact with the dense gas to produce pions which decay into γ-rays with energy of about 500 MeV. Assuming efficient proton acceleration and a distance of 23 Mpc, Chevalier (1983) calculated a γ-ray flux of 1×10^{-9} ($t/20$ days)$^{-6/5}$ photons cm^{-2} s^{-1} for SN 1979C. Current techniques (e.g., COS-B) have a flux limit of about 10^{-6} photons cm^{-2} s^{-1}. A galactic supernova with a dense circumstellar medium should be easily detectable in high-energy γ-rays, providing a test of acceleration me-chanisms for heavy particles (Chevalier, 1983; Berezinsky and Ptuskin, 1989).

5.4. Radiative Interaction

Except for the facts that supernovae are usually discovered relatively late in their evolution and that light curve data on individual supernovae are often sparse, the optical light from supernovae can be directly measured. Kirshner (this volume) reviews the data on supernova light curves. Of particular interest for circumstellar interaction is the ionizing radiation from the supernova, which cannot generally be measured directly. Three sources of energetic radiation of particular importance are the photospheric emission at the time of shock breakout, the Comptonized photo-spheric emission, and the X-rays from the hot interaction shell. Other potential sources of energetic radiation are radioactivity and pulsar activity, but they do not clearly play a role in circumstellar interaction in observed sources.

The acceleration of the shock wave through the outer envelope of the progenitor star leads to high temperatures. The numerical calculation of this process requires high resolution in the outer part of the star; Klein and Chevalier (1978), Falk (1978), and Lasher and Chan (1979) found that the photospheric temperature in the explosion of a red supergiant star reaches $(2-3) \times 10^5$ K. The total energy in ionizing radiation is about 10^{48} ergs and the duration of the burst is about 30 minutes. An important property of the radiation is that electron scattering dominates the opacity so that the emitted photons have their origin relatively deep within the explosion. The result is that the radiation is shifted to higher energies than is expected for a blackbody spectrum (Imshennik and Utrobin, 1977; Klein and Chevalier, 1978), implying that a significant fraction of the radiation is in the soft X-ray range. Klein et al. (1979) searched the data in the HEAO-1 survey for evidence of such a burst without success, but neither HEAO-1 nor any X-ray observatory to date has been well suited for detecting these events. SN 1987A has a smaller radius progenitor star so that less ionizing radiation is expected. Light curve models give a peak photospheric temperature of about 3×10^5 K and a total ionizing energy of about 10^{47} ergs (Arnett, 1988; Woosley, 1988). Dopita et al. (1987) make a similar estimate based on the extrapolation of observed properties. Detailed calculations including the nonblackbody nature of the spectrum have not yet been carried out.

The time during which the photosphere is sufficiently hot to emit ionizing radiation in a blackbody spectrum is relatively brief, typically a few days or less. However, if a hot shell surrounds the supernova due to circumstellar interaction, the photospheric photons can be scattered up in energy by inverse Compton scattering (Fransson, 1982, 1984). For a presupernova $\dot{M} \approx (1-10) \times 10^{-5} \ M_\odot \ \mathrm{yr}^{-1}$ and $v_w = 10 \ \mathrm{km \ s}^{-1}$, the electron scattering optical depth is $\tau_e = (0.5-2) \times 10^{-2}$ near the time of optical maximum light. For $\tau_e \gtrsim 10^{-2}$, the Compton scattering gives an ultraviolet, power-law extension to the photospheric emission; the value of the spectral index α is in the range 1.5–3 (Fransson, 1984). The Compton luminosity can be expressed as

$$L_c = 4.5 \times 10^{40} \left(\frac{R_p}{10^{15} \ \mathrm{cm}}\right)^2 \left(\frac{\tau_e}{10^{-2}}\right) \left(\frac{T_{eff}}{10^4}\right)^{\alpha+3} T_9^{2.35} \ \mathrm{ergs \ s}^{-1},$$

where T_{eff} is the effective temperature of the photosphere. This luminosity is a strong function of T_{eff} and it probably dominates the X-ray luminosity of the hot gas only for $T_{eff} \gtrsim 7000$ K. The X-ray luminosity was discussed in the previous section.

The different character of the three sources of ionizing radiation results in different effects on the surrounding medium. The initial burst is capable of highly ionizing a typical circumstellar medium. However, the recombination time is (Lundqvist and Fransson, 1988)

$$t_{rec} = 0.8 \left(\frac{\alpha_r}{10^{-13} \ \mathrm{cm}^3 \ \mathrm{s}^{-1}}\right)^{-1} \left(\frac{r}{10^{15} \ \mathrm{cm}}\right)^2 \left(\frac{\dot{M}}{5 \times 10^{-5} \ M_\odot \ \mathrm{yr}^{-1}}\right)^{-1}$$

$$\times \left(\frac{v_w}{10 \ \mathrm{km \ s}^{-1}}\right) \mathrm{days},$$

where α_r is the recombination coefficient so that recombination of the gas in the inner circumstellar medium is likely to be important. The cooling time of the gas is even shorter. Thus, while the ionization at large r is effectively frozen in, the gas that is closer in tends to cool and recombine. This gas is strongly affected by the Compton and X-ray flux from the interaction shell.

Fransson (1984) has described the effects of the ionizing radiation on the expanding supernova gas inside the interaction shell. The expanding gas can be divided into four zones depending on the degree to which they absorb radiation from the hot shell. The lower-energy ionizing radiation is absorbed close to the shock front while the more energetic radiation is more penetrating and reaches closer to the photosphere. Starting inward from the reverse shock front, the zones are the corona, the upper chromosphere, the lower chromosphere, and the photosphere. The corona is strongly Compton heated and has a temperature of about 10^5 K. The gas is very highly ionized in this region. The upper chromosphere is optically thin in the Lyman continuum and is heated to a temperature of about 1.5×10^4 K. Ionizations are from the ground state in this region. Ions such as N V, C III, C IV, and Si IV, which have strong ultraviolet transitions, are present. The lower chromosphere is optically thin to the Lyman continuum and ionizations are by collisions and by photoionizations from upper states of hydrogen. Balmer line and Mg II line emission is likely here. The photospheric region is relatively little affected by the circumstellar interaction.

The circumstellar gas outside of the interaction shell is also ionized by the shell radiation (Fransson, 1982; Lundqvist and Fransson, 1988). The gas close to the blast wave initially is heated to a temperature $\gtrsim 10^5$ K and then experiences two rapid drops in temperature, corresponding to the increased cooling by low ionization ions and the drop in the relative electron abundance. At large radii ($r \gtrsim 0.1$ pc) the structure is frozen-in and determined by the outburst and the early Compton luminosity. For high mass loss rates ($\dot{M} \gtrsim 5 \times 10^{-5}~M_\odot~\mathrm{yr}^{-1}$ for $v_w = 10~\mathrm{km~s}^{-1}$) the wind within a few shock radii has sufficient time to recombine. Thus low ionization stage ions are expected to be abundant within a few months after the explosion. For low density winds ($\dot{M} \lesssim 5 \times 10^{-5}~M_\odot~\mathrm{yr}^{-1}$), the dominant metal ions are in the hydrogen- and helium-like ionization stages, while hydrogen and helium are almost fully ionized throughout the evolution. Potentially observable ions having strong resonance lines with $\lambda \gtrsim 912$ Å are: C III $\lambda 977$, C IV $\lambda\lambda 1548$–51, H I $\lambda 1216$, N V $\lambda 1239$–43, O VI $\lambda\lambda 1032$–38, Si IV $\lambda\lambda 1394$–1403, and Mg II $\lambda\lambda 2796$–2803. Absorption lines due to these ions are expected.

As noted in the previous section, the radio emission from SN 1979C shows good evidence for strong circumstellar interaction in this case. The ultraviolet emission from this supernova did show evidence of circumstellar radiative interaction. First, an ultraviolet excess was observed near maximum light with the International Ultraviolet Explorer (IUE) (Panagia et al., 1980). This could be attributed to Compton scattering in the hot interaction shell (Fransson, 1982). Second, broad emission lines of N V $\lambda\lambda 1239$–43, S IV $\lambda\lambda 1394$–1403, N IV] $\lambda 1486$, C IV $\lambda\lambda 1548$–51, He II $\lambda 1640$, N III] $\lambda 1750$, and C III] $\lambda 1909$ were observed (Fransson et al., 1984). The line widths implied velocities in the line-forming region $\gtrsim 8400~\mathrm{km~s}^{-1}$, i.e., velocities comparable to that of gas in the supernova photosphere. The line profile implied formation in a narrow region at less than 1.3 times the photospheric radius. The line intensities gave an electron density $\sim 4 \times 10^9$ cm^{-3} and a nitrogen to carbon ratio, N/C ~ 8. This high nitrogen to carbon ratio suggests that there was substantial mass loss from the progenitor star so that layers that underwent carbon–nitrogen–oxygen processing were uncovered. Finally, a narrow, transient emission line of N III] $\lambda 1750$ was observed that may have been from circumstellar gas external to the interaction shell.

The relative line intensities from SN 1979C and their evolution can be well modeled by emission from the upper chromosphere region discussed above if $\dot{M} \approx 10^{-4}\ M_\odot$ yr^{-1} for $v_w = 10$ km s^{-1} (Fransson, 1984). Circumstellar ultraviolet absorption lines may have been present, but could not be distinguished from strong interstellar absorption along the line of sight; the velocity resolution of the IUE is about 1000 km s^{-1}. The Space Telescope will be valuable for investigating this aspect of circumstellar interaction. Convincing evidence for this phenomenon will be the time evolution of the absorption features.

While the wind velocity for a red supergiant star is expected to be low, there is the possibility of radiative acceleration of the preshock gas. Mechanisms for acceleration are through electron scattering (Fransson, 1982) and scattering in ultraviolet resonance lines (Chevalier, 1981; Fransson, 1986a). For a radiated energy of 10^{49} ergs, Fransson (1986a) estimates that the gas can be radiatively accelerated to a velocity

$$v(r) = 1 \times 10^3 \left(\frac{r}{10^{15}\ \text{cm}}\right)^{-3.3} \text{km s}^{-1}.$$

The Type II supernova SN 1984E did show strong, relatively narrow Hα emission (Dopita *et al.*, 1984). The observed lines width ($\lesssim 3000$ km s^{-1}) indicates that the high density region was considerably preaccelerated.

All of the above discussion applies to radiative interaction with a dense red supergiant wind. SN 1987A is an interesting case because the circumstellar medium immediately surrounding the supernova appears to be of low density, but there probably is dense gas at some distance out that is the remnant of a previous red supergiant phase. Since the distance to the gas is not much less than ct, where t is the age, light travel time effects are important in the interpretation of the emission. These effects were first discussed in detail by Couderc (1939) who considered the scattered light echoes observed around Nova Persei 1901. A pulse of light from the source illuminates an ellipsoid with the source and the observer at the foci. Because the distance to the source is generally large, the source illuminates a paraboloidal region in its vicinity. The intersection of the paraboloid with a shell of radius R_s surrounding the source gives a circle with radius

$$x = \{c(t - t_e)[2R_s - c(t - t_e)]\}^{1/2}$$

from the line of sight, where t_e is the time of emission of the pulse. From the observer's point of view, only the part of the shell close to the line of sight is initially illuminated, but other parts of the shell become visible with time.

Narrow emission lines have been observed from SN 1987A that are likely to be from such a surrounding shell. Lines of He II, C III, N III, N IV, N V, and O III have been observed in the ultraviolet with the IUE (Fransson *et al.*, 1989) and lines of O III and H I in the optical (Wampler, 1988). High-resolution observations with the IUE show that the widths of the lines are less than ~ 30 km s^{-1} (FWHM). The source of ionizing radiation for this gas is probably the photospheric emission at the time of shock breakout because the low density surrounding the progenitor star does not lead to significant ionizing radiation from the hot shell. Thus the ionizing radiation is effectively emitted in a pulse and the light travel time effects can be estimated (Chevalier, 1988). If the shell density is high and the recombination and emission times are short compared to the age, a thin ring on the shell is illuminated. The flux in a line remains constant until time $2R_s/c$ and the line velocity relative to the supernova is

$$V_1 = -V_s\left(1 - \frac{ct}{R_s}\right)$$

where V_s is the velocity of the shell. It can be seen that V_1 shifts to the red linearly with time at a rate that depends of R_s/V_s, which is related to the age of the shell. In the low-density limit, the emissivity of the gas remains approximately constant after it is initially illuminated. Then the flux in a line increases linearly with time until time $2R_s/c$ and the line of sight velocity varies from $-V_s$ to V_1. Observations of the line fluxes show approximately linear increases in some cases and more steady behavior in others (Fransson *et al.*, 1989). It is clear that recombination effects are important and a detailed time-dependent model for the properties of the shell is necessary (Lundqvist and Fransson, 1987; Fransson and Lundqvist, 1989). Lines of N III and N V, which initially showed an approximately linear rise in flux, peaked at an age of 400 days and began to decline in flux (Sonneborn *et al.*, 1988); these observations suggest $R_s \approx 200$ light days $\approx 5 \times 10^{17}$ cm.

An important aspect of the IUE observations is that a nebular analysis of the lines reveals a large nitrogen overabundance with $N/C = 7.8 \pm 4$ and $N/O = 1.6 \pm 0.8$ (Fransson *et al.*, 1989). These are factors of 37 and 12 higher than the solar values, respectively, implying that the gas has undergone substantial carbon–nitrogen–oxygen processing. As described above, a similar situation was observed in the outer layers of SN 1979C. It is plausible that mass loss has played a role in these abundance anomalies.

The winds from red giant stars are known to contain dust often, so that infrared emission from radiatively heated dust might be expected. Such emission has been detected from SN 1979C and SN 1980K (Bode and Evans, 1980; Dwek, 1983). The dust temperature is determined by a balance between the radiative heating and emission from the grain surface. If the supernova radiates approximately like a blackbody of temperature T_{sn} and if the dust absorption efficiency varies as λ^{-n}, the grain temperature is

$$T_g = T_{sn}W^{1/(4+n)},$$

where W is the dilution factor for the supernova radiation. For $n = 1$, we have

$$T_g = 280\left(\frac{T_{sn}}{5000 \text{ K}}\right)^{0.2}\left(\frac{L_{sn}}{10^{42} \text{ ergs s}^{-1}}\right)^{0.2}\left(\frac{r}{10^{18} \text{ cm}}\right)^{-0.4} \text{ K},$$

where L_{sn} is the supernova luminosity.

The dust close to the supernova is evaporated; typical evaporation temperatures for grain materials are in the range 1000–1500 K. The radius out to which dust is evaporated, r_v, is probably determined by the luminosity at the time of shock breakout, when L_{sn} can exceed 10^{44} ergs s^{-1}. For SN 1979C and SN 1980K, Dwek (1983) estimates that r_v is about 3×10^{17} cm. This radius plays an important role in the infrared light curve, which can be calculated from the grain temperature given above. Light travel time effects play a crucial role in the light curve. If the characteristic time that the supernova is bright is short compared to $2r_v/c$, there is a plateau phase until a time $2r_v/c$. The infrared flux then drops as t^{-2} for a highly extended wind and drops more rapidly if there is a cutoff in the wind close to r_v. Thus the length of the plateau phase gives an estimate of r_v, the drop after the plateau phase gives as estimate of the outer extent of the wind, r_w, and the flux yields the dust density in the wind. For SN 1979C and SN 1980K, r_w is somewhat less than 10^{18} cm (Dwek, 1983). The ratio of the infrared emitted energy to the total emitted energy gives an estimate of the optical depth through the dusty wind. The optical depths for SN 1979C and SN 1980K are ~ 0.3 and 0.03, respectively, leading to minimum shell masses of ~ 1–5 $M_\odot$ and ~ 0.1–0.4 $M_\odot$ for the two supernovae (Dwek, 1983). The estimated mass loss rates for $v_w = 10$ km s^{-1}, $> (4$–20$) \times 10^{-5}$ $M_\odot$ yr^{-1} for SN 1979C and $> (0.4$–2$) \times 10^{-5}$ $M_\odot$ yr^{-1} for SN 1980K, are consistent with the results obtained from radio observations. A somewhat higher optical depth than that surrounding SN 1979C would lead to an infrared as opposed to optical supernova; however, the estimated mass loss rate for SN 1979C is at the high end of mass loss rates obtained for galactic red supergiants.

Infrared light echoes are potentially a valuable diagnostic for the circumstellar medium. Dwek (1985) has shown the large differences in the infrared spectra that are expected depending on whether silicate or graphite grains are present. For the carbon-poor composition inferred in the outer layers of SN 1979C, silicate grains are expected. Emmering and Chevalier (1988) showed that asymmetries in the distribution of circumstellar dust can modify the infrared light curves before time $2r_v/c$. This phase is then no longer a plateau, but shows dips and rises that depend on the orientation of the asymmetries. It may eventually be possible to relate such structures to imaging with VLBI techniques.

The same dust grains that give an infrared echo can also give a scattered light echo (Chevalier, 1986). The ratio between the total scattered light and the infrared light just depends on the albedo of the dust grains. Due to the strong forward scattering of typical grains, the scattered light curve is not a plateau before $2r_v/c$, but is a decreasing function of time. Chevalier (1986) failed to find good evidence for scattered light echoes from SN 1979C and SN 1980K, although such echoes might have been observable. The implication may be that the grains have small albedos. On the other hand, Schaefer (1987) has suggested that the light curves of Type II supernovae are frequently affected by circumstellar dust and that the late-time light is dominated by scattered light. This hypothesis needs to be tested by infrared observations of future supernovae.

The apparent evidence for gas from a red supergiant wind in the vicinity of SN 1987A suggests that dust emission might be observable for this nearby supernova. Emmering and Chevalier (1989) have made detailed predictions for the infrared and scattered light echoes on the assumption that silicate grains are present in the shell observed with the IUE. While dust emission from the shell has not been clearly observed, the fading of the supernova has allowed the detection of diffuse optical emission out to 2 arcsec from the supernova (Heathcote, Suntzeff, and Walker, 1989). This component can be plausibly attributed to scattered light emission by dust in an undisturbed red supergiant wind outside the IUE shell (Chevalier and Emmering, 1989). An additional feature is a ring of emission at a radius of about 9 arcsec discovered on January 24, 1989, by Bond *et al.* (1989). If this is light from the supernova emitted near maximum and scattered by surrounding dust, the radial distance from the supernova is about 5 pc. The proximity to a massive star suggests that the dust is circumstellar as opposed to interstellar and exists in a shell surrounding the supernova. Chevalier and Emmering (1989) suggested that the shell is created by the piling-up of the red supergiant wind where its ram pressure becomes equal to that of the surrounding medium. Future observations of this echo feature can be expected to give information on the basic circumstellar grain scattering properties as well as on the spatial distribution of the dust.

5.5. Summary and Future Prospects

The study of circumstellar interaction around supernovae is still in its infancy because observations at wavelengths other than the optical range have been crucial for exploring this aspect of supernovae. The most complete information exists for the Type II supernovae SN 1979C, SN 1980K, and SN 1987A and for these objects it is possible to put together a consistent model that is in accord with expectations from stellar evolution and from observations of stars in our galaxy. However, the mass loss rate inferred for the progenitor of SN 1979C is at the high end of observed stellar mass loss rates. If the outer edge of the dense wind inferred from the infrared echo is correct and if the wind has a velocity of 10 km s^{-1}, the age of the dense wind is about 3×10^4 yr. This is a fraction of the expected time spent as a red supergiant. It may be that stars undergo an unusually high rate of mass loss just before they explode. This phenomenon may also be needed to explain the radio emission from Type Ib supernovae, if they have Wolf–Rayet star progenitors. Analysis of the circumstellar interaction gives valuable clues to the stellar evolution leading up to the explosion. It also places constraints on supernova properties, such as the energetic burst of radiation which occurs at the time of shock break-out, before observations of the event have been initiated.

In addition to basic information on the supernovae, circumstellar interaction can provide a laboratory for studying astrophysical processes. For example, the radio emission gives information on electron acceleration under fairly well-constrained conditions. If high-energy γ-ray radiation is detected from a future nearby event, clues to the origin of cosmic rays may be obtained. On a different subject, the processes of Comptonization and photoionization by an energetic spectrum are similar to processes thought to occur in quasars and in binary X-ray sources. Again, the physical conditions are probably better constrained for the supernova case.

For the future, SN 1987A will continue to be followed closely. Observations of radiative interaction during the next few years should determine many properties of the progenitor mass loss during the red supergiant phase. Space Telescope imaging of the dense shell discovered by the IUE will be especially valuable and will probably illustrate the complexity of the wind interactions. The next major event will occur when the shock front reaches the shell, but this could be 10–20 years in the future. Overall, SN 1987A is a disappointment with regard to circumstellar interaction because the blue supergiant progenitor led to low-density gas in the immediate vicinity of the explosion. We are looking forward to the next normal Type II supernova in our galaxy. Even if the optical emission is greatly reduced by interstellar dust extinction, the circumstellar interaction could give rise to a spectacular source at radio to γ-ray wavelengths.

Acknowledgments

The author's research on supernovae is supported in part by the NSF and NASA.

References

Abbott, D.C. 1982, *Ap. J.*, **263**, 723.
Arnett, W.D. 1987, *Ap. J.*, **319**, 136.
Arnett, W.D. 1988, *Ap. J.*, **331**, 377.
Band, D. 1988, *Ap. J.*, **332**, 842.
Band, D.L. and Liang, E.P. 1988, *Ap. J.*, **334**, 266.
Bandiera, R. 1987, *Ap. J.*, **319**, 885.
Barbon, R., Ciatti, F., and Rosino, L. 1982, *Astr. Ap.*, **116**, 35.
Bartel, N. 1988, in *IAU Symp. 129, The Impact of VLBI on Astrophysics and Geophysics*, eds. M.J. Reid and J.M. Moran (Dordrecht: Reidel), p. 175.
Bartel, N., Rogers, A.E.E., Shapiro, I.I., Gorenstein, M.W., Gwinn, C.R., Marcaide, J.M., and Weiler, K.W. 1985, *Nature*, **318**, 25.
Bartel, N., Rupen, M., and Shapiro, I.I. 1989, *Ap. J. (Letters)*, **337**, L85.
Beall, J.H. 1979, *Ap. J.*, **230**, 713.
Bedogni, R. and D'Ercole, A. 1988, *Astr. Ap.*, **190**, 320.
Berezinsky, V.S. and Ptuskin, V.S. 1989, *Ap. J.*, **340**, 351.
Bode, M.F. and Evans, A. 1980, *M.N.R.A.S.*, **193**, 21P.
Bond, H.E., Panagia, N., Gilmozzi, R., and Meakes, M. 1989, *IAU Circ.*, No. 4733.
Branch, D. 1986, *Ap. J. (Letters)*, **300**, L51.
Canizares, C.R., Kriss, G.A., and Feigelson, E.D. 1982, *Ap. J. (Letters)*, **253**, L17.
Chevalier, R.A. 1981, *Ap. J.*, **251**, 259.
Chevalier, R.A. 1982a, *Ap. J.*, **258**, 790.
Chevalier, R.A. 1982b, *Ap. J.*, **259**, 302.
Chevalier, R.A. 1983, *Ap. J.*, **272**, 765.
Chevalier, R.A. 1984a, *Ann. N.Y. Acad. Sci.*, **422**, 215.
Chevalier, R.A. 1984b, *Ap. J. (Letters)*, **285**, L63.
Chevalier, R.A. 1986, *Ap. J.*, **308**, 225.
Chevalier, R.A. 1987a, *Nature*, **329**, 611.
Chevalier, R.A. 1987b, in *ESO Workshop on SN 1987A*, ed. I.J. Danziger (Garching: European Southern Observatory), p. 481.
Chevalier, R.A. 1988, *Nature*, **332**, 514.
Chevalier, R.A. and Emmering, R.T. 1989, *Ap. J. (Letters)*, **342**, L75.
Chevalier, R.A. and Fransson, C. 1987, *Nature*, **328**, 44.
Chevalier, R.A. and Imamura, J.N. 1983, *Ap. J.*, **270**, 554.
Chiosi, C. and Maeder, A. 1986, *Ann. Revs. Astr. Ap.*, **24**, 329.
Chu, Y.-H., Treffers, R.A., and Kwitter, K.B. 1983, *Ap. J. Suppl.*, **53**, 937.
Couderc, P. 1939, *Ann. d'Ap.*, **2**, 271.

Dopita, M.A., Evans, R., Cohen, M., and Schwartz, R.D. 1984, *Ap. J. (Letters)*, **287**, L69.

Dopita, M.A., Meatheringham, S.J., Nulsen, P., and Wood, P. 1987, *Ap. J. (Letters)*, **322**, L85.

Dwek, E. 1983, *Ap. J.*, **274**, 175.

Dwek, E. 1985, *Ap. J.*, **297**, 719.

Ellison, D.C. and Eichler, D. 1985, *Phys. Rev. Lett.*, **55**, 2735.

Emmering, R.T. and Chevalier, R.A. 1988, *A.J.*, **95**, 152.

Emmering, R.T. and Chevalier, R.A. 1989, *Ap. J.*, **338**, 388.

Epstein, R.I. 1981, *Ap. J. (Letters)*, **244**, L89.

Falk, S.W. 1978, *Ap. J. (Letters)*, **225**, L133.

Fedorenko, V.N. 1984, *Sov. Astr. Lett.*, **10**, 89.

Fransson, C. 1982, *Astr. Ap.*, **111**, 140.

Fransson, C. 1984, *Astr. Ap.*, **133**, 264.

Fransson, C. 1986a, *Highlights of Astr.*, **7**, 611.

Fransson, C. 1986b, in *Radiation Hydrodynamics in Stars and Compact Objects*, eds. D. Mihalas and K.H.A. Winkler, (Berlin: Springer-Verlag), p. 141.

Fransson, C. Benvenuti, P., Gordon, C., Hempe, K., Palumbo, G.G.C., Panagia, N., Reimers, D., and Wamsteker, W. 1984, *Astr. Ap.*, **132**, 1.

Fransson, C., Cassatella, A., Gilmozzi, R., Kirshner, R.P., Panagia, N., Sonneborn, G., and Wamsteker, W. 1989, *Ap. J.*, **336**, 429.

Fransson, C. and Lundqvist, P. 1989, *Ap. J. (Letters)*, **341**, L59.

Heathcote, S., Suntzeff, N., and Walker, A. 1989, *IAU Circ.*, No. 4753.

Herman, J. 1985, in *Mass Loss from Red Giants*, eds. M. Morris and B. Zuckerman (Dordrecht: Reidel), p. 215.

Humphreys, R.M. and Davidson, K. 1979, *Ap. J.*, **232**, 409.

Iben, I. and Tutukov, A.V. 1984, *Ap. J. Suppl.*, **32**, 351.

Imshennik, V.S. and Utrobin, V.P. 1977, *Sov. Astr. Lett.*, **3**, 34.

Itoh, H. and Masai, K. 1989, *M.N.R.A.S.*, **236**, 885.

Jones, E.M. and Smith, B.W. 1983, in *IAU Symp. 101, Supernova Remnants and Their X-Ray Emission*, eds. J. Danziger and P. Gorenstein (Dordrecht: Reidel), p. 83.

Jones, E.M., Smith, B.W. and Straka, W.C. 1981, *Ap. J.*, **249**, 185.

Kennicutt, R.C. 1984, *Ap. J.*, **277**, 361.

Klein, R.I. and Chevalier, R.A. 1978, *Ap. J. (Letters)*, **223**, L109.

Klein, R.I., Chevalier, R.A., Charles, P.A., and Bowyer, S. 1979, *Ap. J.*, **234**, 566.

Kwitter, K.B. 1981, *Ap. J.*, **245**, 154.

Kwitter, K.B. 1984, *Ap. J.*, **287**, 840.

Lamb, S.A. 1978, *Ap. J.*, **220**, 186.

Lasher, G. J. and Chan, K. L. 1979, *Ap. J.*, **230**, 742.

Lundqvist, P. and Fransson, C. 1987, in *ESO Workshop on SN 1987A*, ed. I.J. Danziger (Garching: European Southern Observatory), p. 495.

Lundqvist, P. and Fransson, C. 1988, *Astr. Ap.*, **192**, 221.

Maeder, A. 1981, *Astr. Ap.*, **102**, 401.

Maeder, A. 1987, in *ESO Workshop on SN 1987A*, ed. I.J. Danziger (Garching: European Southern Observatory), p. 251.

Makino, F. 1987, *IAU Circ.*, No. 4336.

McCray, R.A. 1983, *Highlights of Astr.*, **6**, 565.

McKee, C.F. and Hollenbach, D.J. 1980, *Ann. Rev. Astr. Ap.*, **18**, 219.

McKee, C.F., Van Buren, D., and Lazareff, B. 1984, *Ap. J. (Letters)*, **278**, L115.

Nadyozhin, D.K. 1985, *Ap. and Sp. Sci.*, **112**, 225.

Panagia, N. 1973, *A. J.*, **78**, 929.

Panagia, N. *et al.* 1980, *M.N.R.A.S.*, **192**, 861.

Raizer, Yu. P. 1964, *Zh. Prikl. Mat. Tekh. Riz.*, No. 4, 49.

Saio, H., Nomoto, K., and Kato, M. 1988, *Nature*, **334**, 508.

Sakurai, A. 1960, *Comm. Pure Appl. Math.*, **13**, 353.

Schaefer, B. 1987, *Ap. J. (Letters)*, **323**, L51.

Snow, T. P. 1981, *Ap. J.*, **251**, 139.

Sonneborn, G., Kirshner, R., Fransson, C., Cassatella, A., Wamsteker, W., Gilmozzi, R., and Panagia, N. 1988, *IAU Circ.*, No. 4685.
Sramek R.A., Panagia, N., and Weiler, K.W. 1984, *Ap. J. (Letters)*, **285**, L59.
van den Bergh, S. 1988, *Ap. J.*, **327**, 156.
van den Bergh, S., McClure, R.D., and Evans, R. 1987, *Ap. J.*, **323**, 44.
Walborn, N.R., Lasker, B.M., Laidler, V.G., and Chu, Y.-H. 1987, *Ap. J. (Letters)*, **321**, L41.
Wampler, J. 1988, *IAU Circ.*, No. 4541.
Weaver, R., McCray, R., Castor, J., Shapiro, P.R., and Moore, R.T. 1977, *Ap. J.*, **218**, 377.
Weiler, K.W., Sramek, R.A., Panagia, N., van der Hulst, J.M., and Salvati, M. 1986, *Ap. J.*, **301**, 790.
West, R.M., Lauberts, A., Jorgensen, H.E., and Schuster, H.-E. 1987, *Astr. Ap.*, **177**, L1.
Woosley, S.E. 1988, *Ap. J.*, **330**, 218.
Zuckerman, B. 1980, *Ann. Rev. Astr. Ap.*, **18**, 263.

6. Gamma-Rays and X-Rays from Supernovae

PETER G. SUTHERLAND

6.1. Introduction

There are three distinct phases in a supernova when γ-rays and X-rays may be expected to be produced:

(1) When the supernova shock wave breaks through the surface of the star, and/or when this shock encounters any circumstellar material (due to a wind in a presupernova phase of stellar evolution).
(2) When radioactive nuclei, freshly synthesized during the explosion, decay and emit γ-rays.
(3) When the supernova ejecta thin out sufficiently that high-energy radiation from any neutron star/pulsar remnant can penetrate this overlying material.

In the first case, the flux and spectrum of the radiation depend critically on the properties of the stellar atmosphere and wind of the presupernova star, and may or may not ever be easily detected. This source will be discussed, briefly, in Section 6.2; much more on this topic will be found in Chapter 5 by Chevalier. Also included in this section is a short discussion of the high-energy γ-rays produced when cosmic rays interact with the supernova ejecta. In the second case, X-rays are expected as well, because of the energy degradation of the γ-rays through repeated Compton scattering by free and bound electrons in the ejecta. Indeed, these X-rays are expected to emerge first, when the ejecta are still relatively optically thick. The fluxes to be expected from Type I supernovae in this case are orders of magnitude greater than those to be expected for Type II supernovae. The calculations of γ-ray and X-ray fluxes from radioactive decay is the subject of Section 6.3. The observations of γ-rays and X-rays from SN 1987A, and the important clues that they provide about the supernova, will be taken up in Section 6.4. Apart from the X-ray flux below 20 keV, these observations bear exclusively upon the radioactive material and its distribution in the ejecta. The issue of when and how the high-energy emission from any central pulsar emerges will not be discussed here, except in passing. This is because such radiation reflects more on the properties of the pulsar than the supernova, and calculations of the expected fluxes are highly uncertain, as they depend on a number of properties of the newly born neutron star and its electromagnetic spectrum.

6.2. Prompt Emission, Radiation from Interaction with a Wind, and Very High-Energy Radiation

6.2.1. Prompt Emission

The first quantitative prediction of high-energy radiation from supernovae was made by Colgate (1968). This prediction followed on early ideas (Colgate and Johnson, 1960) that supernovae may generate much of the observed cosmic-ray spectrum *hydrodynamically*: As the shock wave runs down the density gradient of the progenitor's atmosphere, it accelerates and a very small mass fraction may become relativistic. As Colgate pointed out, if this is the case, then in a brief time interval after the shock breaks through the surface, the radiation from an optically thin layer is relativistically boosted to γ-ray energies (as high as 2 GeV). Colgate estimated that this γ-ray burst might have an energy as high as 5×10^{47} ergs and Petschek (1967) estimated its duration to be $\sim 1 \times 10^{-5}$ s.

More detailed studies of the prompt emission to be expected when the supernova shock wave approaches and breaks through the progenitor's surface were presented by Klein and Chevalier (1978), Falk (1978), and Chevalier and Klein (1979), for Type II supernovae. These analyses were founded in numerical hydrodynamic calculations, which included radiation pressure and transport (in a flux-limited diffusion approximation) and cooling and heating by bremsstrahlung and Compton processes. These calculations revealed the importance of a radiative precursor ahead of the ion viscous shock. This is due to photons that are emitted by the hot material behind the shock and which diffuse into the cooler material in front of the shock. The precursor heats and accelerates the surface layers. Significant radiation begins to escape from the surface when the shock is at a depth $\tau \sim 25$. The radiation has an effective temperature $T_e \sim 2 \times 10^5$ K, a luminosity $L \gtrsim 1 \times 10^{45}$ ergs s^{-1} and a burst duration $\sim 1 \times 10^3$ s. The burst is in the hard ultraviolet and soft X-ray bands, and may be strongly absorbed by interstellar matter. The spectrum is non-Planckian, with an excess at high energies. When the ion viscous shock breaks the surface, a small mass fraction, heated to temperatures $T \sim 10^8 - 10^9$ K, gives rise to a brief burst of high-energy X-ray bremsstrahlung radiation ($E_x \sim 100$ keV). (The precise temperature is sensitive to the speed of the ion viscous shock wave, which is accelerating down the atmospheric density gradient, and the speed that the material in front has attained through radiative acceleration.) The bremsstrahlung luminosity is a small fraction, ~ 0.01, of the total luminosity. As Klein and Chevalier (1978) noted, the burst duration is considerably shorter than the dynamical time, $\sim 10^5$ s, and so the *observed* burst will be temporally modified by the differing path lengths from the surface to the observer. Lasher and Chan (1979) have given some semianalytic calculations of the ultraviolet-soft X-ray burst near shock breakout. While they did not include all the physical processes incorporated in the Klein and Chevalier and Falk calculations, they arrived at remarkably similar results for the intensity and duration of the burst.

It is amusing to note (Ruderman, 1974) that the prompt radiation bursts described here are strong enough that the average time between supernovae whose bursts can destroy the Earth's atmospheric ozone layer, and thus drastically affect life, is about the time interval between major evolutionary "catastrophes" such as the disappearance of the dinosaurs.

6.2.2. Radiation from Interaction with a Wind

When the ejecta of a Type II supernova encounter the wind associated with the red giant phase of the progenitor, a shock wave is driven into the wind and a reverse shock develops to decelerate the ejecta (Chevalier, 1981, 1982). The bremsstrahlung radiation from the shocked wind material may be estimated to be (Canizares, Kriss, and Feigelson, 1982; Chevalier, 1982):

$$L_{\text{bremss}} \sim 1.6 \times 10^{39} \dot{M}_{-5}^2 w_1^{-2} r_{15}^{-1} T_9^{1/2} \text{ ergs s}^{-1}, \tag{6.1}$$

where $\dot{M}_{-5} = \dot{M}/(10^{-5} M_\odot \text{ yr}^{-1})$ is the wind mass loss rate, $w_1 = w/(10 \text{ km s}^{-1})$ is the wind velocity, $r_{15} = r/(10^{15} \text{ cm})$ is the radius to which the shock has reached, and $T_9 = T/(10^9 \text{ K})$ is the temperature of the shocked wind material. The nondimensional variables all have typical values $\lesssim 1$ at times $t \sim 10^6$ s if the maximum ejecta velocity is $\sim 10,000$ km s^{-1}. The implied X-ray spectra extend up to energies $E_x \sim 100 T_9$ keV. As Chevalier (1982) has pointed out, this simple estimate must be modified if only a small fraction of the ejecta, in the form of a thin shell, has the maximum velocity (Chevalier estimates $\sim 10^{-3} M_\odot$) and this shell encounters a dense wind that slows it down rapidly once it has swept up more than approximately its own mass. Moreover, there may be more bremsstrahlung radiation, with a lower characteristic temperature, arising from the reverse shock that is slowing the shell.

Canizares, Kriss, and Feigelson (1982) observed the Type II supernova SN 1980K in NGC 6946 with the Einstein Observatory and detected an X-ray luminosity (in 0.2–4 keV) of $L_x \sim 2 \times 10^{39}$ ergs s^{-1} for a distance of 10 Mpc. This X-ray detection was made ~ 35 days after the explosion; the flux appeared to decline by a factor $\gtrsim 2$ in the subsequent 47 days. While a *total* bremsstrahlung luminosity $\sim 2 \times 10^{39}$ ergs s^{-1} is not implausible, most of it was expected at energies well above the Einstein Observatory band (this paper appeared before the modifications suggested by Chevalier (1982)). These authors preferred an explanation in terms of the inverse Compton scattering of optical photons from the supernova by relativistic electrons, believed to be responsible for the radio emission observed at approximately the same epoch.

6.2.3. Very High Energy Radiation

Supernovae are believed to be a main source for the cosmic rays observed in the Galaxy. Berezinsky and Ginzburg (1987) estimate that a typical supernova may inject $\sim 10^{50}$ ergs of cosmic rays. They point out that while this energy estimate is reasonably secure, the instantaneous cosmic ray luminosity is highly uncertain since the various mechanisms that can give rise to cosmic rays operate on time scales varying from weeks (acceleration by a central pulsar) to thousands of years (Fermi acceleration mechanisms in the supernova remnant). No matter what the mechanism for accelerating protons to high energies, subsequent collisions of these protons with supernova matter or circumstellar/interstellar matter will copiously generate charged and neutral pions, and the latter kind will then decay in flight and produce high-energy γ-rays. Detection of such γ-rays and the demonstration of a temporal relationship to a specific supernova would tell us much about the cosmic ray acceleration mechanism and possibly the supernova and its environment. Estimates for the fluxes to be expected from a variety of cosmic ray acceleration mechanisms that might operate in association with SN 1987A have been given by Protheroe (1987), Gaisser, Harding, and Stanev (1987), and Berezinsky and Ginzburg (1987). Protheroe also considers likely sources of attenuation of ultra high-energy γ-rays ($\sim 10^{12}$–10^{15} eV) due to photon–photon pair production. Observations made in November 1987 with a very high-energy γ-ray telescope (Raubenheimer et al., 1988) set a 3σ upper limit of 1.7×10^{38} ergs s^{-1} for SN 1987A, which was less than 2% of the total observed luminosity. If cosmic rays and high-energy γ-rays are produced well inside the ejecta, perhaps by a pulsar, this upper limit may be consistent with the opacity of the ejecta at the time.

6.3. Gamma-Rays and X-Rays due to Radioactive Decay

The fresh synthesis of nuclear isotopes in the high temperature ($T \gtrsim 5 \times 10^9$ K) and high density ($\rho \gtrsim 10^9$, at least in SN I) environment of supernovae has been predicted for at least 40 years (Hoyle, 1946). That some of these might be radioactive, with observable consequences in the *visible* (deriving from the energy deposited in the ejecta by γ-rays that cannot, at least initially, escape, or, alternatively, from the energy of fission fragments) has been suspected for nearly as long. The exponential light curve of SN I (half-life of ~ 60 days) suggested ^{7}Be (half-life of 53.3 days) to Borst (1950) and ^{254}Cf (half-life of 60.5 days) to Burbidge et al. (1957). It was soon clear, however, that, given the rate of SN I, the abundances implied for either ^{7}Li (the daughter nucleus in the decay of ^{7}Be) or for other heavy, neutron-rich nuclides expected to be produced in the same r-process environment responsible for the ^{254}Cf would be incompatible with observed abundances. Nevertheless, Clayton and Craddock (1965) made the important observation that a clean test of the radioactive decay models for SN I would be the detection of γ-rays emitted by the daughter nuclei.

With the completion of parametrized studies of explosive silicon burning (Truran, Arnett, and Cameron, 1967; Bodansky, Clayton, and Fowler, 1968), a much better candidate source for the putative radioactivity in SN I was identified. This was ^{56}Ni, the most tightly bound nucleus composed of equal numbers of neutrons and protons (and as important—the most tightly bound nucleus composed of alpha particles). This nuclide would be abundantly produced whenever the state of nuclear statistical equilibrium is attained (by virtue of the high temperatures and densities produced in the explosion) and then quickly frozen out on a dynamical timescale $\lesssim 1$ s. ^{56}Ni first decays to ^{56}Co (half-life of 6.1 days) and then to ^{56}Fe (half-life of 78.5 days). In each step the daughter nuclei are formed in excited states and rapidly decay to their respective ground states through the emission of γ-rays; the important γ-ray lines are given in Table 6.1. While arguments still rage over the last factor of ~ 2 about the implications for the galactic iron abundance, if SN I are not too common in the galaxy (~ 1 per century) and $\lesssim 0.5\, M_\odot$ of ^{56}Ni is produced per event, this radioac-

Table 6.1. Gamma-ray line list.[a]

Energy (MeV)	f	Energy (MeV)	f
0.158	1.00	1.238	0.6758
0.270	0.36	1.360	0.0428
0.480	0.36	1.443	0.0020
0.750	0.50	1.772	0.1600
0.812	0.87	1.811	0.0048
1.562	0.14	1.964	0.0072
		2.015	0.0309
0.511	0.3800	2.035	0.0795
0.734	0.0021	2.213	0.0063
0.788	0.0030	2.598	0.1672
0.847	0.9998	3.010	0.0100
0.978	0.0144	3.202	0.0303
1.038	0.1408	3.254	0.0743
1.140	0.0015	3.273	0.0176
1.175	0.0224	3.452	0.0086

[a] The first six entries are for γ-rays emitted in the decay of the parent nucleus ^{56}Ni and the remaining 23 entries are for γ-rays emitted in the decay of the parent nucleus ^{56}Co. The parameter "f" is the probability of emission for a given line per decay of the parent nucleus.

tive decay model for SN I is probably acceptable on nucleosynthesis grounds. Clayton, Colgate, and Fishman (1969) were the first to perform calculations of the γ-ray fluxes to be expected for this model. At about the same time, Colgate and McKee (1969) calculated (apparently from a suggestion of Truran) the visible light curve for an SN I based upon the energy deposited by the (initially) trapped γ-rays and showed how the ^{56}Co decay (half-life of 78.5 days) could be moderated by expansion and losses in the direction of the observed light curve (half-life of ~ 60 days).

The role of radioactive nuclei in SN II was not as clear. Their light curves are less homogeneous, and few have been followed long enough to reveal the presence, if any, of an exponential phase. Certainly the early light curve, through the first ~ 100 days, is believed to be determined by the release of thermal energy deposited by the supernova shock in the extended, massive envelope of the supergiant progenitor. SN 1969L does show an exponential phase following a plateau phase. Similarly, SN 1980K entered an exponential phase for which it was estimated by Uomoto and Kirshner (1986) that 0.1 $M_\odot$ could provide the necessary energy input. Simulated explosions in very massive progenitors (Weaver and Woosley, 1980a) gave rise to only a modest amount of ^{56}Ni (~ 0.1 $M_\odot$) deeply buried, at low velocity, in the ejecta. SN 1987A has substantially revised and clarified our understanding of all aspects of SN II. We know now from observations in the visible band, at γ-ray energies, and in the infrared that the production of 0.075 $M_\odot$ of freshly synthesized ^{56}Ni occurred during the explosion. The observations and predictions for this supernova will be discussed at length in Section 6.4.

6.3.1. Type II Supernovae

6.3.1.1. Gamma-Ray Lines

We may qualitatively estimate the emergence and characteristics of the γ-ray flux from SN II with the following considerations (Clayton, 1974). To be specific, we will focus on the γ-ray lines associated with ^{56}Co decay, since this nuclide's parent (^{56}Ni) should be the most abundantly produced in the explosion, and the lines are unlikely to emerge until the parent has for the most part disappeared. The flux that remains in a given line is

$$\varphi_i = \frac{(M_{56}/56m)}{4\pi D^2(t_{Co} - t_{Ni})}(e^{-t/t_{Co}} - e^{-t/t_{Ni}})f_i P_{esc}$$

$$= 2.0 \times 10^{-2} \text{ photons cm}^{-2} \text{ s}^{-1} \frac{M_{56}/M_\odot}{D_{Mpc}^2}(e^{-t/t_{Co}} - e^{-t/t_{Ni}})f_i P_{esc}. \quad (6.2)$$

Here M_{56} is the mass of radioactive material (including daughters), m is the atomic mass unit, D_{Mpc} is the distance to the supernova in Mpc, and $t_{Ni} = 8.80$ days $= 7.60 \times 10^5$ s and $t_{Co} = 113.6$ days $= 9.82 \times 10^6$ s are the mean lifetimes of ^{56}Ni and ^{56}Co, respectively. The probability of emission of the ith line, per ^{56}Ni decay, is f_i (often called the branching ratio): for the two strongest lines associated with ^{56}Co decay this is 1.00 (847 keV line) and 0.68 (1238 keV line). The escape probability, P_{esc}, is the probability that a γ-ray will escape *without* scattering at all: to a first approximation, even a single Compton scattering will take the γ-ray out of the line profile (which is determined by the velocity distribution of the radioactive ejecta) due to energy loss to the electron. If we further assume, as is justified by the spherically symmetric model calculations, that the radioactive material is confined to a thin, low-velocity shell, then there is essentially a single optical depth to the surface so that $P_{esc} = e^{-\tau}$. Shortly after the explosion, and long before the γ-rays are likely to be observable, the ejecta "coast" and this optical depth satisfies $\tau = \tau_d(t_d/t)^2$ where t_d is some fiducial time. The γ-rays will only be observable at $t \gg t_{Ni}$ and so the first exponential in the equation above will dominate. Thus the controlling factors in determining the light curve for a given line are

$$e^{-\tau}e^{-t/t_{Co}}f_i \equiv A \exp\left[\frac{-(t - t_m)^2}{2\delta t_m^2}\right]f_i. \quad (6.3)$$

The Gaussian approximation defined here has the time for maximum flux in the line at t_m, a light curve width δt_m, and an amplitude A given by

$$t_m = (2\tau_d)^{1/3}t_{Co}, \quad (6.4)$$

$$\delta t_m = \left(\frac{2^{2/3}}{\sqrt{6}}\right)\tau_d^{1/6}t_{Co}, \quad (6.5)$$

$$A = e^{-1.5t_m/t_{Co}}. \quad (6.6)$$

The fiducial time at which the optical depth τ_d has been evaluated is t_{Co} itself. If the overlying envelope has total mass M_{env} and characteristic velocity v_e then

$$\tau_d = \frac{3\beta\kappa M_{env}}{4\pi v_e^2 t_{Co}^2}; \quad (6.7)$$

β is a dimensionless parameter near unity which depends upon the structure of the envelope. If the radioactive shell is relatively massive or has a very low velocity ($\lesssim 1000$ km s^{-1}), then it may make a large contribution to τ_d: What counts for any shell of material is the combination M/v^2. The opacity is that for Klein–Nishina scattering by all bound and free electrons. (The ratio of the Klein–Nishina cross section to the Thomson cross section is 0.344 at 847 keV and 0.285 at 1238 keV.)

Some plausible values for an SN II are (Weaver and Woosley, 1980b) $v_e \sim 4000$ km s^{-1} and $M_{\rm env} \sim 10\ M_\odot$, in which case $\tau_d \sim 20$ and maximum light in the 847 keV line occurs near 400 days after the explosion, and the light curve width is approximately 120 days. The optical depth in the line at maximum light is 1.8, and the maximum flux in the 847 keV line is 3×10^{-5} photons cm^{-2} s^{-1}, for an SN II at a distance of 1 Mpc with a radioactive mass $M_{56} = 0.3\ M_\odot$. The analysis above, while simple, shows that the characteristics of the γ-ray line light curves are not sensitive to the details of the supernova. Although the 1238 keV line has a smaller branching ratio than the 847 keV line, there will be a greater flux in this line if the Thomson optical depth to the radioactive shell exceeds 6.7. This is true for the representative model considered here at times just before the light curve maximum for the 847 keV line. The light curves for different lines can be used, in principle, to determine the Thomson optical depth of the envelope.

Another consequence of the moderate optical depths to be expected when the lines are strongest is the presence of a strong underlying Compton continuum due to those photons that are scattered. If the differentiation of real background (instrumental and otherwise) from continuum arising in the source is problematical, then the interpretation of line flux measurements may be difficult.

Further information about the supernova can be obtained from the line profiles, if the detectors have sufficient energy resolution and there is good signal-to-noise. The line profile width is, of course, just due to the Doppler shift of γ-rays from nuclei streaming outward. At late times, when the optical depths for all γ-rays are small, the line profiles are symmetrical and centered on the rest energy. At earlier times, if the radioactive shell makes negligible contribution to κ, then the profiles are symmetrical but have a central depression because γ-rays emitted from the limb of the radioactive shell (and hence with radial velocities near zero) have longer paths to traverse through the envelope. Gehrels, Leventhal, and MacCallum (1987) have calculated line fluxes and line profiles for a variety of SN II models in the spirit described above but with a fuller treatment of the envelopes and their opacity distribution. They have neglected the possibly significant γ-ray opacity arising within the radioactive shell. If this is important, then the optical depth to the surface on the near side of the ejecta is less than for the far side, and this leads to a net blueshift of the line, with no central depression. The effect is more pronounced at earlier times but decreases as time goes by. One thing that we have learned from SN 1987A is that the models arising from spherically symmetric explosion calculations are too simple: the radioactive material is nowhere near so centrally concentrated as predicted by the calculations. It is either more-or-less uniformly mixed out to larger velocities, or some instability has caused an inhomogeneous, perhaps "clumpy" distribution. Thus the calculations for model SN II have to be taken as suggestive rather than definitive. Nevertheless, in principle, high-resolution studies of the γ-ray line profiles as a function of time could provide important information about the distribution of radioactive material and the distribution of opacity sources in the supernova. However, given the relative unlikelihood of a sufficiently nearby SN II, even the instruments to be flown on the Gamma-Ray Observatory (see, e.g., Fishman, 1985) are unlikely to provide this degree of detail.

Some 19% of the time ^{56}Co decay proceeds through positron emission. Thus, in addition to the γ-ray lines produced directly through electron capture decay, a pair of 511 keV γ-rays should be produced when the positron subsequently slows down and annihilates. Since there are many other likely astrophysical sources of pair annihilation radiation, the 511 keV line is unlikely to be a useful diagnostic of supernovae. However, the *kinetic* energy (660 keV on average) given up within the ejecta as the positron slows down may play an important role in the energetics of the ultraviolet through infrared display of supernovae at times $\gtrsim 1$ year after the explosion (Axelrod, 1980a, b). This is because, in this so-called "supernebular" phase, there are no other significant sources of energy for the ejecta: the material is increasingly transparent and less and less γ-ray energy is deposited there. (In the unlikely event that there is no small scale magnetic field in the ejecta and the positrons can stream freely away without annihilating, then the supernova should dim rapidly in this phase.)

The above discussion of γ-ray lines has concentrated on ^{56}Ni $\rightarrow$ ^{56}Co $\rightarrow$ ^{56}Fe decay. As Clayton (1974) pointed out, if ^{57}Ni is produced in supernova in the same proportions to ^{56}Ni as in the solar abundance ratio ^{57}Fe : ^{56}Fe $= 1/42$, then lines from ^{57}Co decay may also be detectable. ^{57}Ni decays to ^{57}Co through electron capture and positron emission with a half-life of a mere 36 hours, so the associated γ-rays are unlikely ever to be seen. ^{57}Co decays with a half-life of 270 days and thus, even if its abundance is less, the associated γ-rays will ultimately dominate those from ^{56}Co decay. Almost all (99.8%—see Lederer and Shirley, 1978) ^{57}Co decays end in the second excited state of ^{57}Fe, leading to a line at 136 keV ($f = 0.11$) and a pair of lines at 122 keV plus 14 keV ($f = 0.89$).

There are other trace radioactive isotopes that may be synthesized in the super-nova explosions of massive stars in sufficient quantities to be observable from relatively nearby (10 kpc) events on the times scale of $\sim$ 10 years after the explosion (Woosley, Axelrod, and Weaver, 1981). The most promising of these is ^{44}Ti with a half-life of 48 years. Longer-lived isotopes such as ^{26}Al with a half-life of 7.2×10^5 years are likely to be produced in supernovae and should contribute narrow γ-ray lines to the general interstellar γ-ray background.

6.3.1.2. X-Ray Continuum

In the previous section we have seen that when γ-rays first emerge from an SN II, the ejecta are likely to have optical depths in the lines of approximately a few, and an even larger optical depth for electron scattering. Thus at even earlier times what will emerge from the supernova are hard X-rays due to energy degradation of the γ-rays by multiple Compton scattering. This was first analyzed in some detail by McRay, Shull, and Sutherland (1987) for the case of SN 1987A, and the following argument was presented there and in Sutherland *et al.* (1988).

To estimate the characteristics of the emergence of the X-ray flux we need to balance the effects of energy degradation through multiple-scattering and photoelectric absorption. A photon of energy E_0 has its energy reduced to

$$E_1 = \frac{E_0}{1 + (E_0/mc^2)(1 - \cos \theta)} \tag{6.8}$$

by Compton scattering off a cold electron. For purposes of estimation, we may set $\cos \theta \sim 0$ and then the energy loss equation may be iterated for n scatterings to yield

$$E_n \sim \frac{mc^2}{n} \sim \frac{mc^2}{\tau_s^2} \tag{6.9}$$

independent of the initial γ-ray energy. For ejecta in the homologous expansion phase, $\tau_s = \tau_{s,0}(t_0/t)^2$ where $\tau_{s,0}$ is a fiducial value for the scattering depth at time t_0. The photoelectric optical depth, as a function of X-ray energy, is

$$\tau_a = 1.5\zeta \left(\frac{E}{10 \text{ keV}}\right)^{-3} \times \tau_s, \tag{6.10}$$

where ζ is the metallicity relative to solar (see, e.g., Morrison and McCammon, 1983). At early times the characteristic energy of the degraded photons, as given by eq. (6.9), is very low, and these photons are absorbed with little chance of escape. The critical epoch for emergence of the X-ray flux occurs when the effective absorption optical depth $\tau_{a,eff} \sim [\tau_a(\tau_s + \tau_a)]^{1/2}$ falls below unity; τ_a is evaluated at the characteristic energy of eq. (6.9). We readily find that at this epoch $\tau_{s,crit} \sim 4\zeta^{-1/8}$ and the escaping X-rays have characteristic energy $\sim 30\zeta^{1/4}$ keV. The critical epoch is $t_{crit} \sim [\tau_{s,0}t_0^2/\tau_{s,crit}]^{1/2}$. Essentially no flux *ever* appears below this characteristic energy, and thus only hard X-ray detectors ($E \gtrsim 20$ keV) are predicted to observe these X-rays from SN II.

Taking the representative SN II model previously discussed and assuming a solar metallicity, one has a Thomson scattering depth ~ 60 at the fiducial time of 114 days, and the X-rays would be expected to emerge at a time near 400 days with a flux of $\sim 5 \times 10^{-3}$ photons cm^{-2} s^{-1} for a distance of 1 Mpc, assuming that about half of all γ-rays have been degraded in energy and absorbed.

This estimate of the X-ray flux is very crude for the following reasons:

(1) At the time of X-ray emergence the γ-ray optical depth is only a few and the first few scatterings before the energy is degraded below $\sim mc^2$ are at the smaller, forward-peaked Klein–Nishina cross section, while the above analysis is done with the symmetrical, larger, Thomson cross section.

(2) In any realistic SN II model there is a central concentration of heavier elements that may have profound consequences for the photoelectric absorption.

Furthermore, it is beyond the analysis to calculate the spectral shape and the overall normalization with any accuracy. Reliable calculations of the X-ray spectra to be expected for radioactive SN II models must be done with Monte Carlo techniques. This is even more true for models of SN I, and in the next section the elements of the Monte Carlo approach will be summarized.

6.3.2. Type I Supernovae
Much progress has been made in the last decade in the study of classical Type I supernovae (SN Ia: see Wheeler and Levreault (1985), Uomoto and Kirshner (1985),

and Branch (1986) for a discussion of other subclasses and see Chapters 1, 2, 3, and 8 in this book for a general review of both types; only the γ-ray and X-ray spectra of SN Ia will be discussed here). Arnett (1979) and Colgate, Petschek, and Kriese (1980) discussed the consequences for the light curve of energy released by the radioactive decay of ^{56}Ni produced in the thermonuclear explosion of a white dwarf. Axelrod (1980a, b) showed that the late-time spectra and light curves are well accounted for by the continuing energy input by γ-rays from the radioactive debris. The observation and analysis of the optical spectra of the Type Ia SN 1981B (Branch et al., 1982, 1983) showed that the radioactive material must be blanketed by a substantial mantle of intermediate mass elements (O, Mg, Si, S, and Ca). For these reasons, the model for SN Ia that has received the greatest attention has involved a carbon–oxygen white dwarf in a binary system driven to the point of central ignition by accretion of matter from its companion. The ignition, which occurs as the white dwarf mass approaches the Chandrasekhar limit, is believed to lead to the propagation of a subsonic deflagration wave (Müller and Arnett, 1982; Nomoto, Thielemann, and Yokoi, 1984; Sutherland and Wheeler, 1984; Jeffery and Sutherland, 1985) that incinerates $\sim 1\,M_{\odot}$, with a large central portion fully burned to nuclear statistical equilibrium that subsequently freezes out as, primarily, ^{56}Ni. Uncertainties in the details of the deflagration do not permit specification of precise values for the total amount of incinerated material or the fraction that becomes ^{56}Ni. The explosion and energy input by radioactive decay give very satisfactory explanations for the observed velocities and light curves. Radiative transfer calculations of the thermalized decay energy give maximum light spectra in good accord with the observations, particularly if the partially burned material is mixed (Branch et al., 1985; Harkness 1985, 1986).

Despite these successes, objections to the deflagration model have been raised because of difficulties with the nature and evolution of the progenitor binary system (Fujimoto and Taam, 1982; Iben and Tutukov, 1984; MacDonald, 1984; Sutherland and Wheeler, 1984; Webbink, 1984). Nevertheless, alternative models and their likely γ-ray and X-ray spectra will not be pursued here.

The definitive test of the deflagration model will be the observation of the large γ-ray fluxes that are expected when the ejecta are sufficiently transparent that many of the γ-rays escape with few interactions, typically beginning ~ 30 days after the explosion. If such spectra are observed and confirm the basic model, then comparison of the observations with the theoretical results may permit a determination of the amounts of material fully incinerated (and hence radioactive) and only partially burned. If the model can be thus completely characterized, a comparison of the observed and theoretical fluxes will yield an independent estimate of the distance to the event.

6.3.2.1. Monte Carlo Methods

The fundamental reasons that the calculation of accurate γ-ray spectra, especially for SN Ia, requires a Monte Carlo approach are these:

(1) The γ-ray optical depth (τ_{γ}) at most epochs of interest is approximately a few, and thus one is in a multiple-scattering regime, and each scattering can lead to significant energy loss.

(2) The scattering cross section is energy-dependent and forward-peaked at γ-ray energies.
(3) The radioactive source nuclei are distributed throughout a sizeable fraction of the ejecta.
(4) The ejecta have significant composition and density gradients.
(5) The radioactive material has a sufficiently large velocity dispersion ($\sim$ 10,000 km s^{-1}) that it is *not* a good approximation to assume that a single scattering takes a γ-ray out of the line (thus even the line profiles require a Monte Carlo approach, except perhaps at late times).

The Monte Carlo code described in the following paragraphs was developed by Ambwani and Sutherland (1988) and modifications of it have been used by several authors in the calculation of spectra for SN 1987A.

A model of a partially incinerated white dwarf is selected and then the exploded, homologously expanding model is scaled to the epoch desired. Photons are released from the radioactive zones in proportion to the zone masses and with energies as given by the line list of Table 6.1. The rates, for the production of lines whose parent nuclei are ^{56}Ni and ^{56}Co, respectively, are $R_{Ni} = (1/t_{Ni}) \exp(-t/t_{Ni})$ and $R_{Co} = (\exp(-t/t_{Co}) - \exp(-t/t_{Ni}))/(t_{Co} - t_{Ni})$ when the material is initially pure ^{56}Ni; here $t_{Ni} = 8.80$ days and $t_{Co} = 113.7$ days (Lederer and Shirley, 1978). The radius, r, of the photon's release point in a zone (with inner and outer radii a and b) is set by a random number x in [0, 1): $r = [a^3 + x(b^3 - a^3)]^{1/3}$. The photon is emitted randomly in the local *rest* frame of the ejecta and then its energy and direction are transformed to the observer's frame. Those photons for a given line that escape without scattering will then yield a line profile appropriate to the distribution of velocity with respect to mass in the ejecta.

To determine whether a γ-ray can escape or how far within the ejecta it can travel before interacting, an optical depth τ is chosen according to $\exp(-\tau) = x$, and x random in [0, 1). The photon is then allowed to propagate a distance s until

$$\tau = \int_0^s n_e(s')\sigma(s') \, ds' \tag{6.11}$$

whereupon it suffers an interaction. If the integral of eq. (6.11) is less than τ when the surface is reached, then the photon escapes. The integral along the ray of eq. (6.11) is done with allowance for expansion (dilution) of the material, although in all cases this effect is negligible as the expansion is quite nonrelativistic. In eq. (6.11), n_e is the local electron density and σ is the total cross section, per electron, comprised of:

(1) the total Klein–Nishina cross section at the photon's energy;
(2) the cross section for photoproduction of electron–positron pairs off nuclei for photons with energy above $2m_e c^2$ (see, e.g., Hubbell, 1969); and
(3) the cross section for photoelectric absorption.

Photoelectric absorption is important only when the photon has been scattered down in energy to $\sim 10^2$ keV. Useful tables of absorption coefficients are given by Veigele (1973) and by Henke *et al.* (1982). For energies above 10 keV, it is usually sufficient to determine for each zone an effective absorption coefficient, at a specific energy (say 100 keV), from the elemental abundances for that zone and then to assume the absorption scales as E^{-3} at other energies; this facilitates rapid calculation. Below 10 keV, where absorption edges appear, a more elaborate treatment may be necessary, especially if we are concerned (as in the case of SN II) with calculating accurate fluxes for fluorescence lines.

When it is determined that a photon suffers an interaction, the specific fate is chosen at random in proportion to the ratio of the specific cross section to the total cross section. If scattering is the fate, the photon energy and momentum are transformed to the local rest frame, the polar scattering angle is chosen by solving $\sigma(\theta)/\sigma_{tot} = x$, where $\sigma(\theta)$ and σ_{tot} are, respectively, the integrated (from 0 to θ) and total Klein–Nishina cross sections and x is random in $[0, 1)$, and the azimuthal scattering angle is chosen randomly in $[0, 2\pi)$. The energy and momentum of the scattered photon are transformed back to the observer's frame and the photon is then further followed. The consequences of the Lorentz transformations between the observer's frame and the local rest frame of the scattering electron are in general negligible when compared with the energy change due to the Compton scattering itself. If photoelectric absorption is the fate, a fluorescence photon may be created and then its fate is followed. If pair-production is the fate then it is assumed that the positron annihilates nearby, nearly at rest. Thus two new γ-rays, each of energy 0.511 MeV, are generated at the point of pair-production and their subsequent histories followed.

A photon that escapes is labeled by its energy and its time of escape. The latter information is accumulated so as to determine if the delays due to multiple scattering within the ejecta were significant. In principle, a γ-ray spectrum should be constructed from those photons that escape at the same time, with an appropriate distribution of original times of release, rather than from those photons which were all released at the same epoch. In practice, this is unimportant because the ejecta velocities are nonrelativistic and a γ-ray very rarely scatters more than approximately five times before its energy is lowered to the point where photoelectric absorption is very likely.

The above algorithm is not particularly effective for calculating the *full* spectrum when the mean optical depth for γ-rays is either small or large. In the former case, only rarely do photons scatter enough to build up the low-energy (X-ray) portion of the spectrum. In the latter case, too many photons scatter down to energies where they are photoelectrically absorbed and thus do not escape to contribute at all to the spectrum. This is particulary wasteful! An escape probability algorithm devised by Pozdnyakov, Sobol, and Sunyaev (1983) corrects these deficiencies and has been used in most modifications of the code of Ambwani and Sutherland. In this algorithm a photon is never allowed to fully leave or to be fully absorbed. When a photon starts along a chosen ray (either as a consequence of being emitted or as the outcome of a scattering event), the optical depth along the ray to the surface, τ_s, is then calculated and used to define the escape probability $P_{esc} = e^{-\tau_s}$. A fraction, P_{esc}, of the photon is allowed to escape and thus contributes to the observable spectrum. The remaining fraction of the photon, $1 - P_{esc}$, is allowed to travel along the ray until it suffers an interaction at the point specified by the optical depth

$\tau = -\ln(1 - x(1 - P_{esc}))$ where x is random in $[0, 1)$—this guarantees an interaction before the surface is encountered. A photon begins its life with a "weight," $W = 1$. Its weight is reduced by the fraction that escapes from the start of every new ray, and its life ends whenever that weight drops below some specified threshold. Likewise, only a fraction of a photon is ever absorbed (photoelectrically or in pair-production) and the weight is reduced accordingly. In this algorithm essentially no information is ever thrown away. One cautionary note: In coding Monte Carlo algorithms, it is frighteningly easy to create a program that more-or-less does the job, at least it does not "crash" or generate patently spurious results; but it is in some subtle way flawed. It may even pass simple semianalytic tests, as these may only reflect limiting cases that the flaw does not sample. Thus, the more complex the algorithm (and the escape probability algorithm gets messy when dealing with the photons created by pair-production) the greater is the chance for coding error.

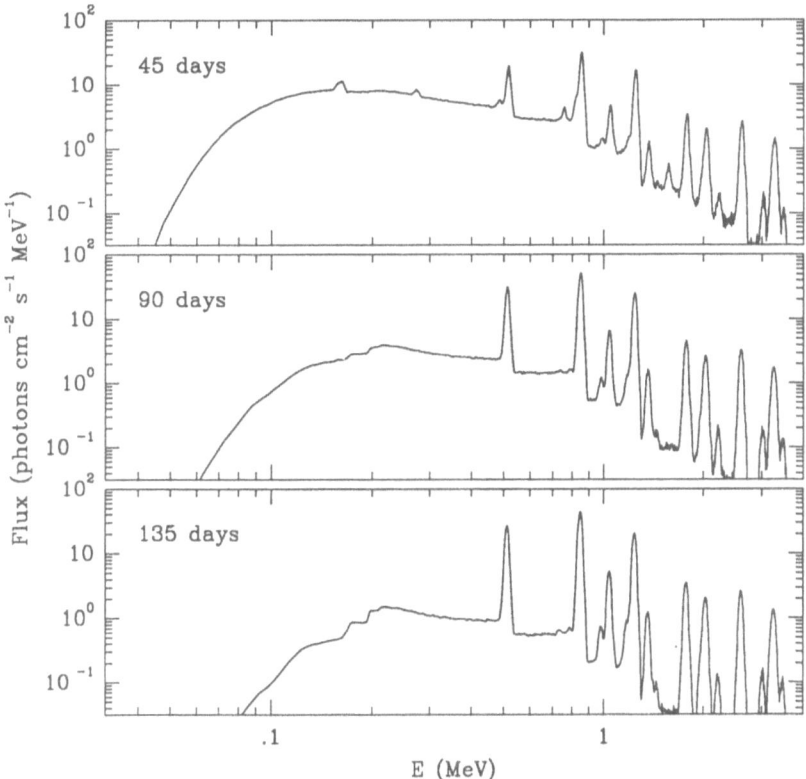

Figure 6.1. Simulated Monte Carlo spectra for a representative model of a Type I supernova. The model consists of a 1.4 $M_\odot$ white dwarf (initial composition was a homogeneous mixture of equal mass fractions of carbon and oxygen) for which the inner 1.2 $M_\odot$ was incinerated to ^{56}Ni and intermediate mass elements. Of this material 0.6 $M_\odot$ started as ^{56}Ni. The outermost 0.1 $M_\odot$ of the ^{56}Ni was (arbitrarily) mixed out uniformly into the partially incinerated matter. This mixing helps to further broaden the γ-ray lines. Compton down-scattering from the decay lines of ^{56}Co forms an X-ray continuum cut off at energies below 100 keV by photoelectric absorption. The distance has been taken as 50 kpc—compare with the results for a SN 1987A model given in Figure 6.9.

6.3.2.2. *Results for SN Ia Models*

A set of three spectra are shown in Figure 6.1 for a representative carbon deflagration model of a SN Ia. The spectra were calculated with the Monte Carlo algorithm described above (with the escape probability refinements incorporated) at times of 45, 90, and 135 days after the explosion. The modeled explosion corresponds to the incineration of the central 1.2 $M_\odot$ of a homogeneous carbon–oxygen white dwarf at the Chandrasekhar limit. Because of our uncertain knowledge of the deflagration mechanism, the fraction of incinerated material that passes through nuclear statistical equilibrium and freezes out as ^{56}Ni was taken as a free parameter: For the results presented here, this fraction was set to be the central 50%. The outer 0.1 $M_\odot$ of the ^{56}Ni was arbitrarily mixed uniformly with the partially incinerated material. The basis for this decision was the kind of improvements that are seen in the synthetic ultraviolet and optical spectrum calculated at maximum light when such mixing is invoked (Branch *et al.*, 1985). (This choice of parameters for a model of an SN Ia would lead to an optical light curve and a velocity profile for the ejecta consistent with observed SN Ia properties. See, e.g., Sutherland and Wheeler (1984).) The spectra reveal the following features, characteristic of these models:

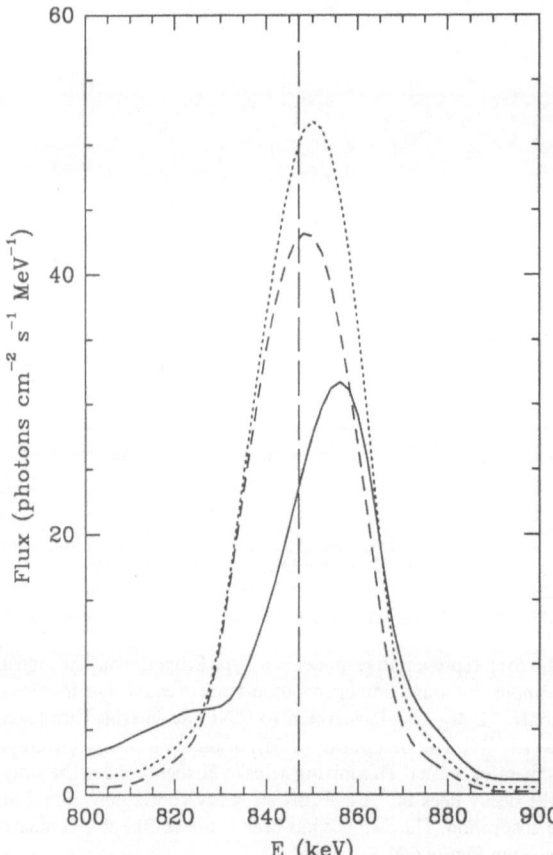

Figure 6.2. An expanded view of the ^{56}Co decay 847 keV line, based upon the simulated spectra shown in Figure 6.1. Note the pronounced blueshift in the line at early times. The 45-day line has a shoulder due to the 812 keV line from ^{56}Ni decay.

(1) the lines associated with ^{56}Ni decay are visible, until at least 45 days, although they are quite weak (the line at 812 keV contributes a shoulder to the ^{56}Co decay line at 847 keV);

(2) the X-ray spectra are very hard, get progressively weaker, and the low-energy cutoff (due to photoelectric absorption) steadily moves to higher energies;

(3) there is never significant flux below 50 keV; and

(4) the line centroids are more blueshifted at earlier times (Figure 6.2 gives an expanded view of the 847 keV line).

The line widths and fluxes for the ^{56}Co 847 and 1238 keV lines are given in Table 6.2. The fluxes are obtained from the spectra by subtracting the average Compton scattering continuum on either side of the lines. It is clear from the table that the line fluxes peak near 90 days after the explosion. This was not the case for the calculations reported by Gehrels, Leventhal, and MacCallum (1987) where the maxima occur nearer 55 days. Their calculations made use of an approximate, semianalytic treatment of the γ-ray escape probability, and necessarily lacked the verisimilitude of the Monte Carlo approach. Indeed, the spectra of Figure 6.2 show that a significant fraction of the photons that are scattered stay within the line profile because of the large velocity range of the radioactive material.

Ambwani and Sutherland (1988) have calculated SN Ia models that span a two-dimensional grid with the incinerated mass, M_{inc}, ranging from 0.90 $M_\odot$ to 1.29 $M_\odot$ and the radioactive mass, M_{rad}, ranging from 0.40 $M_\odot$ to 1.20 $M_\odot$. In their calculations no outward mixing of the radioactive material was considered. They summarized the results in tables of X-ray versus γ-ray luminosities (the division between the two bands was arbitrarily set at 100 keV) and equivalent widths and continuum fluxes for the strong lines at 511, 847, and 1238 keV, at two epochs—30 days and 60 days after the explosion. They concluded that if the underlying paradigm (the carbon deflagration model) was correct and if detailed X-ray and γ-ray observations of a specific event could be made in conjunction with simple optical measurements (Doppler widths of the P Cygni profiles) then a *unique* choice of a model from this grid could be made. In that case, the mass of radioactive material would be known, and a comparison of the observed and calculated γ-ray fluxes would permit a determination of the distance to the event. This method of distance determination would be quite distinct from, and independent of, other methods that have been

Table 6.2. Line fluxes for a model of an SN I at 1 Mpc.[a]

	847 keV	1238 keV
FWHM (keV)	27	40
45 days	2.1	1.7
90 days	3.6	2.7
135 days	3.1	2.2

[a] The line fluxes are in units of 1×10^{-3} photons cm^{-2} s^{-1}. The initial mass of ^{56}Ni is 0.6 $M_\odot$.

applied to supernovae. In light of what we have come to know (or rather, *not* know) about mixing during and after the deflagration phase (not to mention the necessary, but ill-understood, invocation of some degree of outward mixing of the radioactive material in the modeling of SN 1987A), the restriction of the models to the above two-dimensional grid may be naive. Nevertheless, it may be true that the theoretical characterization of SN Ia models will remain relatively simple and thus the comparison of simulated γ-ray spectra with observed spectra may permit a direct determination of the radioactive mass fraction. One characteristic that suggests itself is the X-ray cutoff energy at any epoch, which is sensitive to the mass fractions of radioactive, partially incinerated, and unincinerated material (the energy below which the photoelectric cross section exceeds the Thomson cross section is ~ 100, ~ 50, and ~ 25 keV in these materials, respectively).

6.4. SN 1987A

6.4.1. X-Ray Observations

6.4.1.1. *Ginga Satellite*

The Ginga satellite, with its set of proportional counters with total effective area of 4000 cm^2 and full width field of view of $2° \times 4°$, became operational at about the same time as the SN 1987A explosion. It was used immediately to search for X-ray emission from the supernova, and success was achieved in July 1987 (Dotani *et al.*, 1987; Tanaka, 1988). The major obstacle to observing the supernova is the presence nearby (only 0.6° away) of the very powerful X-ray source LMC X-1. This source, which below 10 keV is *at least* ten times more powerful than SN 1987A ever became, is also known to be time-variable and has a high-energy tail. Two modes of observation were employed to monitor the supernova:

(1) a scanning mode along the line between LMC X-1 and SN 1987A; and
(2) a pointing mode offset by 1° from SN 1987A on the side *away* from LMC X-1.

The former mode led to the initial detection of SN 1987A. The latter mode reduces counting statistics through longer net integration times, although the offset reduces the supernova counting rate by a factor ~ 2 in order to greatly reduce the contribution from LMC X-1 (however, in this mode there is in effect no monitoring of this source). In addition to LMC X-1, other sources in the LMC (including two supernova remnants) and a background (based upon measurements before and after each observation of SN 1987A) are subtracted.

X-ray light curves based upon the Ginga data reported by Makino (1988) and covering the period July 1987–January 1988 are given in Figure 6.3. The data have been divided (by the Ginga team) into two energy bands: (1) the soft X-ray band, from 6 to 16 keV; and (2) the hard X-ray band, from 16 to 28 keV. The contributions from all known non-SN 1987A sources are believed never to exceed 40% in the soft X-ray band and 10% in the hard X-ray band (Tanaka, 1988) except for the initial detection in July 1987. The hard X-ray band light curve stays remarkably steady

Figure 6.3. X-ray light curves for SN 1987A based upon the Ginga data reported by Makino (1988) and covering the period from July 1987 to January 1988. The data have been divided (by the Ginga team) into two energy bands: (1) the soft X-ray band, from 6 to 16 keV; and (2) the hard X-ray band, from 16 to 28 keV.

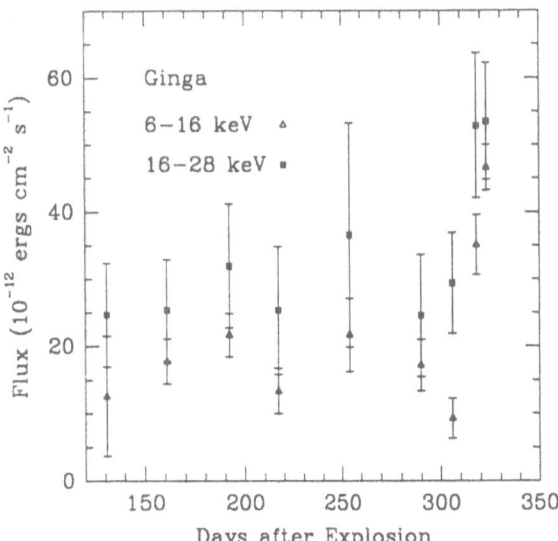

over the interval from 150 to 350 days after the explosion, at a mean level $\sim 3 \times 10^{-11}$ ergs s^{-1} cm^{-2}. The soft X-ray band light curve is quite variable, with variations of a factor ~ 2 on time scales of days to weeks (Tanaka, 1988); its mean level is $\sim 2 \times 10^{-11}$ ergs s^{-1} cm^{-2}. In January 1988 a dramatic flux increase (the "January flare") occurred in the soft X-ray band following a slight dip in late December 1987: the flux increased by a factor ~ 4. The hard X-ray band also show an increase (although less pronounced) in January 1988 that appears to be statistically significant. The apparent correlated behavior between the two bands may be an important clue.

Tanaka (1988) discusses the energy spectrum of SN 1987A. While the spectrum appears to be smooth through the interval 10–20 keV, there appears to be a change of slope near 10 keV, perhaps indicative of two components. As will be discussed below, all of the X-ray data from Ginga and other experiments for energies above 20 keV, and the γ-ray observations, can be explained in terms of the Compton scattering of γ-rays emitted in the decay of ^{56}Co. On the other hand, the detection of significant continuum radiation below 20 keV necessitates another explanation, such as the interaction of supernova ejecta with circumstellar material. Tanaka also notes that the iron K_α line at 6.4 keV was observed when the counting rate was high, i.e., during the January flare. This line appeared with an equivalent width of 200 eV. Tanaka argues that this line may be produced by thermal processes. Alternatively, this line is also expected at some level as a fluorescence line due to the absorption of X-rays with energy above the iron K edge at 7.1 keV.

6.4.1.2. *Mir Space Station*

The Mir Space Station, with its Kvant–Rontgen observatory containing a suite of three X-ray detecting systems, first observed SN 1987A on 1987 August 10 (day

168) and the results for data collected over the following 36 days have been reported by Sunyaev *et al.* (1987). The three experiments were:

(1) the High Energy X-Ray Experiment (HEXE) instrument, comprised of Na I/Cs I phoswich scintillators and sensitive over the range 15–250 keV;
(2) a somewhat similar scintillator instrument, Pulsar X-1, sensitive from 50 to 1000 keV; and
(3) the Coded Mask Imaging Spectrometer (TTM) telescope with a xenon gas proportional counter as detector, sensitive from 2 to 32 keV, but especially so at the lower end.

The supernova was first detected by HEXE, and then Pulsar X-1 was used to obtain the higher-energy spectrum. Only upper limits were obtained from the TTM instrument, at energies below 20 keV. These upper limits are in conflict with the early Ginga observations. The TTM and early Ginga observations were nearly, but not precisely, simultaneous. Moreover, the TTM was never adequately calibrated and thus the data it provided remain somewhat suspect. The mean spectrum above 30 keV was approximately given by $dN/dE = 1.2 \times 10^{-4}(E/30 \text{ keV})^{-1.4}$ photons $\text{cm}^{-2} \text{ s}^{-1} \text{ keV}^{-1}$.

6.4.1.3. *Other X-Ray Observations*

A sounding rocket experiment was conducted by Aschenbach *et al.* (1987) on 1987 August 24, with sensitivity to soft X-rays, 0.2–2.1 keV. No positive detection was made, although an upper limit of 1×10^{-12} ergs $\text{s}^{-1} \text{ cm}^{-2} \text{ keV}^{-1}$ at 1 keV was set at the 95% confidence level. This limit can be used to constrain models involving:

(1) inverse Compton scattering of optical photons by relativistic electrons produced at a shock of the ejecta with circumstellar material; or
(2) thermal emission from such shocks (see below in Section 6.4.3).

The Burst and Transient Source Experiment (BATSE) prototype (for the Gamma Ray Observatory) was used in an observation on day 249 by Wilson *et al.* (1988) and revealed strong continuum flux in their best band from 45 to 200 keV. This band is expected to be free from other contaminating sources. (Below 45 keV, LMC X-4 and/or SMC X-1 may well contribute almost all of their observed excess flux.) The flux at 100 keV was approximately 5×10^{-5} photons $\text{cm}^{-2} \text{ s}^{-1} \text{ keV}^{-1}$ with a spectral index slightly steeper than -1.6. The spectrum observed at this time was considerably more intense and somewhat harder than that detected by Mir/Kvant near day 180 (Sunyaev *et al.*, 1987).

Observations performed with the Caltech GRIP (Gamma-Ray Imaging Payload) by Cook *et al.* (1988a, b) on day 268 detected significant continuum fluxes in three broad bands (a 4.1σ excess in 40–303 keV, a 2.5σ excess in 303–1286 keV, and a 3.0σ excess in 1608–3752 keV). They report a very hard spectrum, with a spectral index -0.9 for the full range 51–1614 keV. It would appear that this determination of the index is dominated by the data above 500 keV. Below 500 keV the data seem consistent with a Compton scattering continuum, with a softer spectral index near -1.6. The flux below 500 keV appears to be in accord with the BATSE prototype observations made at approximately the same time.

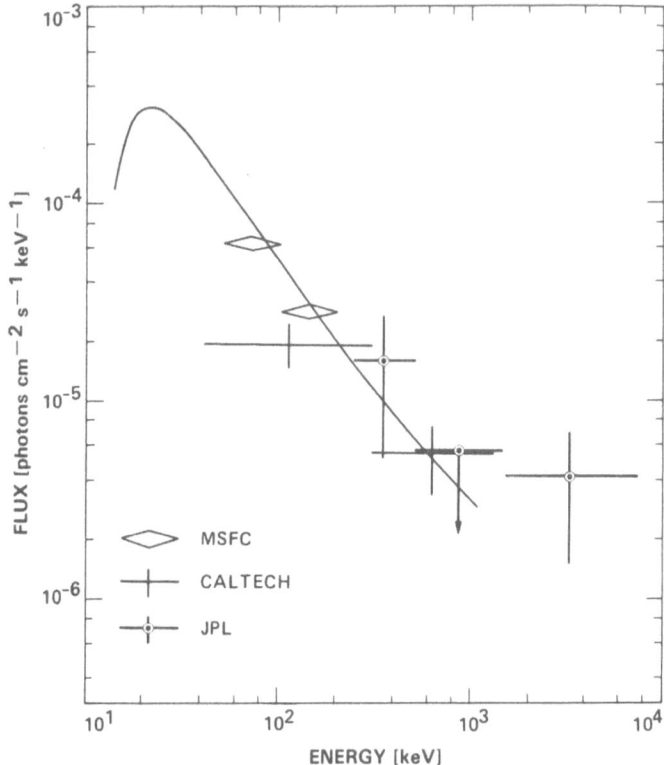

Figure 6.4. Continuum emission above 242 keV for SN 1987A, as measured by te JPL γ-ray spectrometer (Mahoney *et al.*, 1988) on day 286 after the explosion. Observations by Sandie *et al.* (1988) on day 250 and Cook *et al.* (1988b) on day 268 and a theoretical spectrum due to Pinto and Woosley (1988a, b) are also included. This figure has been taken, with permission, from the paper by Mahoney *et al.* (1988).

A JPL high-resolution γ-ray spectrometer, designed to look primarily for γ-ray line emission, was flown on a balloon mission on day 286 (Mahoney *et al.*, 1988). After careful analysis of the continuum background, this experiment has provided continuum flux values for SN 1987A in three broad energy bands spanning 0.242–8.1 MeV. These values together with the continuum results from Sandie *et al.* (1988) and Cook *et al.* (1988a, b) are given in Figure 6.4.

6.4.2. γ-Ray Observations

The γ-ray detectors used to date to observe SN 1987A have employed either Na I/Cs I phoswich scintillators or germanium-based (ionization) detectors. The former have considerably less energy resolution than the latter but the latter are more expensive to operate as they must be cryogenically cooled. When γ-ray lines are being sought, and when these are expected to lie on a Compton continuum generated in the source which cannot be cleanly separated (partly because of limited signal-to-noise) from nonsource background, then the comparisons made among different observations and with theoretical models can be problematical. Adding further to

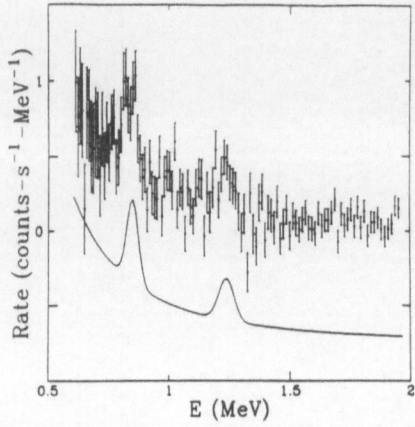

Figure 6.5. The accumulated background-subtracted spectrum for SN 1987A as measured with the SMM satellite, for the period August 1, 1987, to April 1, 1988. The residual continuum is at least in part atmospheric. The solid curve represents the expected response to γ-ray lines at 847 and 1238 keV superimposed on a power-law continuum. This figure is taken, with permission, from the paper by Matz, Share, and Chupp (1988).

the problems of the background is the fact that near the strongest ^{56}Co line at 847 keV is a strong ^{27}Al line at 844 keV, induced in the spacecraft or balloon payload by cosmic rays that produce neutrons in the atmosphere which scatter inelastically off the ^{27}Al. Murphy (1988) has suggested that germanium detectors may be additionally plagued by a similarly excited line at 847 keV in ^{76}Ge.

The NASA Solar Maximum Mission (SMM) satellite was employed indirectly to look for ^{56}Co lines from SN 1987A (Matz, Share, and Chupp, 1988; Matz *et al.*, 1988). The SMM satellite has a wide field of view γ-ray spectrometer (GRS) usually pointed at the Sun, so SN 1987A was observed through the side Cs I shield. Background spectra were accumulated every orbit during Earth occultation of the supernova. Then the background-subtracted spectra were fitted with a power-law continuum (two parameters) and six lines of fixed width, including the sought after ^{56}Co lines at 847 and 1238 keV. The additional lines and continuum used in the fit were necessary to account for residuals in the difference spectra that arise from variations during individual orbits. The resulting mean spectrum for observations from 1987 August 1 to 1988 April 1 is shown in Figure 6.5. The average fluxes in the two lines for data summed from day 159 to day 408 are: at 847 keV— $6.6 \pm 2.0 \times 10^{-4}$ photons cm^{-2} s^{-1}; and at 1238 keV—$6.4 \pm 1.7 \times 10^{-4}$ photons cm^{-2} s^{-1}. There is little available evidence for time-dependence in these data as data sets for intervals five weeks long rarely showed excesses in these lines at greater than the 2σ level. It does, however, appear that the flux in the 1238 keV line peaked near day 240.

Sandie *et al.* (1988) used a high resolution (nominally 2.5 keV), cryogenically cooled, germanium detector to observe SN 1987A for two transits on day 250. The total exposure for the two transits was 5.17×10^5 cm^2 s and 4.17×10^5 cm^2 s at 847 and 1238 keV, respectively. The source spectrum (summed over both transits), the background spectrum, and the difference spectrum are shown in Figure 6.6. This figure clearly shows the problematical ^{27}Al line at 844 keV. The mean flux observed for the two transits in a 5 keV band centered on the 847 line was $5.1 \pm 1.7 \times 10^{-4}$ photons cm^{-2} s^{-1}, and in the broader band 838–850 keV the mean flux was

Figure 6.6. Spectra taken on two transits of SN 1987A on day 250 by Sandie *et al.* (1988) with a high-resolution germanium detector. The first panel is for a pointing on the supernova, the second gives data for pointings ±40° away. The third panel gives the net source spectrum. This figure clearly shows the problematical ^{27}Al line at 844 keV. This figure has been taken, with permission, from Sandie *et al.* (1988).

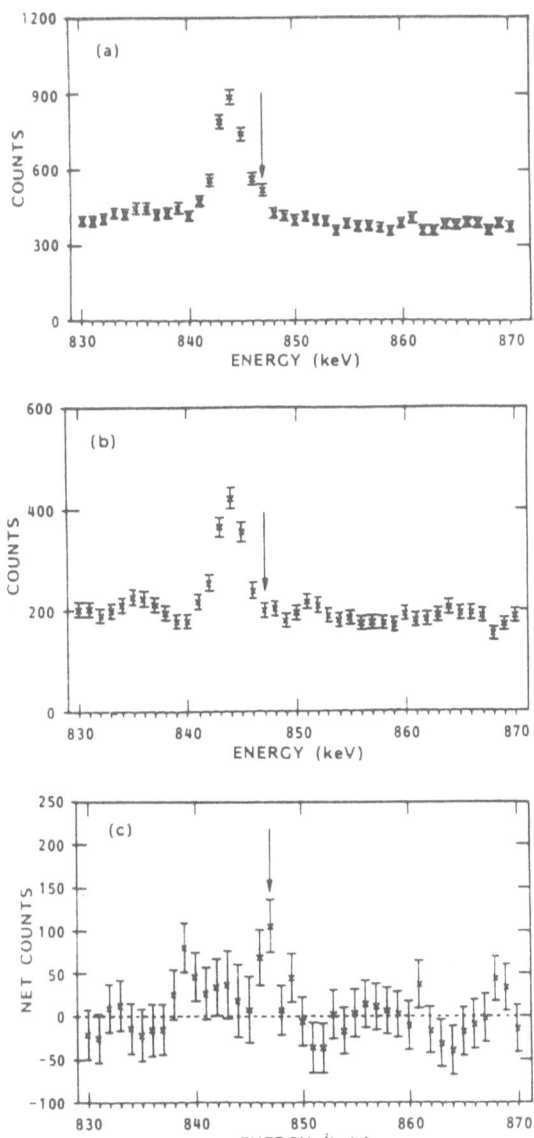

$10 \pm 2.8 \times 10^{-4}$ photons cm^{-2} s^{-1}. The excess counts here could be due to:

(1) kinematical broadening of the line in the source;
(2) a contribution from the Compton scattering continuum in the source;
(3) residual contamination from the ^{27}Al line.

No statistically significant feature appeared near 1238 keV, although the statistical fluctuations in the band 1235–1241 keV would permit a line ratio 1238/847 of 0.4 ± 0.3.

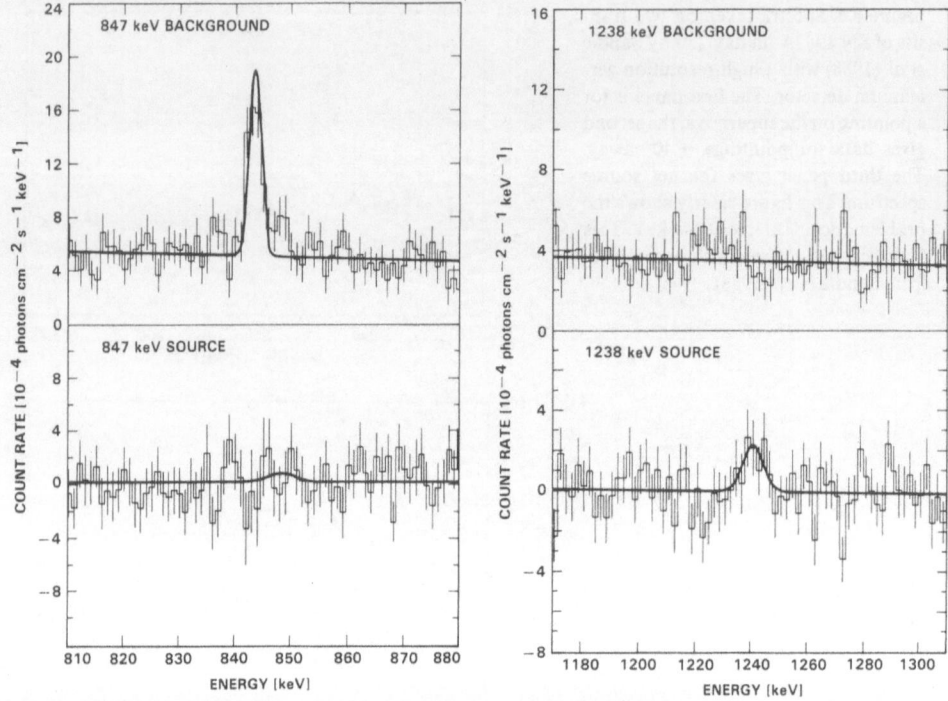

Figure 6.7. Results for SN 1987A from the JPL high resolution germanium spectrometer on day 286 (Mahoney *et al.*, 1988). The solid curves in the upper panels are the best fits to the data with a linear continuum plus known background lines (especially the ^{27}Al line at 844 keV). There is no clear sign of the ^{56}Co line at 847 keV but the 1238 keV line is detected. This figure has been taken, with permission, from the paper by Mahoney *et al.* (1988).

The GRIP detector (Cook *et al.*, 1988a, b) broad band observations (day 268) in the energy range 545–1614 keV were also fitted with a model consisting of a straight-line continuum and two Gaussian profiles centered at 847 and 1238 keV, with widths equal to the instrumental energy resolution (approximately 7%). The resulting line fluxes were found to be, respectively, $0.5 \pm 0.6 \times 10^{-3}$ photons cm^{-2} s^{-1} and $1.3 \pm 0.8 \times 10^{-3}$ photons cm^{-2} s^{-1}.

A JPL high-resolution γ-ray spectrometer with a germanium detector was flown on day 286 by Mahoney *et al.* (1988). The background and source spectra around the ^{56}Co 847 and 1238 keV lines are shown in Figure 6.7. There is no clear sign of the line at 847 keV (the flux in a Gaussian "line" profile of nominal width 5.6 keV at 847 keV was $0.5 \pm 0.7 \times 10^{-3}$ photons cm^{-2} s^{-1}). The 1238 keV line was detected, with the following characteristics: the line was centered at 1240.8 ± 1.7 keV with an intrinsic width of 8.2 ± 3.4 keV FWHM and a flux of $2.1 \pm 0.7 \times 10^{-3}$ photons cm^{-2} s^{-1}. After correcting for the recessional velocity of the LMC, the line center corresponds to a blueshift of approximately 900 ± 400 km s^{-1}. The line width indicates a velocity dispersion in the source ~ 2000 km s^{-1}, which is consistent with that expected for the bulk of the ^{56}Co in many models. These observations cannot tell us much about the small amount of ^{56}Co that is expected to have been mixed out to much higher velocities.

On day 319 the Gamma-Ray Advanced Detector (GRAD) was launched on a balloon flight from Antarctica. Preliminary analysis (Rester *et al.*, 1988) of the data revealed two features at approximately the 2σ level near 847 and 1238 keV. Both lines were broad and showed possible splitting, although the interpretation of the 847 keV line is obscured by the background (arising in the payload) ^{27}Al line at 844 keV.

On day 433 the GRIS collaboration (Barthelmy *et al.*, 1988) flew a high-resolution germanium spectrometer from a balloon. Their preliminary analysis clearly revealed the presence of the 1238 keV line with a flux of $8.1 \pm 1.7 \times 10^{-4}$ photons cm^{-2} s^{-1}. The line had no discernible Doppler shift and had a width of 15–25 keV (3600–6000 km s^{-1}). The preliminary analysis did not yield a separation between the instrumental ^{27}Al line at 844 keV and the ^{56}Co line at 847 keV. A strong continuum flux in the 120–700 keV was observed.

The γ-ray line observations are summarized in Figure 6.8, adapted from Matz, Share, and Chupp (1988). There are no irreconcilable differences, except possibly for the upper limit set by Sandie *et al.* (1988) on the 1238 keV line on day 250. This limit, which was set for a *narrow* feature, would presumably increase and be more in accord with the other observations if the line is broadened.

6.4.3. Modeling of the X-Rays and Gamma-Rays from SN 1987A

There have been three phases in the modeling of the X-ray and γ-ray fluxes from SN 1987A. In the first phase, little was known about the supernova, and the importance of ^{56}Ni $\rightarrow$ ^{56}Co $\rightarrow$ ^{56}Fe decay was unclear. The first published predictions concerning the hard X-ray fluxes that might be expected were made by McCray, Shull, and Sutherland (1987). They considered two possible primary sources for high-energy radiation:

(1) the γ-rays associated with the fresh synthesis of ^{56}Ni; and
(2) energy input by a pulsar with a hard, Crab-like, spectrum extending up to energies $\gtrsim 1$ MeV.

The competition between energy degradation through repeated Compton scattering and the increasing probability for photoelectric absorption at low energies were discussed, and it was shown how these, together with a model for the supernova, could be used to determine the time of emergence and the low-energy cutoff of the X-radiation. Chan and Lingenfelter (1987) at almost the same time estimated light curves for the 847 keV line of ^{56}Co decay. Their curves were characterized by a single parameter, M_{ej}^2/E_{ej} (M_{ej} and E_{ej} are the total mass and kinetic energy of the ejecta after the explosion). (These calculations were at the same level of approximation as those presented in Section 6.3.1 above.) Gehrels, MacCallum, and Leventhal (1987), using an elaborate electron and photon propagation code (but one that lacks the full realism of a Monte Carlo code), calculated complete γ-ray and X-ray spectra from ^{56}Co decay for three simple models that seemed appropriate at the time. They emphasized that the hard X-rays would emerge before the γ-rays attained maximum intensity, by $\gtrsim 100$ days in the case of the massive envelope model. The estimates presented in these three papers were intended primarily to alert the X-ray and γ-ray observing communities to the possibilities that SN 1987A might provide; the estimates were made before a consensus had developed on a

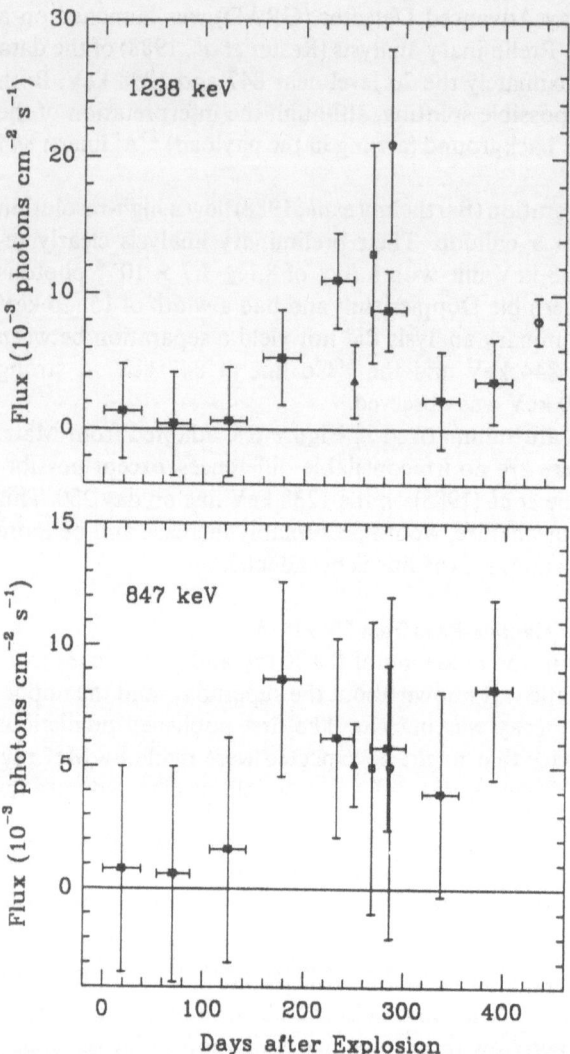

Figure 6.8. Summary of the γ-ray line observations. The data for SMM (solid circles) are from Matz, Share, and Chupp (1988); the points at days 250, 268, 286, and 433 are from Sandie *et al.* (1988), Cook *et al.* (1988b), Mahoney *et al.* (1988), and Barthelmy *et al.* (1988), respectively. The point at day 250 for the 1238 keV line is a converted upper limit, and is in mild conflict with the other observations taken at nearby times.

realistic model for the supernova. It was clear, however, that if the ejecta were massive ($M_{ej} \sim 10\ M_\odot$), the energy was conventional ($E_{ej} \sim 1 \times 10^{51}$ ergs), and only $\sim 0.1\ M_\odot$ of ^{56}Ni was produced (and at the inner edge of the ejecta), then no detectable fluxes were expected until $\gtrsim 500$ days after the explosion and only the most sensitive γ-ray detectors were likely to be successful.

In the second phase of the modeling, as a consequence of ultraviolet through optical and infrared (UVOIR) observations that spanned by then ~ 150 days, a consensus was achieved as to many of the properties of the supernova:

(1) the progenitor had a helium core mass very near 6 $M_\odot$ (of which presumably 1.4 $M_\odot$ collapsed to become a neutron star);

(2) $\sim 0.08\ M_\odot$ of ^{56}Ni must have been synthesized in the explosion;
(3) at the instant of the explosion there was a hydrogen envelope of $3-10\ M_\odot$; and
(4) a corresponding total kinetic energy of $\sim(0.5-1.5) \times 10^{51}$ ergs.

Predictions made at this time by Xu *et al.* (1988) and Ebisuzaki and Shibazaki (1988a) were that X-rays would not be observable until late 1987 and γ-rays would become detectable at an even later date. These predictions were confounded by the early X-ray detection by the Ginga and Mir satellites near day 175 and by the nearly contemporaneous detection of a γ-ray signal. The early onset of the high-energy fluxes could be reconciled by mixing some of the ^{56}Co outward, to lower optical depths (Itoh *et al.*, 1987; Leising, 1988; Pinto and Woosley, 1988a; Ebisuzaki and Shibazaki, 1988b; Shibazaki and Ebisuzaki, 1988; Shull and Xu, 1988; Sutherland *et al.*, 1988). Apart from the need to explain the observations, there was a clear motivation for considering some sort of mixing: When the supernova shock wave passes from the helium core into the low density hydrogen envelope a reverse shock wave develops and moves inward to decelerate the core material. This causes a density inversion which *must* be prone to the Rayleigh–Taylor instability (see, e.g., Chevalier and Klein, 1978). In the stellar evolution and hydrodynamics codes that assume spherical symmetry, such instabilities cannot be readily incorporated and the ^{56}Co shell on the inside of the core is fully decelerated to $\lesssim 1200$ km s^{-1}. (It was, of course, the low velocity and high optical depth of this shell that led to the earlier erroneous predictions.) The exact consequences of the Rayleigh–Taylor instability are unknown, and include the possibilities of essentially uniform mixing down to small angular scales or the development of "fingers" or "clumps." The nature of the instability is also complicated by the fact that, initially, sufficient energy is released by the ^{56}Ni and trapped *locally* that there must be dynamical consequences (Woosley, Pinto, and Ensman, 1988): in the absence of overlying material the ^{56}Ni shell would accelerate to ~ 4000 km s^{-1}. Modelers working with current Monte Carlo codes are essentially obliged to assume spherically symmetrical mixing.

The third phase of the modeling (which continues into 1989) is characterized by:

(1) Increasingly tight constraints due to UVOIR observations, for the bolometric light curve (Catchpole *et al.*, 1987; Menzies *et al.*, 1987; Hamuy *et al.*, 1988) and for the velocity profile of the matter in the hydrogen envelope and helium core (Phillips *et al.*, 1988).
(2) The need to explain the long-lasting (at least through day 400) and approximately constant γ-ray and X-ray signals.

The modeling of the UVOIR bolometric light curve (especially in the interval when the hydrogen recombination wave moves inward through the envelope and before the radiative diffusion timescale becomes shorter than the dynamical time) and the material velocity profile demand detailed models. The two groups most actively engaged in this are those of Nomoto and Woosley, and recent summaries of their reseach programs have been given by Shigeyama, Nomoto, and Hashimoto (1988) and Woosley (1988). It is natural that the most complete studies of the high-energy emission from SN 1987A have been conducted by these groups. Indeed, the X-ray and γ-ray observations have provided important constraints on certain properties of the underlying model. The following remarks are based upon the current (late 1988) understanding expressed by Pinto and Woosley (1988a, b) and Kumagai *et al.* (1988), who come to very similar conclusions.

Figure 6.9 gives simulated spectra for SN 1987A for model 10 HMM of Pinto and Woosley (1988b) from 200 to 3500 days after the explosion. In this model the mass of the hydrogen envelope overlying the helium core is 10 $M_\odot$ and the total kinetic energy of the ejecta is 1.3×10^{51} ergs. These numbers are in part constrained by the bolometric UVOIR light curve and by the lowest observed velocity of hydrogen (Phillips *et al.*, 1988). (See Woosley (1988) for an extensive discussion of how the models are put together. In general, model 10HMM meets more of the observational constraints—both UVOIR and high-energy—than any other.) The mass of radioactive material is accurately known to be 0.075 $M_\odot$ (for a distance of 50 kpc or distance modulus of 18.5) from the exponential behavior of the bolometric light curve after July 1987 and the fact that for ~ 200 days thereafter less than 1% of the γ-ray energy is calculated to escape. The radioactive material has been partially mixed out through what was the helium core and even into the hydrogen envelope in order to explain the early and prolonged detection of X-rays and γ-rays. (Model 10HMM is essentially the same as model 10HM of Pinto and Woosley (1988a) except for the more extensive mixing). The mixing is *not* assumed to be homogeneous; there is a gradient in the radioactive mass fraction and most of the ^{56}Co is to be found at low velocities. The calculated X-ray fluxes are consistent with the observations. The comparison of the γ-ray line fluxes with the observations, while generally exhibiting consistency, points out some of the difficulties in interpreting results from experiments with different detectors and a strong background (both instrumental and from the source, in the Compton continuum). For example, Pinto and Woosley report the following 847 keV "line" fluxes for model 10HMM at days 200, 250, and 400:

(1) 3.3, 5.4, and 6.8×10^{-4} photons cm^{-2} s^{-1} in an 8 keV band as appropriate to a germanium detector; and

(2) 6.4, 9.4, and 12×10^{-4} photons cm^{-2} s^{-1} in a 56 keV band as appropriate to a Na I detector.

The factors of ~ 2 between the two types of detector are due to the jump in the Compton continuum on the low-energy side of the line and to the line "wings" arising from the relatively small amount of ^{56}Co mixed out to velocities $\gtrsim 1500$ km s^{-1}. Similar results are obtained for the 1238 keV line. However, many observers (G. Share (1988), private communication) argue that their analysis procedures account adequately for these effects (typically they fit a Gaussian line shape above a continuum *and* allow for the Compton shelf) and thus their quoted line fluxes are reliable. In addition to these possibly detector-dependent fluxes, the germanium detectors should also see a blueshift in the line centroid at early times by ~ 4 keV for the 847 keV line and ~ 6 keV for the 1238 line. It is my belief that we may not see a definitive resolution of the apparent differences among experiments or between experiments and models.

Other features of note to be seen in Figure 6.9 are the emergence of ^{57}Co lines after approximately day 600 and the iron, cobalt, and nickel fluorescence lines at

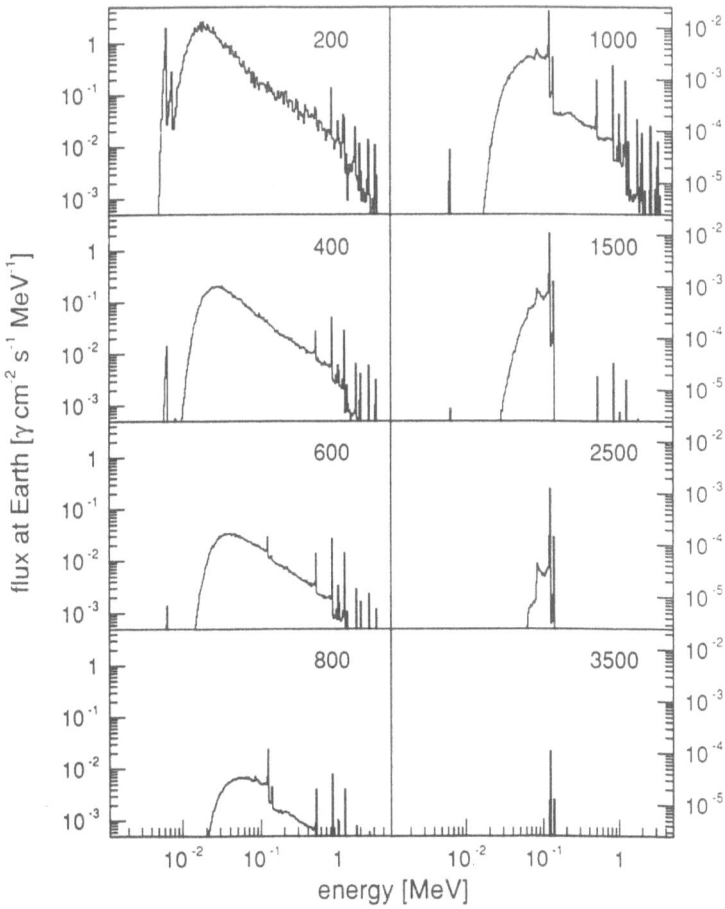

Figure 6.9. Simulated Monte Carlo spectra from Model 10HMM of SN 1987A of Pinto and Woosley (1988b); the distance to the LMC is taken to be 50 kpc. Compton down-scattering from the decay lines of ^{56}Co form a power-law spectrum cut off at energies below 20 keV by photoelectric absorption. Note the K_α fluorescence lines of nickel at 8.2 keV, of Co at 7.6 keV, and of iron at 6.4 keV. After day 600, the 122 and 136 keV lines from ^{57}Co decay being to appear, gradually dominating the lines of ^{56}Co. The panels are labeled by the time (in days) after the explosion.

6.4, 7.6, and 8.2 keV, respectively. The strengths of the ^{57}Co decay lines at 122 and 136 keV are predicated on the solar abundance ratio of ^{57}Fe : ^{56}Fe $= 1/42$; if these lines are observed we will learn if SN 1987A provided the right kind of site for the synthesis of an isotope sensitive to the neutron excess and freeze out conditions. The strength of the fluorescence lines is directly proportional to the metallicity of the hydrogen envelope.

Kumagai *et al.* (1988) have reported very similar results for the X-rays and γ-rays to be expected from their models. These models have a helium core similar to Woosley's, but a less massive hydrogen envelope (6.7 $M_\odot$) and a correspondingly lower total energy of 1×10^{51} ergs. In their earlier calculations (Itoh *et al.*, 1987) ^{56}Co was mixed out to at most (model HY) 6.0 $M_\odot$ or 2400 km s^{-1}, slightly penetrating the hydrogen envelope, and the mixing was homogeneous. Their most satisfactory results now come from a model wherein the mixing has a gradient and extends out to 10 $M_{\odot}$ or 4200 km s^{-1}. They predict that the γ-ray lines and hard X-rays should be detectable until early 1989, but that the X-rays in the hard Ginga band, 16–30 keV, ceased to be detectable by approximately day 400.

Ironically (hardly any pun intended), the most complete information about the distribution of freshly synthesized ^{56}Ni and its decay products may come from infrared observations. Rank *et al.* (1988) have reported the detection of lines from low ionization states of nickel and especially cobalt by a detector flown on the Kuiper Airborne Observatory in November 1987. The detailed interpretation of these lines requires some modeling of the ionization equilibrium within the ejecta.

An indirect signature of the X-rays and γ-rays has been suggested by Graham (1988). The infrared 1.083 μm line of He I dramatically strengthened from May 1987 to July 1987. The upper level of this transition is populated by recombination of He II. Graham argues that the most plausible agent for maintaining the required level of ionization of helium in the cool envelope is the flux of scattered decay radiation.

Soon after the explosion of SN 1987A, it was suggested by many (see, e.g., Ostriker, 1987) that a young, rapidly rotating pulsar might be providing a substantial power input to the supernova ejecta, perhaps even enough to power most of the light curve. McCray, Shull, and Sutherland (1987) considered the consequences of this hypothesis in terms of the X-ray spectrum that might be observed if a substantial fraction of the power output of the putative pulsar was in high-energy radiation, as is the case for the Crab pulsar. From July 1987 to late 1987 the UVOIR light curve tracked within 1% the ^{56}Co decay exponential, consistent with the very small fraction of γ-ray energy that was calculated to escape (Pinto and Woosley, 1988a, b). By day 345, the UVOIR light curve was observed to have dropped by 7% below the ^{56}Co exponential (Catchpole and Whitelock, 1988), but again this is consistent with the increasing escape probability for γ-rays (Pinto, Woosley, and Ensman, 1988). The implied limit on any power input from a pulsar is thus $L_{\text{pulsar}} \lesssim 1 \times 10^{39}$ ergs s^{-1}. Mastichiadis, Ögelman, and Kirk (1988) have calculated when X-rays from a pulsar might shine through the ejecta and exceed in intensity the X-rays due to Compton scattering of radioactive decay γ-rays. They adopted a Crab-like spectrum with a luminosity of 1×10^{36} ergs s^{-1} per decade between 1 keV and 1 GeV, and performed Monte Carlo simulations for this radiation and for radioactive decay γ-rays for a simple model of the supernova. The emergence of the X-rays depends, as in the γ-ray case, on photoelectric absorption and Compton scattering, with differences arising from the presumed constancy of the pulsar radiation and its assumed spectrum which extends down to low energies where the energy degradation through scattering is less significant. They find that if the pulsar has the luminosity of the present-day Crab, breakout would occur at approximately 20 months after the explosion and might be detectable, at first in the 6–16 keV band of Ginga and later in the 20–45 keV band of the HEXE detector on Mir. Their results are readily scaled to more, or less, powerful pulsars, but in the latter case the expected fluxes will be below current sensitivities.

The low-energy Ginga band (6–16 keV) observations have remained very difficult to model. The progenitor of SN 1987A almost certainly went through two wind phases, when it was a red supergiant (RSG) and then later, prior to the explosion, as a blue supergiant (BSG). Thus the simple discussion of Section 6.2.1 does not apply.The early radio observations (Turtle *et al.*, 1987) and lack of early detectable X-ray emission support the notion of a low-density, medium-velocity (~ 500 km s^{-1}) BSG wind (Chevalier and Fransson, 1987). The duration of the BSG phase is uncertain, and may range from 10^4 yr to $\gtrsim 10^5$ yr (Woosley, 1988). The RSG wind has made its presence known through narrow ultraviolet emission lines (see, e.g., Panagia *et al.*, 1987) that were lit up by the original ultraviolet flash at shock breakout. The evolution of the flux in these lines suggests that the inner edge of the RSG wind lies at $\gtrsim 2.4 \times 10^{17}$ cm (Chevalier, 1988). Itoh *et al.* (1987) did detailed calculations of the X-ray emission to be expected when the supernova shock strikes the RSG wind (the BSG wind was ignored), *before* the Ginga and Mir detections. Masai *et al.* (1987) revised the calculations in light of the observations. Unfortunately, they seem to require the inner edge of the RSG wind to be near 1×10^{16} cm, a distance incompatible with the current understanding of the narrow ultraviolet lines (Chevalier, 1988). In a later paper, in part to counter this objection and in part to interpret the variable Ginga low-energy band observations (especially the "January flare"), Masai *et al.* (1988) have developed a more elaborate, "clumpy" model of the circumstellar environment. They argue that during the transition from the RSG phase to the BSG phase, when the wind was accelerating and its density decreasing, Rayleigh–Taylor instabilities develop and dense clumps of slow-moving material may be found in the BSG wind. When these are shocked by the supernova ejecta X-rays are emitted (primarily by the reverse shock that develops in the ejecta). Bandiera, Pacini, and Salvati (1988) have instead proposed that the low-energy X-ray flux, and its variability, may be attributed to radiation from a central pulsar and its synchrotron nebula that is penetrating fragmented supernova ejecta.

6.5. Concluding Remarks

SN 1987A has allowed the seedling of γ-ray line astronomy to blossom. We now have direct confirmation of fresh nucleosynthesis in supernovae. It is unfortunate that NASA's plans for the Gamma-Ray Observatory have had to be rescheduled from the mid 1980s to the early 1990s. (On the other hand, if the GRO had been launched in 1985, with its planned 2-year mission, it might have missed SN 1987A on the early side!) The GRO, specifically its OSSE detector (see Kurfess *et al.*, 1983), will have a sensitivity of $(2-3) \times 10^{-5}$ photons cm^{-2} s^{-1} for 10^6 s observations, and would have generated very accurate γ-ray line light curves for SN 1987A. However, the Na I detectors would still not have afforded the energy resolution necessary to get the distribution with velocity of the radioactive material. The GRO should detect SN Ia out to 10 Mpc, and SN II to perhaps 1 Mpc. It will not be until the next generation of high-resolution, high-sensitivity germanium detectors (see, e.g., the discussion of the Nuclear Astrophysics Explorer by Matteson, Teegarden, and Mahoney, 1988) that γ-ray observers will be able to sample the rich delights of the Virgo cluster at 20 Mpc. As is the case in the rest of supernova astronomy, we eagerly await the next galactic event to really provide a challenge to observers and theorists.

References

Ambwani, K. and Sutherland, P.G. 1988, *Ap. J.*, **325**, 820.

Arnett, W.D. 1979, *Ap. J. (Letters)*, **230**, L37.

Aschenbach, B., Briel, U.G., Pfeffermann, E., Bräuninger, H., Hippmann, H., and Trümper, J. 1987, *Nature*, **330**, 232.

Axelrod, T.S. 1980a, Ph.D. thesis, University of California, Santa Cruz.

Axelrod, T.S. 1980b, in *Type I Supernovae*, ed. J.C. Wheeler (Austin: University of Texas), p. 80.

Bandiera, R., Pacini, F., and Salvati, M. 1988, *Nature*, **332**, 418.

Barthelmy, S., Gehrels, N., Leventhal, M., MacCallum, C.J., Teegarden, B.J., and Tueller, J. 1988, *IAU Circ.*, No. 4593.

Berezinsky, V.S. and Ginzburg, V.L. 1987, *Nature*, **329**, 807.

Bodansky, D., Clayton, D.D., and Fowler, W.A. 1968, *Ap. J. Suppl.*, **16**, 299.

Borst, L.B. 1950, *Phys. Rev.*, **78**, 807.

Branch, D. 1986, *Ap. J. (Letters)*, **300**, L51.

Branch, D., Buta, R., Falk, S.W., McCall, M.L., Sutherland, P.G., Uomoto, A., Wheeler, J.C., and Wills, B.J. 1982, *Ap. J. (Letters)*, **252**, L61.

Branch, D., Doggett, J.B., Nomoto, K., and Thielemann, F.-K. 1985, *Ap. J.*, **294**, 619.

Branch, D., Lacy, C.H., McCall, M.L., Sutherland, P.G., Uomoto, A., Wheeler, J.C., ad Wills, B.J. 1983, *Ap. J.*, **270**, 123.

Burbidge, E.M., Burbidge, G.R., Fowler, W.A., and Hoyle, F. 1957, *Rev. Mod. Phys.*, **29**, 547.

Canizares, C.R., Kriss, G.A., and Feigelson, E.D. 1982, *Ap. J. (Letters)*, **253**, L17.

Catchpole, R.M. *et al.* 1987, *M.N.R.A.S.*, **229**, 115.

Catchpole, R.M. and Whitelock, P.A. 1988, *IAU Circ.*, No. 4544.

Chan, K.W. and Lingenfelter, R.E. 1987, *Ap. J. (Letters)*, **318**, L51.

Chevalier, R.A. 1981, *Ap. J.*, **251**, 259.

Chevalier, R.A. 1982, *Ap. J.*, **259**, 302.

Chevalier, R.A. 1988, *Nature*, **332**, 514.

Chevalier, R.A. and Fransson, C. 1987, *Nature*, **328**, 44.

Chevalier, R.A. and Klein, R.I. 1978, *Ap. J.*, **219**, 994.

Chevalier, R.A. and Klein, R.I. 1979, *Ap. J.*, **234**, 597.

Clayton, D.D. 1974, *Ap. J.*, **188**, 155.

Clayton, D.D., Colgate, S.A., and Fishman, G.J. 1969, *Ap. J.*, **155**, 75.

Clayton, D.D. and Craddock, W.L. 1965, *Ap. J.*, **142**, 189.

Colgate, S.A. 1968, *Can. J. Phys.*, **46**, S476.

Colgate, S.A. and Johnson, M.H. 1960, *Phys. Rev. Lett.* **5**, 325.

Colgate, S.A. and McKee, C. 1969, *Ap. J.*, **157**, 623.

Colgate, S.A., Petschek, A.G., and Kriese, J.T. 1980, *Ap. J. (Letters)*, **237**, L81.

Cook, W.R., Palmer, D.M., Prince, T.A., Schindler, S.M., Starr, C.H., and Stone, E.C. 1988a, in *Nuclear Spectroscopy of Astrophysical Sources*, eds. N. Gehrels and G. Share (New York: American Institute of Physics), AIP Conference Proceedings, No. 170.

Cook, W.R., Palmer, D.M., Prince, T.A., Schindler, S.M., Starr, C.H., and Stone, E.C. 1988b, *Ap. J. (Letters)*, **334**, L87.

Dotani, T. *et al.* 1987, *Nature*, **330**, 230.

Ebisuzaki, T. and Shibazaki, N. 1988a, *Ap. J.*, **328**, 699.

Ebisuzaki, T. and Shibazaki, N. 1988b, *Ap. J. (Letters)*, **327**, L5.

Falk, S.W. 1978, *Ap. J. (Letters)*, **226**, L133.

Fishman, G.J. 1985, in *Proceedings of Workshop on Cosmic Ray and High-Energy Gamma-Ray Experiments for the Space Station Era*, eds. W.V. Jones and J.P. Wefel (Baton Rouge: Louisiana State University Press), p. 400.

Fujimoto, M.Y. and Taam, R.E. 1982, *Ap. J.*, **260**, 249.

Gaisser, T.K., Harding, A., and Stanev, T. 1987, *Nature*, **329**, 314.

Gehrels, N., Leventhal, M., and MacCallum, C.J. 1987, *Ap. J.*, **322**, 215.

Gehrels, N., MacCallum, C.J., and Leventhal, M. 1987, *Ap. J. (Letters)*, **320**, L19.

Graham, J.R. 1988, *Ap. J. (Letters)*, **335**, L53.

Hamuy, M., Suntzeff, N.B., Gonzalez, R., and Martin, G. 1988, *Astron. J.*, **95**, 63.

Harkness, R.P. 1985, in *Supernovae as Distance Indicators*, ed. N. Bartel (Berlin: Springer-Verlag), p. 183.

Harkness, R.P. 1986, in *Radiation Hydrodynamics in Stars and Compact Objects*, eds. D. Mihalas and K.-H.A. Winkler (Berlin: Springer-Verlag), p. 166.

Henke, B.L., Lee, P., Tanaka, T.J., Shimabukuro, R.L. and Fujikawa, B.K. 1982, *Atomic Data and Nuclear Data Tables*, **27**, 1.

Hoyle, F. 1946, *M.N.R.A.S.*, **106**, 343.

Hubbell, J.H., 1969, *Photon Cross Sections, Attenuation Coefficients, ad Energy Absorption Coefficients from 10 keV to 100 GeV*, NSRDS-NBS 29 (Washington: National Bureau of Standards).

Iben, I., Jr. and Tutukov, A.V. 1984, *Ap. J. Suppl.*, **54**, 335.

Itoh, M., Kumagai, S., Shigeyama, T., Nomoto, K., and Nishimura, J. 1987, *Nature*, **330**, 233.

Jeffery, D. and Sutherland, P.G. 1985, *Astrophys. & Space Sci.*, **109**, 277.

Klein, R.I. and Chevalier, R.A. 1978, *Ap. J. (Letters)*, **223**, L109.

Kumagai, S., Itoh, M., Shigeyama, T., Nomoto, K., and Nishimura, J. 1988, *Astron. and Astrophys.*, **197**, L7.

Kurfess, J.D., Johnson, W.N., Kinzer, R.L., Share, G.H., Strickman, M.S., Ulmer, M.P., Clayton, D.D., and Dyer, C.S. 1983, in *Gamma-Ray Astronomy in Perspective of Future Space Experiments*, eds. G. Vedrenne and K. Hurley (Oxford: Pergamon); *Adv. Space Res.*, **3**, No. 4, 109.

Lasher, G. and Chan, K.L. 1979, *Ap. J.*, **230**, 742.

Lederer, M.C. and Shirley, V.S. 1978, *Table of Isotopes*, 7th ed., (New York: Wiley) pp. 160–163.

Leising, M.D. 1988, *Nature*, **332**, 516.

MacDonald, J. 1984, *Ap. J.*, **283**, 241.

Mahoney, W.A., Varnell, L.S. Jacobson, A.S., Ling, J.C., Radocinski, R.G., and Wheaton, W.A. 1988, *Ap. J. (Letters)*, **334**, L81.

Makino, F. 1988, *IAU Circ.*, No. 4532.

Masai, K., Hayakawa, S., Inoue, H., Itoh, H., and Nomoto, K. 1988, *Nature*, **335**, 804.

Masai, K., Hayakawa, S., Itoh, H., and Nomoto, K. 1987, *Nature*, **330**, 235.

Mastichiadis, A., Ögelman, H., and Kirk, J.G. 1988, *Astron. and Astrophys.*, **201**, L19.

Matteson, J.L., Teegarden, B.J., and Mahoney, W.A. 1988, in *Nuclear Spectroscopy of Astrophysical Sources*, eds. N. Gehrels and G. Share (New York: American Institute of Physics), p. 417.

Matz, S.M., Share, G.H., and Chupp, E.L. 1988, in *Nuclear Spectroscopy of Astrophysical Sources*, eds. N. Gehrels and G. Share (New York: American Institute of Physics), p. 51.

Matz, S.M., Share, G.H., Leising, M.D., Chupp, E.L., Vestrand, W.T., Purcell, W.R., Strickman, M.S., and Reppin, C. 1988, *Nature*, **331**, 416.

McCray, R., Shull, J.M., and Sutherland, P. 1987, *Ap. J. (Letters)*, **317**, L73.

Menzies, J.W. *et al.* 1987, *M.N.R.A.S.*, **227**, 39.

Morrison, R.L., and McCammon, D. 1983, *Ap. J.*, **270**, 119.

Müller, E. and Arnett, W.D. 1982, *Ap. J. (Letters)*, **261**, L109.

Murphy, M.J. 1988, *Ap. J. (Letters)*, **334**, L95.

Nomoto, K., Thielemann, F.-K., and Yokoi, K. 1984, *Ap. J.*, **286**, 644.

Ostriker, J.P. 1987, *Nature*, **327**, 287.

Panagia, N. *et al.* 1987, *IAU Circ.*, No. 4514.

Petschek, A.G. 1967, *Science*, **156**, 239.

Phillips, M.M., Heathcote, S.R., Hamuy, M., and Navarette, M. 1988, *Astron. J.*, **95**, 1087.

Pinto, P.A. and Woosley, S.E. 1988a, *Ap. J.*, **329**, 820.

Pinto, P.A. and Woosley, S.E. 1988b, *Nature*, **333**, 534.

Pinto, P.A., Woosley, S.E., and Ensman, L.M. 1988, *Ap. J. (Letters)*, **331**, L101.

Pozdnyakov, L.A., Sobol, I.M., and Sunyaev, R.A. 1983, *Soviet Sci. Rev., Ap. and Space Phys.*, **2**, 189.

Protheroe, R.J. 1987, *Nature*, **329**, 135.

Rank, D.M., Pinto P.A., Woosley, S.E., Bregman, J.D., Witteborn, F.C., Axelrod, T.S., and Cohen, M. 1988, *Nature*, **331**, 505.

Raubenheimer, B.C., de Jager, O.C., Nel, H.I., North, A.R., and van Urk, G. 1988, *Astron. and Astrophys.*, **193**, L11.

Rester, A.C., Eickhorn, G., Coldwell, R., Trombka, J., Starr, B., and Lasche G.P. 1988, *I.A.U. Circ.*, No. 4535.

Ruderman, M.A. 1974, *Science*, **184**, 1079.

Sandie, W.G., Nakano, G.H., Chase, L.F., Jr., Meegan, C.A., Wilson, R.B., Paciesas, W.S., and Lasche, G.P. 1988, *Ap. J. (Letters)*, **334**, L91.

Shibazaki, N. and Ebisuzaki, T. 1988, *Ap. J. (Letters)*, **327**, L9.

Shigeyama, T., Nomoto, K., and Hashimoto, M. 1988, *Astron. and Astrophys.*, **196**, 141.

Shull, J.M. and Xu, Y. 1988, in *Supernova 1987A in the Large Magellanic Cloud: Proceedings of the Fourth George Mason Astrophysics Workshop*, eds. M. Kafatos and C.A. Michalitsianos (Cambridge: Cambridge University Press), p. 371.

Sunyaev, R. *et al.* 1987, *Nature*, **330**, 227.

Sutherland, P.G. and Wheeler, J.C. 1984, *Ap. J.*, **280**, 282.

Sutherland, P.G., Xu, Y., McCray, R., and Ross, R.R. 1988, in *IAU Colloq. 108, Atmospheric Diagnostics of Stellar Evolution: Chemical Peculiarity, Mass Loss, and Explosion*, ed. K. Nomoto (Berlin: Springer-Verlag), p. 394.

Tanaka, Y. 1988, in *IAU Colloq. 108, Atmospheric Diagnostics of Stellar Evolution: Chemical Peculiarity, Mass Loss, and Explosion*, ed. K. Nomoto (Berlin: Springer-Verlag), p. 399.

Truran, J.W., Arnett, W.D., and Cameron, A.G.W. 1967, *Can. J. Phys.*, **45**, 2315.

Turtle, A.J. *et al.* 1987, *Nature*, **327**, 38.

Uomoto, A. and Kirshner, R.P. 1985, *Astron. Ap.*, **149**, L7.

Uomoto, A. and Kirshner, R.P. 1986, *Ap. J.*, **308**, 685.

Veigele, W.J. 1973, *Atomic Data Tables*, **5**, 51.

Weaver, T.A. and Woosley, S.E. 1980a, *Ann. N.Y. Acad. Sci.*, **336**, 335.

Weaver, T.A. and Woosley, S.E. 1980b, in *Supernovae Spectra*, eds. R. Meyerott and G.H. Gillespie (New York: American Institute of Physics), p. 15.

Webbink, R.F. 1984, *Ap. J.*, **277**, 355.

Wheeler, J.C. and Levreault, R. 1985, *Ap. J. (Letters)*, **294**, L17.

Wilson, R.B., Fishman, G.J., Meegan, C.A., Paciesas, W.S., and Pendleton, G.N. 1988, in *Nuclear Spectroscopy of Astrophysical Sources*, eds. N. Gehrels and G. Share (New York: American Institute of Physics), p. 73.

Woosley, S.E. 1988, *Ap. J.*, **330**, 218.

Woosley, S.E., Axelrod, T.S., and Weaver, T.A. 1981, *Comments Nucl. Part. Phys.*, **9**, 185.

Woosley, S.E., Pinto, P., and Ensman, L. 1988, *Ap. J.*, **324**, 466.

Woosley, S.E. and Weaver, T.A. 1986a, *Ann. Rev. Astr. Ap.*, **24**, 205.

Woosley, S.E. and Weaver, T.A. 1986b, in *Radiation Hydrodynamics in Stars and Compact Objects*, eds. D. Mihalas and K.-H.A. Winkler (Berlin: Springer-Verlag), p. 91.

Xu, Y., Sutherland, P., McCray, R., and Ross, R.R. 1988, *Ap. J.*, **327**, 197.

7. Neutrinos from Supernovae

ADAM S. BURROWS

7.1. Introduction

More than half a century ago, in a prescient paper, Baade and Zwicky (1934) first connected neutron star formation with supernovae. To these authors there seemed to be a natural correspondence between the large binding energies of neutron stars ($GM^2/R \sim 10^{53}$ ergs), whose existence had only just recently been postulated (Landau, 1932), and the large explosion energies that seemed to be associated with the newly identified supernovae. Such were the ambiguities in supernova light curves, spectra, and distances and such was the novelty of the new supernova concept that as much as 10^{53} ergs in supernova light and debris kinetic energy seemed perfectly consistent with the observations. The binding energy that must be shed to form the ultracompact neutron star, naturally, was to escape as light and kinetic energy in some combination. Indeed, how else?

No doubt, Baade and Zwicky would be gratified to learn that the existence of neutron stars is now amply verified and that a connection between neutron star birth and supernovae is a firm tenet of modern astrophysics, though one does not always imply the other. However, they might be astonished to learn that when neutron star birth does accompany a supernova explosion, the kinetic energy of the debris and its optical emissions amount to only $\sim 1\%$ and $\sim 0.01\%$, respectively, of the $\geq 10^{53}$ ergs reservoir. Fully 99% of the energy of neutron star formation is now known to emerge as *neutrinos* in a prodigious burst that, during its few *seconds*, is as "bright" as the entire observable universe. This neutrino burst announces both the birth of a neutron star and the death of a massive star ($M \geq 8\,M_\odot$) and precedes, by only hours, the optical pyrotechnics of the supernova itself. Curiously then, from the energetic point of view, the supernova is only a sideshow to the main event: neutron star birth.

The theory that implicates neutrinos with neutron stars and supernovae has a distinguished pedigree that spans several decades and has involved at various times hundreds of researchers from around the world. Its evolution has paralleled and relied upon developments in nuclear and neutrino physics, hydrodynamic and transport techniques, stellar theory, and, above all, the weak interactions that have occurred in fits and starts since the 1950s. Every decade has transformed the field. Whether we now possess all of the physics necessary to resolve the outstanding issues of the "supernova" puzzle definitively by the 1990s remains to be seen. What is clear is that we now know a great deal about the progenitors, the astronomy, the statistics, and the characteristics of supernovae and neutron stars, and we possess detailed models of implosion and explosion dynamics and the attendant neutrino emissions.

The recent (Feb. 23.316 (UT), 1987) epochal detections by the IMB (Bionta *et al.*, 1987) and the Kamiokande II (Hirata *et al.*, 1987, K II) collaborations of a neutrino burst from the LMC supernova, SN 1987A, have abruptly wrenched supernova modelers out of 30 years of theoretical isolation. That the standard model of neutrino bursts fits these data has been demonstrated by numerous authors (Arafune and Fukugita, 1987; Bahcall, Press, and Spergel, 1987; Bahcall, Dar, ad Piran, 1987; Bludman and Schinder, 1988; Bruenn, 1987; Burrows, 1987a–f; Burrows and Lattimer, 1987; Hari Dass *et al.*, 1987; Kahana, Cooperstein, and Baron, 1987; Krauss, 1987; Lamb, Melia, and Loredo, 1987; Lattimer, 1987; Lattimer and Yahil, 1989; Mayle and Wilson, 1987; Sato and Suzuki, 1987; Schaeffer, Declais, and Jullian, 1987; Schramm, 1987; Spergel *et al.*, 1987; Burrows, 1988a, b). However, given the rocky history of our picture of the neutrino signature of collapse, it is remarkable and certainly gratifying that theory and experiment confronted one another when they did. Had we detected a supernova burst in the early 1960s, we would not have been able to claim victory, as we, perhaps a bit too overconfidently, do now, though it was in the 1960s that the connection between neutrinos and supernovae was forged. Therefore, at this juncture in the history of supernova studies, it would seem useful to step back and reflect on the evolution over the last three decades of the standard neutrino burst and supernova models. In Section 7.2 I conduct such a review. As it is neither possible nor desirable to be comprehensive, I present a brief, idiosyncratic, and nontechnical critical bibliography and rely on a rather corpulent literature to fill in the gaps. As a complete history is not being attempted, I apologize in advance for any omissions and oversights. These paragraphs are meant only to give the reader a flavor for the evolution of the field.

Peppered throughout the historical discussion is the modern view of collapse dynamics, supernovae, and the character of the neutrino burst. In Section 7.3 the theory of core collapse supernovae is reviewed. Because of the profound opacity to light of the dense core that experiences collapse, we "see" this core directly only through its neutrino signature. Every bump and wiggle echoes the internal convulsions of the event and can provide clues about both the supernova mechanism and the neutron star that remains.

Section 7.4 discusses the only neutrino observations of a supernova so far, SN 1987A. While the agreement with calculations has been gratifying, there remain, of course, plenty of outstanding issues in supernova theory to be tested. These are highlighted throughout the text. Since neutrinos give us the only real access to the physics inside the collapse, it is important that observation of these particles continue. In an appendix I describe some of the available or contemplated neutrino detectors capable of good time resolution and therefore of shedding light on supernova mechanisms.

This paper is not a detailed listing of the techniques and equations of neutrino transport, but is rather a survey of personal observations and thoughts on the neutrino signatures of supernovae, in general, and SN 1987A, in particular.

7.2. A Short History of Supernova Neutrino Theory

7.2.1. 1957–1969

Our modern view of stellar collapse and nucleosynthesis dates back to the seminal paper by Burbidge *et al.* (1957, B^{2}FH) in which the "onionskin" model of late, massive star evolution was developed and popularized. It was shown that a massive star made of hydrogen would evolve in stages by thermonuclear burning into a structure of nested shells of progressively heavier elements. An outer shell of hydrogen surrounded, in turn, shells of, for example, helium, oxygen, and silicon, and a dense ($> 10^8$ g cm^{-3}) iron core developed in its center. Since iron is at the peak of the nuclear binding energy curve, as the iron core grew and its temperature exceeded $\sim 7 \times 10^9$ K, the iron would *endo*thermically photodissociate into alphas and neutrons. The effective adiabatic index thereby decreased below the critical value of $\frac{4}{3}$ and the iron core became unstable to *stellar collapse*.

In the B^{2}FH model, later developed more fully by Hoyle and Fowler (1960, HF), the supernova arises from the thermonuclear explosion of the oxygen zone that momentarily follows the iron core in collapse to higher temperatures and densities. The researchers B^{2}FH did not directly associate the supernova with the collapsing iron core itself. The explosion lifts the processed shells of heavy elements into space to contaminate subsequent generations of stars.

This paper was predominatly concerned with the synthesis of the elements, but developed the connection between nuclear physics, supernovae, nucleosynthesis, massive stars, iron cores, and stellar collapse that is the essence of the modern theory. However, it was off-the-mark on a number of particulars. We now associate only Type II (and perhaps Type Ib; see Harkness and Wheeler, this volume) supernovae with massive stars and core collapse. The progenitors of Type Ia's are thought to be population II, carbon-oxygen white dwarfs pushed by some binary process past the Chandrasekhar limit into complete thermonuclear disruption. Radioactive ^{56}Ni, the predominant product of a carbon–oxygen thermonuclear runaway, and its daughter ^{56}Co (not ^{254}Cf, as B^{2}FH suggested), are thought to power the optical light curve of Type Ia's (see Woosley, this volume).

The high entropy of the B^{2}FH cores resulted in a broad range of large core radii and masses, unconnected with the Chandrasekhar mass that figures so prominently today. The researchers B^{2}FH said little about the fate and dynamics of these unstable iron cores and did not develop the connection with neutron stars. Furthermore, a "massive" star to B^{2}FH could be as light as $\sim 1.5~M_\odot$. It is now thought that stars lighter than $\sim 8~M_\odot$ shed their mantles quiescently and leave white dwarfs, while most stars that, when on the main sequence, are heavier than $\sim 8~M_\odot$ become Type II or Type Ib supernovae and leave neutron stars. Prodigious neutrino bursts ($> 10^{53}$ ergs) accompany the core collapse of these $\geq 8~M_\odot$ stars, but not the death of $< 8~M_\odot$ stars or Type Ia supernovae. Perhaps only $\sim 10^{49}$ ergs of few MeV electron-type (ν_e) neutrinos from compression-induced electron capture issue from Type Ia's (Nomoto, 1984a). Even if the K II water Cherenkov detector had an electron detection threshold of 0 MeV, such a neutrino trickle would be invisible to it beyond ~ 100 pc. Realistic thresholds greater than 6.0 MeV render even the above detection range moot.

Nevertheless, though B^{2}FH laid the foundations for all subsequent work on stellar evolution and collapse supernovae, they were unaware of the important role of neutrinos in both.

Chiu (1961) was the first to suggest that the neutrino cooling of the cores of massive stars would leave them not extended and nondegenerate, as B^{2}FH had thought, but compact and degenerate. He showed that at the high temperatures and densities reached during advanced burning stages, plasmon decay, e^+/e^- pair annihilation, and the photoneutrino process dominate photon loss as a sink for the thermonuclear energy. Though neutrinos are produced in far smaller numbers than photons in the core, due to their notoriously small interaction cross section, they can escape, whereas the strongly coupled photons cannot. Neutrinos are important because, of all the known particles, they do not participate in either the electromagnetic or the strong interaction, but do have a low production threshold (low mass). Since they are indeed produced (through the weak interaction), they can and do dominate the cooling.

Not only does neutrino loss accelerate the later evolution of massive stars, but it implies "core convergence." Since degenerate cores become unstable at roughly the same Chandrasekhar mass (M_c), allowing for variations in the electron fraction (Y_e) and entropy profiles, the mass of the core, when it becomes unstable, is roughly independent of the total mass of the star, which could range from $\sim 8\ M_\odot$ to ~ 60 $M_\odot$. The critical masses taking these profiles into account are in the narrow range 1.2–2.0 $M_\odot$ near the classical Chandrasekhar result of 1.44 $M_\odot$ (Woosley and Weaver, 1986a, b). Though it took more than a decade for the transformation wrought by Chiu's neutrinos to be generally appreciated, the picture that has emerged is one of the accumulation of a distinct, degenerate "white dwarf" core in the center of a loosely coupled massive envelope of shells of progressively lower atomic number. While we now believe that the cores of ~ 8–10 $M_\odot$ stars are not iron, but O–Ne–Mg (Nomoto, 1984b), and that electron capture can have a hand in initiating collapse, the B^2FH concept of core collapse is still the focus of modern research. Core convergence established the important connection between the Chandrasekhar mass and the masses of the neutron stars formed in Type II supernova. However, even though a star's thermal neutrino luminosity just before core instability might dwarf its photon luminosity by as much as ten orders of magnitude, such emission is still much too weak to be detected by any foreseeable terrestrial detector.

Though it was Chiu (1964) who was instrumental in the gradual realization that neutrinos were even more important after collapse ensued, it was with the paper by Colgate and White (1966, CW) that the field of supernova neutrinos was truly born. They conducted what were not only the first numerical simulations of stellar collapse, but also were the first hydrodynamic computations of any kind in astrophysics. They concluded that the oxygen burning model of B^2FH and HF would not work because in spherical symmetry such burning could not reverse supersonic infall. However, Fowler and Hoyle (1964) suggested a combination of rotation and oxygen burning that still stands in the wings, if necessary (Bodenheimer and Woosley, 1983). Colgate and White (1966) contained the basic ingredients of any modern calculation: current progenitor models, neutrino processes, and an equation of state that ranged to and beyond nuclear densities ($\rho_N \sim 2.6 \times 10^{14}\ \mathrm{g\ cm^{-3}}$) and included a mix of particles (e.g., neutrons, protons, electrons, photons, nuclei, ...). Three progenitor models were used, whose structures were basically polytropic: a 10 $M_\odot$ star on a "high adiabat" that was shown to collapse by the iron–helium phase transition of B^2FH, and two "low adiabat" stars, one 1.5 $M_\odot$ and the other 2.0 $M_\odot$, that were shown to collapse by electron capture. The quoted masses are total stellar masses, as the focus of research had not yet shifted to the central core. In the collapse calculations of CW, before the center had reached nuclear densities, the electrons of the initially symmetric matter ($Y_e = Y_p (\sim Y_n)$, in nuclei) had captured away ("$e^- + p \rightarrow n + v_e$"), liberating electron neutrinos. The completely neutronized core continued to collapse until ρ_N was reached, at which point matter in an inner core, which comprised $\sim 5\%$ of the total stellar mass, stiffened (due to the high neutron degeneracy pressure at and above ρ_N), rebounded, and finally stopped with a thud. The rest of the neutronized inner star continued to rain down on the static inner core, now bounded by a hot accretion shock. In the model of CW, the gravitational energy of this infalling material, which because of the compactness of the inner core was the canonically high neutron star value of ~ 100 MeV baryon^{-1}, was radiated at the shock as thermal neutrino–antineutrino pairs (v_e–$\bar{v}_e$) with a high blackbody temperature (T_v) of ~ 40 MeV.

What was the role of these neutrinos in the supernova? Colgate and White's answer was the following. They noticed that the high-energy shock neutrinos had mean-free-paths in the infalling material that were not small. In a rather ad hoc fashion, they deposited approximately a half of the shock neutrino energy in the infalling mantle and allowed the rest to escape. The deposited energy amounted to $\sim 10^{53}$ ergs, and a violent explosion ensued. The essence of this model is the transfer of gravitational energy from the inner regions to the more loosely bound outer regions by the neutrino intermediary. According to the CW prescription, energy deposition took 10–14 ms, after which time explosion halted accretion and neutrino emission ceased. This neutrino-mediated energy deposition mechanism along with the direct, hydrodynamic core bounce mechanism of Colgate and Johnson (1960) set the agenda for all subsequent studies of core collapse supernovae, even though we now know that supernova energies are not $\sim 10^{53}$ ergs, but 10^{51} ergs, only a small fraction of the "available" gravitational energy. Here CW were following in the footsteps of Baade and Zwicky (1934). In addition, their prescription for neutrino transport was crude and not internally consistent. In particular, if we can deposit energy in the outer regions by neutrino absorption or scattering, it is much easier to do so in the inner regions nearer the source. And this done, the neutrinos are degraded to lower energies and the rate of energy transfer is decreased. The fact that the transfer of energy (or momentum) via neutrinos from a dense, inner region to a less dense, outer region requires anomalous opacity distributions has stood in the way of all such supernova mechanisms down to the present. In short, the conditions that are required for such a success usually ensure failure. Finally, the observed SN 1987A neutrino signal looked nothing like the CW signal. $T_v = 40$ MeV neutrinos (antineutrinos) emitted over a period of ~ 10 ms stands in marked contrast to $T_v = 3$–5 MeV antineutrinos observed over ~ 10 s. Also, because of the much larger interaction cross sections in water at $T_v = 40$ MeV, the integrated CW signal would have been greater than or equal to ten times larger in K II and greater than or equal to thirty times larger in IMB than has been actually reported.

Arnett (1966, 1967) followed CW with much more consistent neutrino transport and a more sophisticated equation of state. He recognized the need to solve the neutrino *diffusion* equations and incorporated them into his finite-difference scheme. The neutrinos (again, v_e and $\bar{v}_e$) were thermal with zero chemical potential, and they interacted via v–e^- and v–v scattering and v_e(n, p)e^- or $\bar{v}_e$(p, n)e^+ absorption. With Bahcall's (1964) old v–e^- scattering cross sections and no knowledge of neutral currents, v–e^- scattering was thought to dominate. Arnett concluded that the direct bounce mechanism of Colgate and Johnson (1960) did not work, and that the CW neutrino transfer mechanism was too simplistic, but that it could be made to work via neutrino energy *diffusion* from inside to outside. In his calculations, neutrino energy transport out of the inner regions allowed the core to achieve higher densities than otherwise and thereby "release" more gravitational energy, which by neutrino diffusion on dynamical times could reach the outer regions and lead to explosion. The explosion energies of $(1.4$–$4.0) \times 10^{52}$ ergs were high, but more reasonable, and importantly, the neutrino emission temperatures were ≤ 10 MeV. Mu-type neutrinos (v_μ and $\bar{v}_\mu$) were also considered, but since the weak interaction physics of the time require the presence of rare, high-mass muons ($m_\mu \sim 100$ MeV), v_μ's were thought to be of little consequence. In Arnett's work, we see the first serious back-of-the-envelope calculation of neutrino burst detectability. However, the duration (Δt) of the burst is still not much more than a few hundred milliseconds. Nevertheless, the trend to larger Δt's and smaller T_α's was in the right direction. The reader should note that the reciprocal nature of Δt and T_v is a universal feature of supernova models. Since the total energy radiated is a constant and the size of the radiating surface is set by neutron star characteristics ($R \geq 10$ km), small Δt's always imply high T_v's, and long Δt's always imply low T_v's. The downscattering that yields a low T_v implies a long diffusion time (Δt). T_v and Δt stem from the same physics.

Arnett, like CW, still neutronized on infall and regarded the v_e–$\bar{v}_e$'s as thermal. However, his more sophisticated equation of state and neutrino transport resulted in a bounce at, not ρ_N, but $\rho \sim 10^{13}$ g cm^{-3}. Indeed, this thermal bounce is what modern calculations would obtain either if neutronization were accomplished during infall or if initial core entropies were high ($s > 2 \times$ Boltzmann's constant per baryon). Early, rapid capture on nuclei would leave them too neutron-rich to avoid neutron drip. High entropies would dissociate nuclei at moderate densities. The nonrelativistic thermal neutrons thereby released are a stiff ($\gamma = \frac{5}{3}$) gas which would reverse collapse when abundant.

These low-density, thermal bounces dominated the field for a decade. It was not until it was recognized that v_e's were trapped (see below), that net neutronization ceased *during* collapse, and that initial core entropies were low ($s \leq 1$) that our modern view of collapse and the neutrino burst emerged. Nevertheless, CW and Arnett established once and for all the centrality of neutrino and weak interaction physics in collapse supernovae. Subsequent studies consist of refinements and extensions, however important, of this fundamental work.

7.2.2. 1970–1979
In the 1960s, it seemed that neutrino energy transport on dynamical times (less than or equal to tens of milliseconds after bounce) was responsible for the supernova explosion and that the supernova problem had been solved. The residue of collapse was a nascent neutron star, and ample gravitational energy was available to unbind the rest of the progenitor. However, the 1970s were to challenge this view. Wilson (1971) constructed a code with a multienergy group neutrino transport scheme and

thermal ν_μ losses and found that, far from aiding explosion, the neutrinos actually damped it. He concluded that it was the prompt hydrodynamic bounce-shock mechanism of Colgate and Johnson (1960) that obtained, and that due to neutrino losses, it obtained only barely. Nevertheless, his explosion energies were in the realistic range of $\sim 10^{51}$ ergs. Bruenn (1975), with a more sophisticated neutrino transport scheme that did not treat ν–e^- scattering as absorption, as Wilson had done, seconded Wilson's conclusion concerning neutrino transfer and the bounce shock. These first salvos of caution began battles between the prompt mechanism and the neutrino-mediated mechanism and between explosion and fizzle that persist to this day. The fundamental problem lies in the marginality of supernovae. Only $\sim 1\%$ of the gravitational energy of a neutron star is required to power the observed explosions. Only one tenth of that ($\sim 10^{50}$ ergs) is required to unbind the outer progenitor shells. This unfortunately makes the supernova problem one of *details* and subtle additions and subtractions. The community was slow to realize that much of the $\geq 10^{53}$ ergs was not directly associated with the supernova and was emitted in neutrinos on timescales that may have little to do with those of the explosion.

Nuclear physics, the weak interaction, and hydrodynamics with gravity are synergistically coupled in the supernova problem. The 1970s saw a multiplication of interest in these components, as they relate to supernova theory, and major advances in neutrino transport technique and practice (Imshennik and Nadyezhin, 1973, 1979; Lamb and Pethick, 1976; Yueh and Buchler, 1976; Arnett, 1977; Bludman and Van Riper, 1978; Lichtenstadt *et al.*, 1978; Nadyezhin, 1978; Tubbs, 1978; Sawyer and Soni, 1979). However, our picture of the behavior of neutrinos in "collapse" and, as a consequence, of collapse itself was importantly altered by two revelations.

The first was the realization by Mazurek (1974, 1975, 1977) and Sato (1975) that the ν_e's liberated by electron capture during collapse are trapped in the flow long before bounce. They concluded that most of the neutronization of the "iron" cores does not occur, as had been assumed during the previous decade, during infall. The "optical" depth (τ_{ν_e}) of the increasingly dense collapsing core to the progressively higher energy ($\varepsilon_\nu \propto \mu_e \propto \rho^{1/3}$) capture neutrinos, whose interaction cross sections scale with $\sim \varepsilon_\nu^2$, rapidly exceeds unity near $\rho \sim 10^{11}$ g cm^{-3} ($\ll 10^{13}$ g cm^{-3} and ρ_N). In short, the neutrino escape timescale (τ_{esc}, free streaming or scattering) exceeds the characteristic dynamical timescale ($\tau_d \sim 1$ ms/$\rho_{12}^{1/2}$) before significant electron capture and ν_e loss takes place. A *degenerate* sea of ν_e's builds up and establishes a chemical (beta) equilibrium between ν_e and e^- capture ($e^- + p \rightleftarrows n + \nu_e$) at a constant electron lepton fraction ($Y_l = Y_e + Y_{\nu_e}$; $Y_e \sim 4Y_\nu$) between 0.33 and 0.40. This value is near "iron's" original value of ~ 0.5 and is far from a neutron star's 0.05.

Trapping profoundly affects the evolution of cores. Maintaining high electron fractions during infall allows nuclei to be proton-rich enough to avoid significant neutron drip during collapse. With few free neutrons, the soft ($\gamma \sim \frac{4}{3}$), relativistic leptons (e^-'s and ν_e's) continue to dominate the gas pressure, and collapse is unimpeded. Infall proceeds beyond 10^{13} g cm^{-3} and is eventually halted near ρ_N, only because near nuclear densities, the nonrelativistic nucleons ($\gamma > \frac{5}{3}$) are finally squeezed out of nuclei (Lamb *et al.*, 1978; Bethe *et al.*, 1979). This phase transition to stiff matter leads to the rebound of the inner "homologous" core into the imploding outer core. At the interface, a strong shock wave is generated that may be the supernova in its infancy. The inner core and the material that subsequently rains down upon it achieve hydrostatic equilibrium within milliseconds. This "protoneutron" star is only marginally bound ($O(10^{51}$ ergs)) as trapping throttles off significant energy loss during infall.

Trapping implies that both lepton loss and energy loss are retarded and must proceed *after* the dynamical phase of collapse during the protoneutron star phase. The gravitational energy of collapse is converted, not into the thermal energy of free neutrons, but into the degeneracy energy of trapped leptons. Compression to high densities raises the Fermi energies (μ_i) of the electron and the electron neutrino to ~ 300 MeV and to ~ 200 MeV, respectively, from their initial values of ~ 10 MeV and "0 MeV." At bounce, the central temperature (T_c) is only ~ 10 MeV. The higher ρ_c and the higher average ν_e energies that result from trapping lead to greater neutrino opacities and depths ($\tau_{\nu_e} \sim 10^{4-5}$). As a consequence, the few *tenths*-of-a-second neutrino signal durations of earlier calculations are transformed into the current theoretical durations of *seconds* (see Burrows and Lattimer, 1986, BL) that seem to have been verified by K II (5.6 s) and IMB (12.4 s) on Feb. 23.316(UT), 1987. The long duration of the LMC burst can be seen as indirect confirmation of the basic concept of lepton trapping.

The second revelation of the 1970s was the weak neutral current (Tubbs and Schramm, 1975; Freedman, Schramm, and Tubbs, 1977). The new electroweak theory implied new neutrino/matter couplings and cross sections. Via neutral currents, ν_μ and ν_τ pairs (hereafter, "ν_μ's") can be produced directly at high temperatures and densities without the presence of a muon or a tauon (newly introduced in that decade). Plasmon decay, $e^+ - e^-$ annihilation, etc., yield not only $\nu_e - \bar{\nu}_e$ pairs, but ν_μ's in amounts that require their consideration in the protoneutron star and supernova (Wilson, 1974) problems as well. Indeed, in modern theory, ν_μ energy loss figures prominently in the cooling of young neutron stars.

Neutral currents also introduce new neutrino scattering processes, such as $\nu_i(n, n)\nu_i$ and $\nu_i(p, p)\nu_i$, that compare favorably with the important charged-current process, $\nu_e(n, p)e^-$, and whose cross sections are species-independent. The opacities (κ) to ν_e's of 50%–50% free neutron–proton matter at 10^{11} g cm^{-3}, due to the former (ν_e-nucleon) and the latter (charged current) process, are $\sim 10^{-20}$ (ε_ν/MeV)^2 cm^2 g^{-1} and $\sim 2 \times 10^{-20}$ (ε_ν/MeV)^2 cm^2 g^{-1}, respectively. These are approximately one hundred times larger at 10 MeV than the current value of $\nu_e - e^-$ scattering ($\sim 10^{-21}$ (ε_ν/MeV) cm^2 g^{-1}), but, curiously, are comparable to the $\nu_e - e^-$ scattering cross section employed in the 1960s (Bahcall 1964). Freedman (1974) showed that neutral currents allow neutrinos of all species to scatter coherently on nuclei ($\nu_i(A, A)\nu_i$). The corresponding transport opacity for pure iron is $\sim 3 \times 10^{-20}$ (ε_ν/MeV)^2 (A/56) cm^2 g^{-1}. This is larger, but not much larger, than either of the above, and is comparable to their sum. Note that the Thomson scattering ($\gamma - e^-$) opacity is ~ 0.2 cm^2 g^{-1}. During collapse and trapping, coherent Freedman scattering by "iron" is the most important source of opacity. When $\rho = 10^{11}$ g cm^{-3} and $\varepsilon_\nu = 10$ MeV, at which time $R \sim 100$ km, the mean free path is

$$\lambda = \frac{1}{\kappa \rho} \sim \frac{1}{3 \times 10^{-20}(10)^2 10^{11}} \sim 30 \text{ km}.$$

Surprisingly, the old ν_e opacities (with the old ν_e-e^- and $\nu_e(n, p)e^-$ cross sections of either neutron-rich or electron-rich matter) were close to those we now employ in similar environments. Historically, trapping did not require neutral currents (Mazurek, 1974).

With the advent of coherent scattering, a new mechanism was proposed for the still-problematic supernova: neutrino momentum deposition (Schramm and Arnett, 1975; Bruenn, Arnett, and Schramm, 1977). It was thought that the large ν_e luminosity during the first few milliseconds after the shock breakout of the neutrinosphere (see below) might exceed the Eddington luminosity and reverse the undissociated mantle of iron. The large Freedman scattering cross section was the key. However, the Eddington neutrino luminosity is $\sim 10^{55}$ ergs s^{-1}, almost two orders of magnitude greater than current calculations yield. Momentum deposition requires a huge contrast in opacity in the object: low values on the inside and high values on the outside. The high coherent cross sections of iron are not matched by correspondingly lower values inside. Indeed, the large ν_e-nucleon opacities, quoted above, that are relevant in the post-shock matter during breakout, when multiplied by the large matter densities behind the shock, imply mean free paths behind the shock that are approximately five times those in the unshocked outer mantle where coherent scattering obtains. This is the reverse of what is required. In general, neutral currents make both neutrino energy and neutrino momentum transfer mechanisms for supernovae more, not less, difficult on dynamical times.

Trapping and neutral currents have indeed changed our predictions of the average neutrino energies, species mix, and duration of neutrino bursts. While the 1970s did not solve tthe supernova problem, it transformed both it and our view of the neutrino signature in ways not generally appreciated.

7.2.3. 1980–1989

The supernova problem is not yet solved. However, in the 1980s, all the issues involved have been brought into much sharper focus. There has been palpable improvement in massive star models (Nomoto 1984b; Woosley and Weaver 1986a, b), and in our knowledge of the masses, entropy profiles, electron fraction (Y_e) profiles, and statistical mechanics of the unstable "Fe" or O–Ne–Mg cores. We now know that these critical cores can have masses between $\sim 1.2 \, M_\odot$ and $2.0 \, M_\odot$, and that central entropies are in the cold range between 0.6 and 1.0. The core size has a direct bearing on the success or failure of the direct mechanism of collapse supernovae (see Cooperstein, this volume). If the core is large ($\geq 1.3 \, M_\odot$, Burrows and Lattimer, 1983) and the trapped lepton fraction (Y_l) is low (< 0.4), the weak rebound generates a weak shock that must dissociate many nuclei and survive breakout electron neutrino losses (Van Riper, 1978), but cannot. It stalls into a standing accretion shock and the prompt mechanism is aborted (Mazurek, 1982). A small core ($\leq 1.3 \, M_\odot$) with a large trapped lepton fraction ($\gtrsim 0.4$) experiences a strong rebound that drives a strong shock. This shock must plow through a smaller mantle, will suffer fewer breakout neutronization losses (Burrows and Mazurek, 1983), and can succeed (Bowers and Wilson, 1982; Burrows and Lattimer, 1985; Bruenn, 1985, 1986; Myra et al., 1987). A soft supranuclear equation of state can also energize the bounce and increase the chances for direct success (Baron, Cooperstein, and Kahana, 1985; Van Riper, 1988). Unfortunately, there is still disagreement on both the softness of the nuclear equation of state and the value of Y_l (trapped).

Relevant to the outstanding Y_l question are the nature and extent of electron capture on nuclei during infall (Fuller, Fowler, and Newman, 1980, 1982; Fuller, 1982; Zaringhalam, 1983; Cooperstein and Wambach, 1984) and the degree to which v_e–e^- energy downscattering during trapping enhances lepton loss (Bruenn, 1985; Myra et al., 1987). Lepton fractions as low as ~ 0.33 and as high as 0.39 are still quoted, the former being disastrous for the direct mechanism. However, improvements in the neutrino transport algorithms and input physics should settle the issue in the next few years from a theoretical vantage (van den Horn and van Weert, 1981; Schinder and Shapiro, 1982, 1983; Cooperstein, van den Horn, and Baron, 1986; Myra et al., 1987; Bruenn, 1985, 1986, 1987; Cooperstein, 1990; Giovanoni, Ellison, and Bruenn, 1989).

Though there is no consensus on how small cores ($< 1.3\ M_\odot$) explode, there is a general consensus that large cores ($> 1.4\ M_\odot$) cannot overcome the formidable obstacles to the direct mechanism. Their bounce-shocks are sapped by dissociation and neutrino losses and stall into accretion. This is a most unsatisfactory state of affairs. How then do the massive stars with larger cores become supernovae? Wilson (1985) has recently found that a stalled shock can be re-energized by the heating of the shocked mantle by neutrinos from the cooling and accreting core on timescales of, not milliseconds, but hundreds of milliseconds to a second. This is a modification of the old neutrino transport models of CW and Arnett that *requires* initial failure. The "delayed" mechanism has been diagnosed by Bethe and Wilson (1985) and Lattimer and Burrows (1984), among others, and, though still problematic, is a frontrunner, along with the prompt mechanism in the supernova theory race (see Mayle, this volume). The convective enhancement of the neutrino emissions crucial to this mechanism may also play a role (Arnett, 1987; Burrows, 1987g). The neutrino burst from the next nearby core collapse supernova, if it is close enough, will allow us to distinguish the prompt from the delayed mechanism and identify the role, if any, of convection.

7.3. The Neutrino Signature of Core Collapse Supernovae

The white dwarf-like core of a star whose main sequence mass exceeds $\sim 8\ M_\odot$ and whose thermonuclear life lasts $\leq 2.5 \times 10^7$ yr collapses in less than a second to a protoneutron star, whose subsequent transformation into a neutron star proper (or black hole) takes seconds. The supernova shock wave travels from the core to the stellar surface in between an hour (SN 1987A) and a day (red supergiant) and disassembles the star in an explosion that lasts from months to years. Though $\sim 10^{49}$ ergs is radiated in supernova light, and the kinetic energy of the debris is

$\sim 10^{51}$ ergs, more than 10^{53} ergs, or 0.1–$0.2\,M_{\odot}$, is liberated in a neutrino burst that simultaneously announces the massive star's death, the neutron star's birth, and the supernova explosion itself. It was such a burst that was detected on Feb. 23.316(UT), 1987.

Stellar collapse is the only context in which the weak interaction has a direct effect on macroscopic dynamics. This dynamics is stamped on the neutrino signature whose spectra, time behavior, and species mix enable us to diagnose the event. However, since the neutrino cross sections are annoyingly small, proximity (< 10 kpc) is required to test theories in satisfying detail. Only one in 10^{15} of the 1987A $\bar{v}_e$'s that coursed through the IMB detector was actually caught. When the next nearby core collapses, both general and specific predictions concerning its neutrino emissions might be useful and are provided in the following paragraphs.

7.3.1. Infall Epoch: Before Bounce

The central temperature, central density, and the initial radius of a critical Chandrasekhar mass ($\sim 1.5\,M_{\odot}$) in the core of a massive star are 0.6–0.8 MeV, 10^{9-10} g cm^{-3}, and $\sim 4 \times 10^3$ km, respectively. The Fermi energy (μ_e) of relativistic electrons is $\sim 5(\rho_9 Y_e)^{1/3}$ MeV. The luminosity ($\sim 10^{49}$ ergs s^{-1}) and average energy (~ 1 MeV) of the thermal neutrinos radiated during quiescent burning just before collapse are each so low that these neutrinos are not detectable with devices of realistic size or energy threshold. However, the collapse, initiated by a combination of electron-capture and iron photodissociation, drives μ_e to higher and higher values that easily exceed the electron capture thresholds on "iron" and the few ambient free protons. As the density climbs, the electron capture rate ($\propto \mu_e^5$, for capture on free protons) increases dramatically. The electron neutrinos from electron capture during infall constitute the first phase of the neutrino burst. The average energy of capture neutrinos ($\langle \varepsilon_{v_e} \rangle \propto \mu_e$) starts at ~ 10 MeV near $\rho \sim 10^{10}$ g cm^{-3} and increases to 20–30 MeV near $\rho \sim 10^{11-12}$ g cm^{-3}, at which point the capture neutrinos are trapped. Trapping implies not only that Y_e for a given mass zone stops decreasing, but that the average energy of the escaping v_e's has a maximum that reflects the electron Fermi energy at trapping. Without trapping, μ_e (transparent) would climb to 50–100 MeV, even as Y_e fell, and the spectrum of the escaping infall v_e's would be much harder.

Mass zones are constantly entering the 10^{11-12} g cm^{-3} range during infall. Therefore, despite trapping, the capture v_e luminosity (L_{v_e}) actually increases with time. The temperatures in the transparent zone ($\rho < 10^{11}$ g cm^{-3}) are sufficiently small (≤ 1 MeV) that $\bar{v}_e$'s and v_μ's do not contribute significantly at this time. However, the total energy radiated in v_e's during infall, before bounce, is only $\sim 10^{51}$ ergs, though L_{v_e} can approach $\sim 10^{52-53}$ ergs s^{-1}. Infall energy loss would be much higher if trapping did not occur. Representative collapse ($\rho_i \rightarrow \rho_f$) timescales are: $\rho_c = 10^{10} \rightarrow 10^{11}$ g cm^{-3}, $\Delta t_{if} \sim 10^2$ ms; $\rho_c = 10^{11} \rightarrow 10^{12}$ g cm^{-3}, $\Delta t_{if} \sim 25$ ms; $\rho_c = 10^{12} \rightarrow 10^{13}$ g cm^{-3}; $\Delta t_{if} \sim 5$ ms; and $\rho_c = 10^{13} \rightarrow 10^{14}$ g cm^{-3}, $\Delta t_{if} \sim 1$ ms. A good fit to Δt_{if} is $5\rho_{12,i}^{-0.7}$ ms. Therefore, a zone spends ~ 20 ms in the trapping zone where most of the capture, neutronization, and v_e emission occurs. Though the infall phase may last hundreds of milliseconds, the characteristic infall capture flux timescale is only ~ 30 ms.

The dominant nucleus and mix of nuclei near the peak of the nuclear binding energy curve change during collapse in such a way as to make the e^--nuclei capture rate and spectrum highly problematic. In addition, the free-proton fraction is a sensitive function of the initial core entropy. As a result, even the relative contributions of capture on free protons and capture on nuclei to the total capture rate and spectrum can, at present, be estimated only crudely (Fuller, 1982; Cooperstein and Wambach, 1984). Furthermore, these questions have a direct bearing on the precise value of Y_l (trapped). Truly, an observation of the ν_e's from infall can speak volumes about fundamental collapse physics.

During infall, the core breaks into two distinct regions: the subsonic, "homologous" inner core ($M \sim 0.5$–$0.8\ M_\odot$) and the supersonic outer core (mantle). Upon reaching nuclear densities, the homologous core rebounds as a unit. Because the outer core is out of sonic contact with its inner counterpart, the two regions collide violently. A shock wave is formed at the boundary (sonic point, $\rho \sim 10^{13-14}\ \mathrm{g\ cm}^{-3}$, $R \sim 20$ km) and is driven into the mantle by the expanding inner core piston. This shock wave, whose initial speed is $\sim c/4$, moves outward in mass to lower densities and greater radii. Post-shock entropies are high enough (~ 5–9) to dissociate the nuclei encountered. If the shock is energetic enough and the outer core is light enough, this bounce-shock will continue without pause, reverse the direction of the outer zones, and be the supernova explosion in its infancy. This is the prompt mechanism for collapse supernovae. However, the bounce-shock may stall into accretion near $M \sim 1.3\ M_\odot$ ($R \sim 200$ km). If this accretion shock is eventually revived by neutrino heating on long timescales (0.1–1.0 s), the delayed mechanism for collapse supernovae obtains. In either case, the inner core plus subsequent accreta achieve hydrostatic equilibrium within milliseconds. This residue is a protoneutron star.

7.3.2. Prompt Breakout Burst

At bounce, the integrated Rosseland depth, $\tau_{\nu_e}(r) = \int_r^\infty dr/\lambda_{\nu_e}$, to ν_e's from radius r to infinity has at $r = 0$ the high value of $\tau_{\nu_e}(0) \sim 10^4$, but decreases rapidly with increasing radius and decreasing stellar density ($d \ln \rho/d \ln r < 0$). The surface $\tau_{\nu_e}(R) = \frac{2}{3}$ defines the "neutrinosphere," which separates the inner opaque region from the outer transparent region. The neutrinosphere is the effective radiating surface for neutrinos, in perfect analogy with the "photosphere" of normal light-emitting stars. Quite naturally, its position differs from species to species and is a function of time. At bounce, the ν_e-sphere is at $R = 50$–80 km and $\rho \sim 10^{11-12}$ g cm^{-3}. Whichever mechanism obtains, the bounce-shock is formed in the neutrino-opaque region, but quickly (≤ 1 ms) races out to the lower density, transparent region. When the shock hits the ν_e-sphere, there is a sharp burst of ν_e's whose luminosity can reach a peak of $\sim 5 \times 10^{53}$ ergs s^{-1}. This is the so-called prompt "breakout" burst that is supplied by rapid electron-capture on the newly shock-liberated free protons. When the shock is at high densities, the post-shock matter is rich with ν_e's which, because of the high opacities and consequently "long" ($>$ milliseconds) escape times, are trapped in the flow. However, when the shock breaks through the ν_e-sphere near post-shock densities of $\sim 10^{11}$ g cm^{-3}, the characteristic ν_e escape times become comparable to the shock motion (or dynamical) timescales, and v_e's can pour out into the transparent region and to infinity. In dissociated matter at $\rho = 10^{11-12}$ g cm^{-3}, the capture timescales are less than 1 ms and the escape times are ~ 1 ms. Therefore, the outer shocked mantle between $\sim 10^{12.5}$ g cm^{-3} and 10^{11} g cm^{-3} is rapidly neutronized. A trough in Y_e is formed within a few milliseconds, which bottoms out at ~ 0.15 and encompasses a few tenths of solar masses in the periphery of the residue (see Burrows and Mazurek, 1982; and Figure 7.5). The lepton fractions in the inner regions at high τ_{v_e} do not change during this prompt v_e burst.

The total energy of the prompt burst depends on the progenitor structure and on whether the bounce-shock succeeds. If the "iron" core is small and its envelope is tenuous, the shock encounters, and must promptly neutronize, less mass. The burst energy in this case is only $\sim 10^{51}$ ergs, and its half-width is ~ 1 ms. However, if the "iron" core is fat and its envelope is thick, the shock encounters much more mass during breakout, and as much as $\sim 3 \times 10^{51}$ ergs can be radiated during a few milliseconds. The prompt losses are exacerbated in the latter case by the fact that, in it, the direct mechanism is aborted, the shock stalls and sinks back, and more capture v_e's are "squeezed" out. Indeed, the dissociation "losses" and the prompt burst losses are the major obstacles to direct shock success. Therefore, the magnitude of the breakout v_e burst is a good diagnostic of the supernova mechanism and its unambiguous detection will shed a great deal of light on supernova physics. The spectrum of the prompt burst is a mixture of the capture specturm and a thermal spectrum. The shock wave entropizes and puffs out the outer mantle. The post-shock v_e-sphere radius is 80–100 km, and its temperature is ~ 4–5 MeV. The breakout spectrum can be fitted by a v_e blackbody with a temperature of ~ 4.5 MeV and an η_v $(= \mu_v/T_v)$ of ~ 3 during the first post-breakout milliseconds. Whatever the actual breakout v_e spectrum, though its average v_e energy (15–20 MeV) is similar to that during infall, its shape is very different. The time between bounce and breakout is no more than 1 ms. The abrupt transition in the v_e spectrum before and after breakout is a distinctive characteristic of the current theory of stellar collapse.

7.3.3. Cooling and Neutronization of a Protoneutron Star

With good reason, the focus of most modern research into stellar collapse has been the mechanism of Type II supernovae. However, since, as we have explained, much of the gravitational energy is trapped during infall, the neutrino emissions are perforce associated with the quasi-static protoneutron star phase. If the long-term neutrino heating mechanism obtains, this phase and the supernova phase roughly coincide for as long as ~ 1 s. But if the prompt mechanism obtains, there is a clean separation between the launch of the supernova and the phase during which the bulk of the neutrinos are radiated. Therefore, the protoneutron star phase itself deserves special scrutiny as a distinct and fascinating problem in stellar evolution.

There had been some preliminary work on the problem of young neutron stars before the appearance of SN 1987A (Nadyezhin, 1978, 1983; Sawyer and Soni, 1979; Sawyer, 1980; Burrows, Mazurek, and Lattimer, 1981; Goodwin, 1982; Burrows and Lattimer, 1986 (BL)). Fortunately, the IMB and K II detections have stimulated a great deal of new theoretical interest in the problem that will no doubt refine our understanding of the physics of neutron star birth. What follows is a general

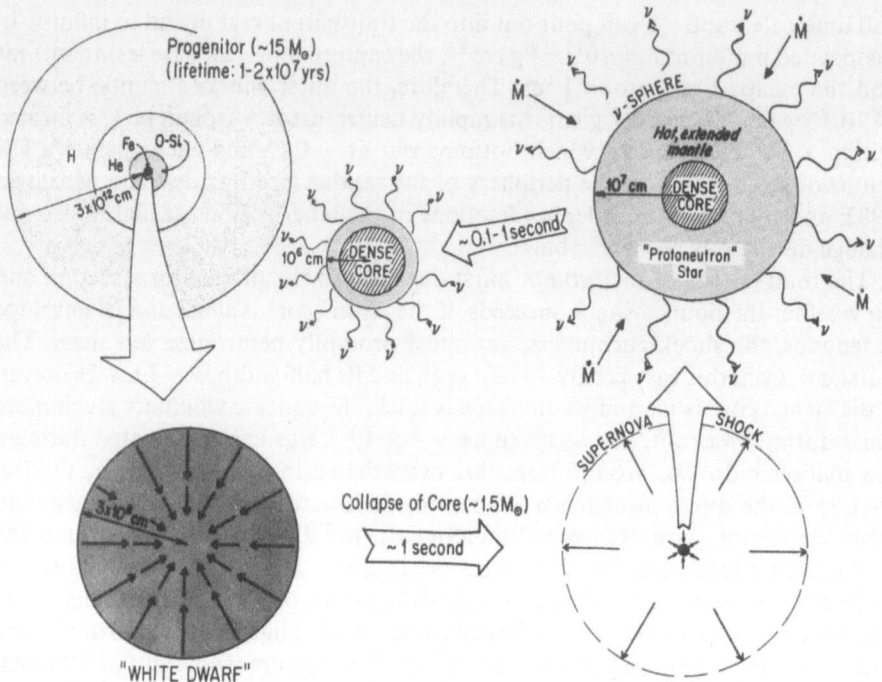

Figure 7.1. A depiction of the transition from progenitor, through collapse to explosion, the early protoneutron star phase, and the late protoneutron star phase. See text for details.

discussion of this phase. For a more complete treatment, the reader is referred to Burrows (1988c, BL), and Mayle and Wilson (1987).

Within 10 ms, the prompt ν_e burst blends into the long-term cooling and neutronization flux of the protoneutron star. The protoneutron star is hot, bloated, and electron- and energy-rich. Though the prompt ν_e burst is dramatic, this next phase of emission involves $> 10^{53}$ ergs and requires seconds. Since the shock wave has heated the peripheral zones, $\bar{\nu}_e$ and "ν_μ" thermal seas are generated within milliseconds by thermal pair, plasmon, and nucleon bremmstrahlung processes. The emission of these neutrinos quickly starts to compete with the ν_e emission, whose advantage rapidly fades as mantle neutronization slows. After no more than 20 ms, all six neutrino species are contributing almost equally to the luminosity. The sequence of prompt ν_e burst followed by $\bar{\nu}_e$ "turn-on" is another characteristic and prediction of the standard model.

Figure 7.1 depicts the transition from core instability to neutron star. The protoneutron star begins as a dense, cold (low s) core ($M \sim 0.5$–$0.8\ M_\odot$, $\rho_c \sim 5 \times 10^{14}$ g cm^{-3}, $R \sim 20$ km, $\langle T \rangle \sim 10$ MeV, $s \sim 1.0$) surrounded by a hot (high s) shocked mantle ($M \sim 0.5$–$0.8\ M_\odot$, $\rho < 10^{13}$ g cm^{-3}, $R \sim 100$ km, $T = 4$–10 MeV, $s \sim 5$–9) that is undergoing accretion at a rate that depends on the progenitor structure and the supernova mechanism. If the prompt mechanism obtains, $\dot{M}$ is initially low (~ 0.1–$0.2\ M_\odot$ s^{-1}) and decreases quickly. If it does not, $\dot{M}$ can be high

($\geq 0.2 \, M_\odot \, s^{-1}$) and will persist at a healthy level for as long as the shock is stalled. However, since the density profile of the inner progenitor structure that is feeding the protoneutron star with its last few tenths of a solar mass is steep, even in the latter case there is a secular decrease in $\dot{M}$, albeit with a longer characteristic time.

The first cooling phase involves the neutronization and cooling of the hot mantle and starts with "mantle collapse" (BL). Within ≥ 100 ms, the lower density mantle radiates a good fraction of its internal energy and shrinks to ~ 10–20 km. Approximately half of the gravitational energy change during the quasi-hydrostatic "deflation" of the mantle is radiated. The other half pumps up the mantle temperatures to ~ 30–50 MeV. This is the normal negative specific heat effect in stars and implies the paradoxical result that energy and entropy loss first leads to a temperature *increase*. All during the mantle collapse phase, a decaying accretion rate is contributing to the total neutrino luminosity $\sim GM\dot{M}/R$). Accretion results in additional compression which further increases the mantle temperature. It has been shown that the shocked mantle is unstable to entropy-driven convective motions that are quite difficult to model. The effect of convection on the neutrino luminosities during the first ~ 500 ms could be significant and may positively affect the viability of the long-term mechanism (Burrows, 1987g). Convection should enhance both L_i and the effective emission temperatures and accelerate the mantle phase. However, barring a possible pivotal role in the supernova itself, it should not qualitatively alter the general features of the cooling phase.

Though the average mantle temperature skyrockets during mantle collapse, the effective temperatures of the ν_e- and $\bar{\nu}_e$-spheres remain roughly constant at ~ 4.0 and 5.0 MeV, respectively, suitably redshifted. The effective $\bar{\nu}_e$-sphere temperature ($T_{\bar{\nu}_e}$) is always a bit higher than the corresponding quantity for ν_e's (T_{ν_e}). The major sources of opacity for the ν_e's and $\bar{\nu}_e$'s are charge-current absorption, $\nu_e(n, p)e^-$ and $\bar{\nu}_e(p, n)e^+$, respectively, and neutral current scattering, $\nu_i + (n, p) \rightarrow \nu_i + (n, p)$. Since the periphery near the neutrinospheres is quite neutronized, the $\bar{\nu}_e$ opacity is smaller than that for ν_e. Therefore, the radiating surface of the $\bar{\nu}_e$'s is deeper, at high temperatures. Since the ν_μ's do not participate in the charged-current reactions, the ν_μ-sphere is deeper still, and temperatures of 7–8 MeV have been assigned to it (Burrows ad Mazurek, 1982; Woosley, Wilson, and Mayle, 1986; Bruenn, 1987; Mayle, Wilson, and Schramm, 1987).

After many hundreds of milliseconds to a second, the heat and electron neutrinos in the dense inner core start gradually to emerge. Diffusion from deep inside now begins to resupply the radiating region and support the neutrino losses. The huge temperature spike in the mantle starts to decay. The timescale for diffusion from the dense interior is sufficiently long (seconds) that the lion's share of the mantle emission phase is over before its losses can be compensated by core sources. Therefore, the neutrino emissions from protoneutron stars naturally break up into two phases: the early mantle collapse, accreting, convecting phase that may last 0.5–1.0 s, during which the supernova is decided, and a long-term core cooling phase with a characteristic time of many seconds. This second phase might persist for many tens of seconds at observable levels. In addition, if the accretion rate is sufficiently high for a sufficiently long time, the protoneutron star will experience a general relativistic instability and collapse dynamically in less than 1 ms to a *black hole*. In this stellar-mass black hole formation scenario, the accretion-enhanced neutrino luminosity would be higher than "normal" and would be truncated abruptly after 1–2 s. The supernova might be aborted. Further, an intermediate protoneutron star phase is required. A more detailed discussion of this scenario for black-hole formation can be found below in Section 7.4.4.

It is useful to note that, in effect, during the protoneutron phase, one high energy (~ 100–200 MeV), degenerate electron neutrino in the interior downscatters and diffuses to the exterior where it is "converted" into approximately ten thermal neutrinos of all species that escape with ~ 10–20 MeV. There is a small excess of ν_e over $\bar{\nu}_e$ emission to effect neutronization. Crude estimates of the characteristic lepton and energy loss timescales can be obtained from simple sphere diffusion arguments (Burrows, 1984). The results are

$$\tau_1 \sim \frac{3R^2}{\pi^2 c \lambda_0} \frac{dY_1}{dY_{\nu_e}} \sim 3 \text{ s},$$

and

$$\tau_{\text{Th}} \sim \frac{3R^2}{\pi^2 c \lambda_0'} \left(\frac{E_{\text{Th}}^0}{2E_\nu^0} \right) \sim 10 \text{ s}.$$

where the symbols have their standard meanings. Though these numbers tell us that neutronization and cooling are simultaneous and that the timescales are not milliseconds or minutes, a real stellar evolution calculation is required in order to go beyond trivialities.

Presumably, the baryon mass of a neutron star is not known *a priori*; but could cover a range that extends many tenths of solar masses around 1.4 $M_\odot$. Table 7.1 shows the model parameters of a subset of the grid of general relativistic (GR) protoneutron star calculations recently performed by Burrows (1988c). This model grid is not comprehensive but is meant to suggest the range of possible behaviors of the protoneutron star family. $\dot{M}_0$ is the initial accretion rate onto a protoneutron star whose initial mass is either 1.2, 1.3, or 1.4 $M_\odot$ (see Table 7.1). τ is the assumed

Table 7.1. Model parameters.

Model	Equation of state	Initial mass → Final mass ($M_\odot$)	$\dot{M}_0$ ($M_\odot$ s^{-1})	τ (s)[a]
52	Stiff	1.2 → 1.2	0.0	—
53	Stiff	1.3 → 1.3	0.0	—
54	Stiff	1.4 → 1.4	0.0	—
55	Stiff	1.3 → 1.5	0.4	0.5
56	Stiff	1.3 → 1.6	0.6	0.5
57	Stiff	1.3 → 1.8	1.0	0.5
69	Stiff	1.3 → 2.3 (BH)[b]	2.5	0.5
72	Stiff	1.3 → 2.3 (BH)	1.0	1.5
59	Soft	1.2 → 1.2	0.0	—
62	Soft	1.3 → 1.5	0.4	0.5

[a] The exponential time constant for the mass accretion rate whose initial value is $\dot{M}_0$. $\dot{M}_0 \tau$ is the total mass added during the calculations if a black hole did not form.
[b] BH indicates a black hole formed.

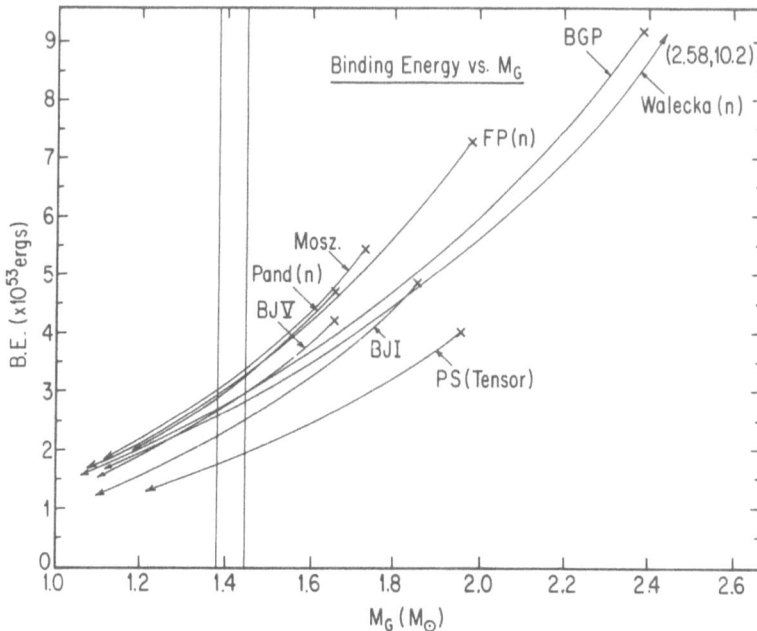

Figure 7.2. Binding energy in units of 10^{53} ergs versus gravitational mass for cold, catalyzed neutron stars with various nuclear equations of state found in Arnett and Bowers (1977). The vertical lines near 1.4 $M_\odot$ mark the masses of the components of PSR 1913 + 16 (1.444 $M_\odot$, 1.384 $M_\odot$).

exponential accretion time constant. "Stiff" refers to a nuclear equation of state like BJI (Bethe and Johnson, 1974) and "soft" refers to a nuclear equation of state like BJV (ibid). The nature of the true nuclear equation of state is, in fact, unknown, but plots of binding energy (BE) versus gravitational mass for neutron stars with various viable equations of state are given in Figure 7.2 for the reader's perusal. The total binding energy radiated during the neutrino burst might be between 10^{53} ergs and 5×10^{53} ergs.

Figure 7.3 depicts the evolution of $T_{\bar{\nu}_e}$ versus time for two representative models of Table 7.1. With the blackbody assumption of these models, $\langle \varepsilon_{\bar{\nu}_e} \rangle \sim 3.15 T_{\bar{\nu}_e}$. The results for models generated by Mayle, Wilson, and Schramm (1987, MWS), Mayle (1987, M), Mayle and Wilson (1987, MW), and Bruenn (1987, BR) are shown for comparison. The quantities for $\bar{\nu}_e$ are highlighted because of their importance in water Cherenkov detectors. We see that all workers agree that during the first 1–2 s, $T_{\bar{\nu}_e}$ is between ~ 4.0–5.0 MeV. As models 59 and 62 show, $T_{\bar{\nu}_e}$ decays only gradually. The abrupt jump in $T_{\bar{\nu}_e}$ seen in the work of MW occurs when the long-term supernova commences and accretion is consequently halted. This jump, though it may not actually be so abrupt, when accompanied by the corresponding *fall* in the various neutrino luminosities due to the cessation of accretion should be a clear signal of the long-term supernova mechanism. Furthermore, Mayle (1985) has suggested that small characteristic oscillations in luminosity might accompany the delay before the supernova.

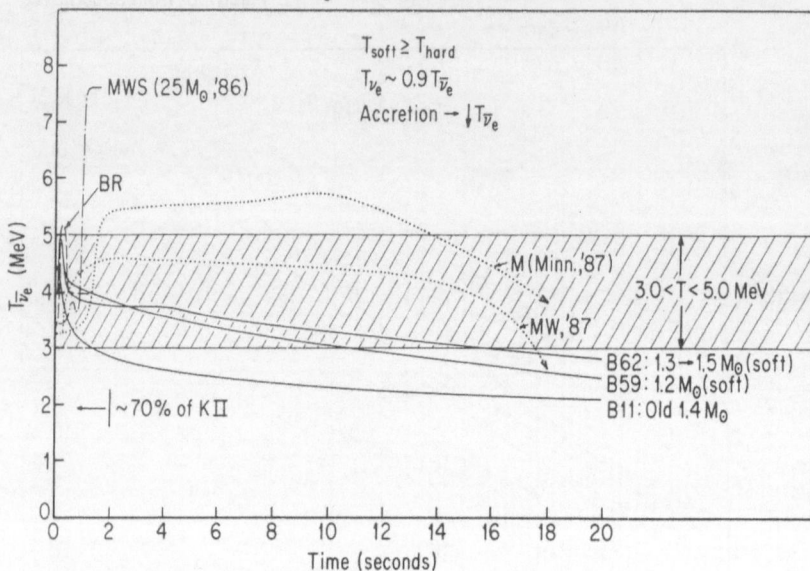

Figure 7.3. Redshifted antielectron neutrino temperature (in megaelectronvolts) versus time (in seconds) for various protoneutron star evolutions, from Burrows (1988c, B11, B59, B62), Bruenn (1987, BR); Mayle, Wilson, and Schramm (1987, MWS), Mayle (1987, M); and Mayle and Wilson (1987, MW). The shaded region covers the wide 3.0–5.0 MeV range allowed by the K II and IMB data. The best fit is 4.0–4.5 MeV. The models B11 and MWS are definitely excluded, while the others agree broadly with one another and the detections (see Table 7.3). See text for details.

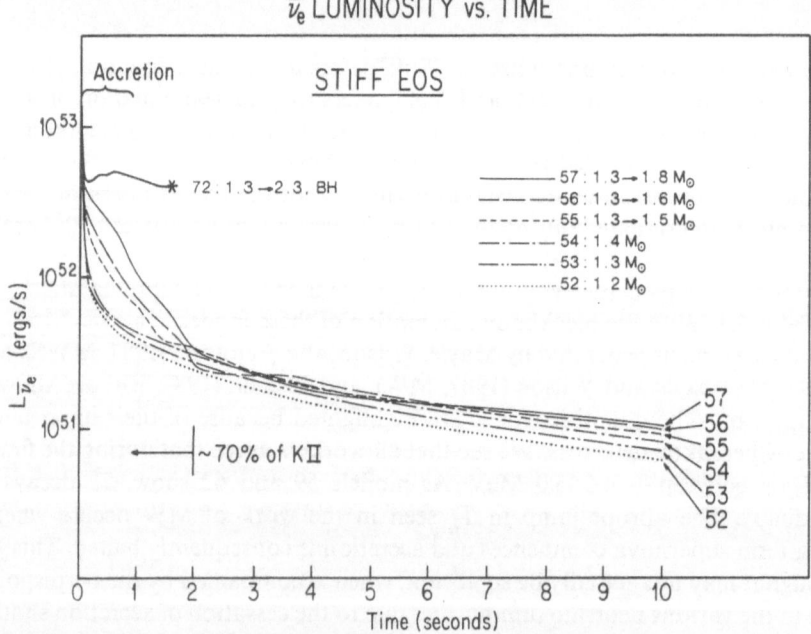

Figure 7.4. Redshifted $\bar{\nu}_e$ luminosity versus time (in seconds) for all the stiff models of Table 7.1, except model 69. The masses of models 52–54 are constant at 1.2, 1.3, 1.4 $M_\odot$ (baryon), respectively, while the others experience various degrees of accretion. After ~ 2 s, due to the high accretion rate, model 72 collapsed to a black hole. The two phases of emission (early, $t < 1$–2 s, and late, $t > 2$ s) are clearly seen. This figure is taken from Burrows (1988c). See text for furter discussion.

Figure 7.4 shows the evolution of the $\bar{\nu}_e$ luminosity after prompt burst versus time for most of the stiff models of Table 7.1. The two-phase nature of the protoneutron star emission is apparent. L_{ν_e} and $L_{\nu_\mu}/4$ are always within $\sim 10\%$ of $L_{\bar{\nu}_e}$ after breakout, and T_{ν_μ} is always only a bit less than two times $T_{\bar{\nu}_e}$. $L_{\bar{\nu}_e}$ starts near 4×10^{52} ergs s^{-1} and can stay above 10^{52} ergs s^{-1} for about 1 s for the heavier models with more accretion. Note the behavior of model 72. After only ~ 2 s of accretion, the baryon mass of the core went critical and a black hole was formed.

The evolutions of $L_{\bar{\nu}_e}$ for the softer equation of state are similar to those in Figure 7.4, for the same baryon mass. However, though the larger binding energy available in soft models is compensated in the calculation of $L_{\bar{\nu}_e}$ by their longer diffusion times (they are denser), after ~ 5 s, the decay in $L_{\bar{\nu}_e}$ slows. Soft models emit more, and longer, though they behave like stiff models in the early seconds (Mayle and Wilson, 1987; Burrows, 1988c). For the digitally minded, Table 7.2 shows the evolution of both $L_{\bar{\nu}_e}$ and $T_{\bar{\nu}_e}$ versus time for the first 20.0 s in the "life" of stiff models 55, 56, and 57. Note that it takes ~ 10 s for $T_{\bar{\nu}_e}$ to decay to 3.0 MeV. This tenacity is necessary to explain the long duration (5.58 s) of the IMB signal, as sensitive as it is to the high-energy tail of the $\bar{\nu}_e$ spectrum (see Section 7.4.3). This spectrum, though blackbody to a good approximation, should deviate from black-

Table 7.2. $L_{\bar{\nu}_e}$ and $T_{\bar{\nu}_e}$ versus time.

Time (ms)	Model 55 $L_{\bar{\nu}_e}$ (10^{51} ergs s^{-1})	$T_{\bar{\nu}_e}$ (MeV)	Model 56 $L_{\bar{\nu}_e}$ (10^{51} ergs s^{-1})	$T_{\bar{\nu}_e}$ (MeV)	Model 57 $L_{\bar{\nu}_e}$ (10^{51} ergs s^{-1})	$T_{\bar{\nu}_e}$ (MeV)
25	126.5	5.36	129.9	5.46	135.2	5.46
50	34.1	5.08	43.5	5.33	53.2	5.36
100	24.5	4.92	29.4	4.66	38.7	4.41
200	17.3	4.35	22.4	4.25	32.8	3.71
500	11.1	4.06	14.0	3.71	23.9	3.68
1000	6.48	3.81	7.34	3.87	10.4	3.49
2000	3.40	3.68	3.00	3.65	3.05	3.46
3000	2.36	3.62	2.27	3.62	2.40	3.55
4000	1.84	3.49	1.97	3.52	1.78	3.43
5000	1.61	3.40	1.71	3.46	1.56	3.30
6000	1.41	3.30	1.50	3.36	1.47	3.27
8000	1.10	3.11	1.20	3.19	1.25	3.14
10000	0.87	2.98	0.97	3.01	1.05	3.01
12000	0.73	2.86	0.83	2.92	0.89	2.89
14000	0.60	2.70	0.67	2.76	0.78	2.79
16000	0.52	2.60	0.57	2.67	0.67	2.70
18000	0.44	2.51	0.49	2.57	0.60	2.63
20000	0.37	2.41	0.45	2.51	0.52	2.54
E_T^a (10^{51} ergs) =	228.4		261.9		326.3	
#K II =	11.46		13.10		15.68	
#IMB =	5.95		6.66		7.38	

[a] E is the total neutrino energy radiated within 20 s. #K II and #IMB are the corresponding total number of $\bar{\nu}_e$ events expected in the K II and IMB detectors, respectively, if the distance to SN 1987A is 50 kpc.

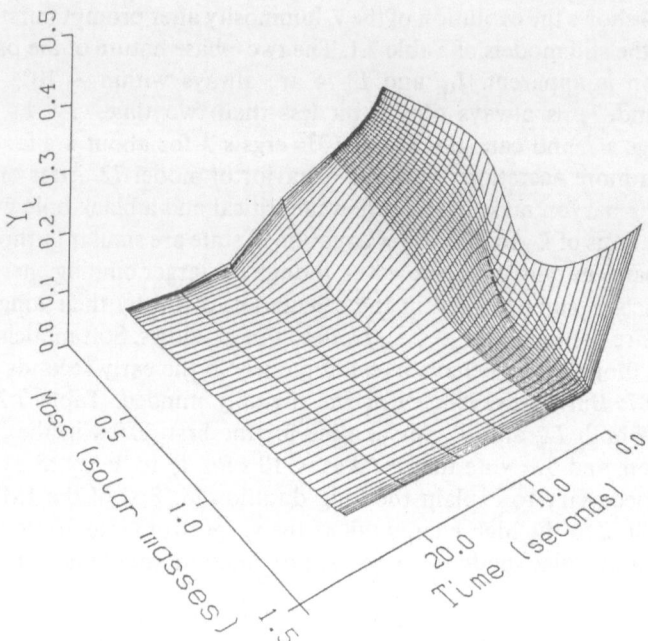

Figure 7.5. A three-dimensional plot of lepton fraction per baryon, Y_l ($= Y_e + Y_{\nu_e}$), versus enclosed baryon mass (in $M_\odot$) versus time (in seconds) for the first 25 s in the life of protoneutron star model 54. The background is the initial Y_l versus mass profile. Note the trough in the mantle. The foreground is the final Y_e ($Y_\nu \sim 0$) profile of the neutronized neutron star. Leptons are lost only over many seconds. The figure is from Burrows (1988c).

body at high energies. At higher energies, neutrinos are more efficiently trapped and, hence, are effectively radiated from further out, at lower temperatures (see Woosley, Wilson, and Mayle, 1986; Mayle, Wilson, and Schramm, 1987; Giovanoni, Ellison, and Bruenn, 1988). The magnitude of this effect has not yet been well calculated and should be a focus of future work.

Though not directly observable, the evolution of the temperature and lepton profiles in protoneutron stars is instructive. A three-dimensional history of the Y_l profile for the first 25 s after breakout of model 54 is depicted in Figure 7.5. In the background is the lepton fraction versus enclosed mass at $t = 0$. Note the mantle lepton trough discussed previously. The foreground is the neutron star electron profile ($Y_e \sim 0.03$–0.05) after 25 s. Figure 7.5 demonstrates that full neutronization proceeds quite slowly. Figures 7.6 and 7.7 depict the first 30 s of temperature profile evolution for stiff model 55 and soft model 62, respectively. The only difference between these two models is the equation of state. The different lines on each plot represent the T-profile at different times, every 100 ms for 2 s, and thereafter, every 2 s. The initial profiles are flat near ~ 10 MeV (the bottom), but within 100 ms, the temperature spike that attends mantle collapse is manifest. Heat diffuses both outward and inward until the peak decays, at which time the monotonic temperature profile begins to decrease uniformly. Higher temperatures are reached in soft model

Figure 7.6. Local matter temperature (in megaelectronvolts) versus enclosed baryon mass ($M_\odot$) at various times for the first 30 s in the life of protoneutron star model 55 (STIFF, 1.3–1.5 $M_\odot$). The bottom line depicts the initial temperature profile. Snapshots are taken every 100 ms for 2 s and then every 2 s until 30 s. Note the peak in mantle temperature that results from mantle collapse. The final central temperature is < 20 MeV.

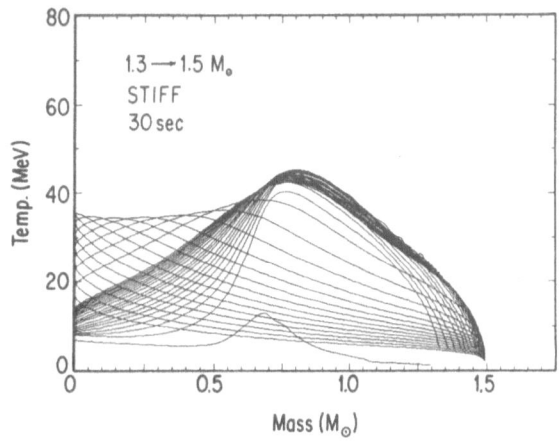

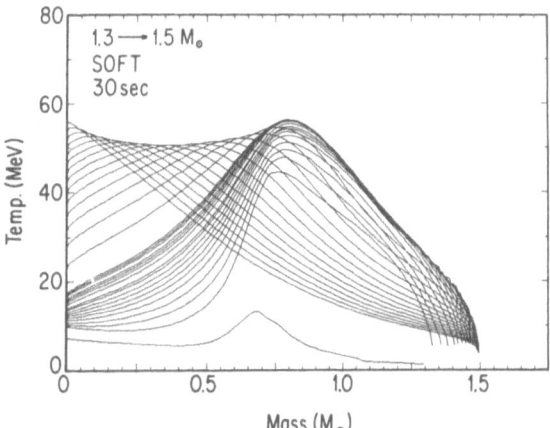

Figure 7.7. Same as Figure 7.6, except for soft model 62. Note the significantly higher temperature achieved here and that the central temperature is still > 50 MeV after 30 s.

62 than in stiff model 55, because softness facilitates compression. Since higher densities and, hence, higher opacities, are achieved in the soft model, the central temperature is high (50 MeV), even after 30 s. The behavior depicted in Figures 7.5, 7.6, and 7.7 is not unique, but represents the generic history of protoneutron stars.

7.4. The Neutrinos from SN 1987A

The epochal neutrino data gathered by the IMB (Bionta *et al.*, 1987) and K II (Hirata *et al.*, 1987) collaborations are shown in Table 7.3 and Figure 7.8. As stated in the introduction, these data have been scrutinized in detail by an army of pundits. Quite a bit of information has been derived about not only "collapse," but also neutrino and exotic particle properties (Krauss, 1988). This paper will not recapitulate the detailed analyses of the author or others that can be found now in many journals.

Table 7.3. Data from the Kamiokande II and IMB detectors.

Event #	Time (s)	Electron energy (MeV)	Angle with respect to Large Magellanic Cloud (degrees)
Kamiokande II			
1	0.000	20.0 ± 2.9	18 ± 18
2	0.107	13.5 ± 3.2	40 ± 27
3	0.302	7.5 ± 2.0	108 ± 32
4	0.323	9.2 ± 2.7	70 ± 30
5	0.507	12.8 ± 2.9	135 ± 23
6[a]	0.685	6.3 ± 1.7	68 ± 77
7	1.540	35.4 ± 8.0	32 ± 16
8	1.728	21.0 ± 4.2	30 ± 18
9	1.915	19.8 ± 3.2	38 ± 22
10	9.219	8.6 ± 2.7	122 ± 30
11	10.432	13.0 ± 2.6	49 ± 26
12	12.439	8.9 ± 1.9	91 ± 39
IMB			
1	0.000	38 ± 7	80 ± 10
2	0.411	37 ± 7	44 ± 15
3	0.650	28 ± 6	56 ± 20
4	1.141	39 ± 7	65 ± 20
5	1.562	36 ± 9	33 ± 15
6	2.683	36 ± 6	52 ± 10
7	5.010	19 ± 5	42 ± 20
8	5.581	22 ± 5	104 ± 20

[a] Excluded by the Kamiokande II collaboration.

Rather, we sketch out the important conclusions concerning protoneutron stars and "supernovae" and make some general remarks about the new status of the field. For further comments, the reader is referred to Burrows (1988c).

Table 7.3 demonstrates that the problems of small-number statistics must play a central role in any but the most general conclusions. However, even those conclusions are quite useful to a field emerging from three decades of data starvation.

The IMB and K II detections demonstrate above all else that neutrinos, or particles just like neutrinos, are indeed radiated from Type II supernovae. That fact in itself is gratifying and a testament to the early pioneers of collapse theory.

7.4.1. General Analysis

The large angles of the events with respect to the LMC (shown in Table 7.3) imply that most were not v–e^- scatterings, which are quite forward-peaked, but that most or all were indeed $\bar{v}_e$ absorptions. Positron emission following $\bar{v}_e$ absorption is almost isotropic. However, the IMB data show a marked preference for the forward hemisphere that may well be a statistical fluke, but is as yet unexplained (Matthews, 1987; van der Velde, 1988). Furthermore, the first K II event is forward-peaked and may indeed be a v_e–e^- event. The angle of the second K II event has recently been revised upward (Suda, 1988) and is no longer a good scattering candidate. Unfortu-

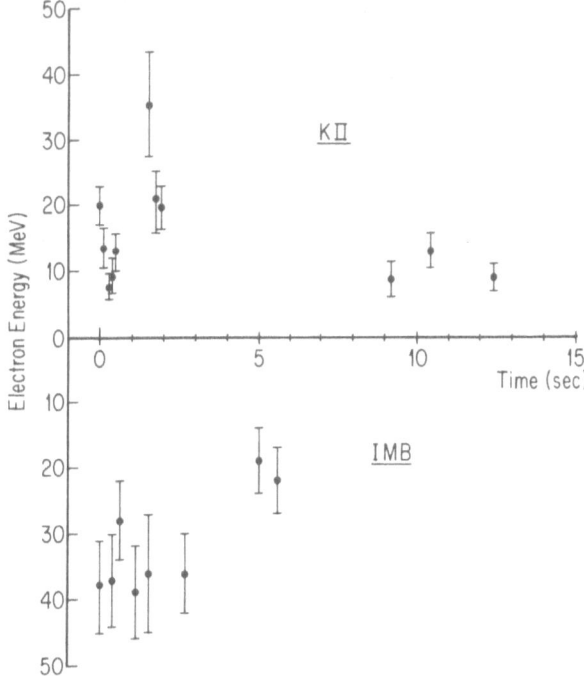

Figure 7.8. Electron energy (in megaelectronvolts) versus time (in seconds) for the K II and IMB data. Note that the IMB scale is inverted. Time = 0 corresponds to the first event. The electron energy refers to the secondary electron or positron that generated the Cherenkov light detected by the phototubes in the detectors. $\varepsilon_\nu \sim \varepsilon_e + 2.0$ MeV if the "electron" is a "positron" from the process $\bar{\nu}_e(p, n)e^+$.

nately, whether any of these events is a scattering event cannot be definitively determined from these sparse data.

The best inferred average source temperature is 3.0–5.0 MeV. There is a slight indication from the data that the source is actually cooling, as one would expect. Figure 7.8 of the "electron" energy versus time shows this more clearly. The late-time events are all lower in energy. The shorter duration of the IMB signal (5.58 s) with respect to the K II signal (12.44 s) also suggests that the source is cooling, since a given decay in temperature implies an even more rapid decay in the high-energy tail of a thermal, or near-thermal, spectrum. The high-energy threshold of the IMB detector (~ 19 MeV) makes it much more sensitive to this tail than the K II detector with its lower threshold (~ 7 MeV) and, hence, one would expect the IMB signal to turn off sooner, as it did.

The evolution of the signal can be fitted reasonably well, not by a single exponential, but by two exponentials, one after the other, with τ's of ≤ 1.0 s and ~ 4.0 s, respectively. In other words, two phases, an early short one and a later long one, seem indicated. These timescales of seconds are comfortably consistent with the discussion in Section 7.3.3. No neutrino mass, source pulsing, or neutrino oscillations are necessary to explain the data, though none of these exotica are absolutely eliminated. Early ($t < 0.5$ s) $\bar{\nu}_e$ luminosities of $\sim 4 \times 10^{52}$ ergs s^{-1} (with wide error bars) and later (seconds) $\bar{\nu}_e$ luminosities of 10^{51} ergs s^{-1} can be derived. These luminosities and the inferred temperatures can be used to obtain radii between 10 km and 50 km. There is a slight indication that the radius evolved from the higher value to 10–20 km within the first second. These radii are near those always quoted for neutron stars, and near those for nothing else. The total $\bar{\nu}_e$ energy radiated is $\sim(3\text{–}5) \times 10^{52}$ ergs, assuming a distance of 50 kpc. If we multiply by 6 to account for the other five neutrino species approximately, a total neutron star binding energy of $(2\text{–}3) \times 10^{53}$ ergs is derived. This is our first "direct" measurement of a neutron star's binding energy and it is surprisingly close to what was expected (Figure 7.2).

7.4.2. The Kamiokande II and IMB Detectors

Both K II and IMB are water Cherenkov detectors with fiducial masses for supernova detection of 2140 t and 6800 t, respectively. They each register light generated in the water volume by means of banks of inward-pointing photomultiplier tubes at the tank boundaries. A relativistic electron or positron (these detectors cannot distinguish which) will generate a Cherenkov cone of light. If the "electron" energy is above threshold, sufficient Cherenkov light will be generated to trigger enough tubes to identify a real event above background. However, the detection efficiency is not 100% above threshold, and is electron energy-dependent. This efficiency curve depends upon phototube coverage, radioactive background, etc., and is different for different detectors.

Both collaborations have published efficiency curves (Bionta *et al.*, 1987; Hirata et al., 1987; Matthews, 1987). Analytic fits to these published curves are

$$E(\varepsilon_e) = 0.93 - e^{-(\varepsilon_e/9.0 \text{ MeV})^{2.5}}, \qquad 7.0 \text{ MeV} < \varepsilon_e; \qquad \text{K II}, \qquad (7.1)$$

and

$$E(\varepsilon_e) = 0.3975x - 0.02625x^2 - 0.59, \qquad 1.9 < x < 7.6,$$
$$= 0.915, \qquad\qquad\qquad\qquad\qquad x > 7.6; \qquad \text{IMB}, \qquad (7.2)$$

where $x = \varepsilon_e/10$ MeV, ε_e is the electron or positron energy, and $E(\varepsilon_e)$ is the detection efficiency. Note that the efficiency of the K II detector at 10 MeV is $\sim 66\%$ and at 15 MeV is a full $\sim 90\%$, but that the IMB detector efficiency is only $\sim 10\%$ at 20 MeV and only $\sim 60\%$ at 40 MeV. In addition, the IMB efficiencies are quoted to only $\pm 5\%$, which results in no small uncertainty in the 20–40 MeV range.

Water is made up of protons, electrons, and oxygen nuclei with number densities (N_i) of 6.7×10^{31}, 3.35×10^{32}, and 3.35×10^{31} particles per kilotonne, respectively. The $\bar{\nu}_e$'s interact with protons via the super-allowed charged-current reaction, $\bar{\nu}_e(p, n)e^+$. The total cross section for this reaction is

$$\sigma_p = 0.941 \times 10^{-43} \left(\frac{\varepsilon_\nu}{\text{MeV}}\right)^2 (1 + \delta_{\text{wm}}) \left(1 - \frac{Q}{\varepsilon_\nu}\right) \left(\left(1 - \frac{Q}{\varepsilon_\nu}\right)^2 - \left(\frac{m_e}{\varepsilon_\nu}\right)^2\right)^{1/2} \text{ cm}^2,$$
$$(7.3)$$

where Q is the neutron–proton mass difference (1.29335 MeV), δ_{wm} is the weak-magnetism correction (Vogel, 1984; Schramm, 1987), and the other terms have their standard meanings. In eq. (7.3) above, we have set ε_ν equal to $\varepsilon_e + Q$. The term δ_{wm}

is approximately equal to $-0.00325 (\varepsilon_v - Q/2)$/MeV. It is small at low anti-neutrino energies, but approaches $\sim 10\%$ in the 30 MeV range. Hence, while not too important for the K II prediction and within current statistics in any event, it cannot be ignored when analyzing or predicting the IMB response.

All the neutrino types scatter off of electrons via the process, $v_i(e^-, e^-)v_i$. While not the dominant process for the $\bar{v}_e$'s in water, at ~ 10 MeV it is the dominant interaction process for all the other species. The total scattering cross section is (Sehgal, 1974)

$$\sigma_i = \tfrac{1}{2}\sigma_0 \Lambda_i \left(\frac{\varepsilon_v}{m_e c^2} + \tfrac{1}{2} \right), \tag{7.4}$$

where $\sigma_0 = (4G^2 m_e^2 \hbar^2)/\pi c^2 = 1.705 \times 10^{-44}$ cm^2 and Λ_i is a constant that depends on the Weinberg angle (θ_W) and the species.

If $\sin^2 \theta_W$ were 0.25, Λ_{v_e}, $\Lambda_{\bar{v}_e}$, Λ_{v_μ}, and $\Lambda_{\bar{v}_\mu}$ would equal $\tfrac{7}{12}$, $\tfrac{3}{12}$, $\tfrac{1}{12}$, and $\tfrac{1}{12}$, respectively. In this case, $\sum_{i \neq v_e} \Lambda_i$ would equal Λ_{v_e} and $\Lambda_{v_e}/\Lambda_{v_\mu}$ would equal 7. With a current experimental value of $\sin^2 \theta_W$ of 0.23, Λ_{v_e}, $\Lambda_{\bar{v}_e}$, $\Lambda_{\bar{v}_\mu}$, are ~ 6–8% lower and Λ_{v_μ} is $\sim 9\%$ higher, but $\Lambda_{v_e}/(\sum_{i \neq v_e} \Lambda_i)$ is still ~ 1 (0.973). We find when $\sin^2 \theta_W = 0.23$ that σ_{v_e} approximately equals 0.92×10^{-44} ε_v/MeV cm^2. Therefore, at 10 MeV, $\sigma_p(\bar{v}_e)/\sigma_{v_e} \sim 100$. Since there are in water five times as many electrons as protons, the v_e scattering signal should be only $\sim \tfrac{1}{20}$ of the $\bar{v}_e$ absorption signal. Therefore, the ratio of absorption signal to the *total* scattering signal should be ~ 10 at 10 MeV. However, when the different detection efficiency corrections and the actual spectra and temperatures are used, this ratio is generally between 15 and 20.

Haxton (1987) has pointed out that at high temperatures ($T_v > 5.0$ MeV), the oxygen absorption processes,

$$v_e + {}^{16}\text{O} \rightarrow {}^{16}\text{F} + e^- \qquad (\varepsilon_{\text{th}} = 15.4 \text{ MeV}), \tag{7.5}$$

and

$$\bar{v}_e + {}^{16}\text{O} \rightarrow {}^{16}\text{N} + e^+ \qquad (\varepsilon_{\text{th}} = 11.4 \text{ MeV}), \tag{7.6}$$

can compete with electron scattering, but never with $\bar{v}_e$ absorption. His results indicate that these processes depend steeply on energy above threshold (ε_{th}) (see Figure 7.13).

With the above, the event rate in terrestrial detectors is

$$\frac{d\mathcal{N}_i}{dt} = \frac{MN_i}{4\pi D^2} F_i \frac{\int f(\varepsilon_v, T(t)) \varepsilon_v^2 (d\sigma_i/d\varepsilon_e) E(\varepsilon_e) \, d\varepsilon_e \, d\varepsilon_v}{\int f(\varepsilon_v, T(t)) \varepsilon_v^2 \, d\varepsilon_v}, \tag{7.7}$$

where F_i is the total neutrino number flux ($\#$/s) of species i from the protoneutron star, f is the neutrino spectral distribution function, D is the distance of the collapse, M is the detector mass (in kt), the top integral is threshold-bounded, but the bottom is not. For $\bar{v}_e$–p absorption, $d\sigma/d\varepsilon_e \sim \sigma_p \delta(\varepsilon_v - \varepsilon_e - Q)$.

7.4.3. Model Results

Using eq. (7.7), the cumulative number of $\bar{v}_e$ events ($\mathcal{N}_{\bar{v}_e}$) versus time expected in either K II or IMB for the stiff models in Table 7.1 are plotted in Figures 7.9 and 7.10. These figures are taken from Burrows (1988c). The model parameters and total energy (E_T) radiated after 20 s are superposed on the graphs for quick reference.

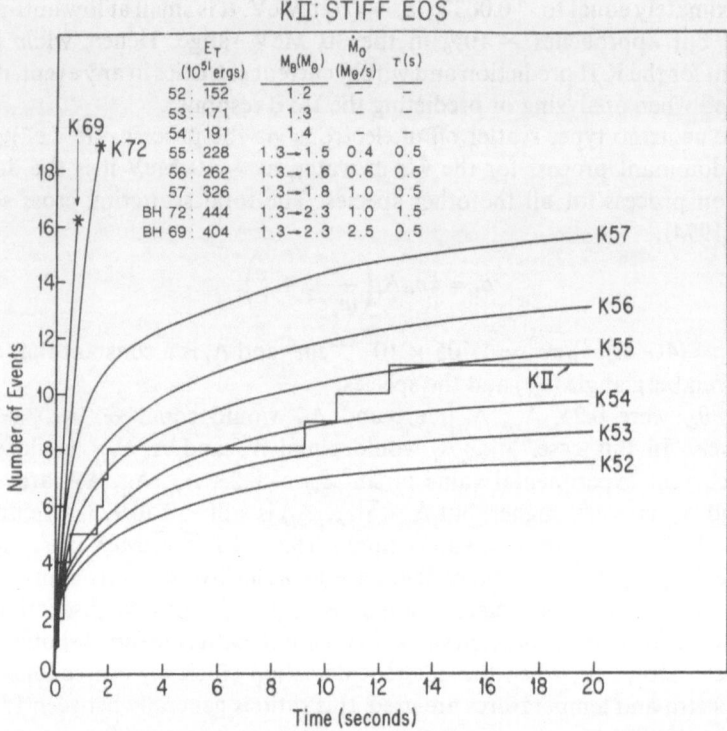

Figure 7.9. The predicted integrated number of $\bar{\nu}_e$ events in K II versus time (in seconds) for the first 20 s for the stiff models of Table 7.1. Efficiency and threshold corrections are included. The inset recapitulates the model parameters of Table 7.1 and includes the total neutrino energy lost during the 20 s. BH for models 69 and 72 indicates that a black hole formed. The stepped line is the K II data for comparison. The K in the line identifier refers to K amiokande and the accompanying number refers to the model number. The figure is from Burrows (1988c). See the text for details.

The intrinsic source signal strength ($S_{\bar{\nu}_e}$), the total number of $\bar{\nu}_e$ events per kilotonne of water at 1 kpc in a perfect detector, is $\sim (1.6 \pm 0.4) \times 10^4$. This measure of burst strength is useful for back-of-the-envelope calculations concerning detectors of arbitrary size and future collapses at arbitrary distances. The prefixes K or I on the model-identifying line labels serve to emphasize the detector under consideration in the figure. The actual accumulated event histogram for either K II or IMB (so labeled) is superposed on the appropriate figure for comparison with the models. Aside from the assumption that the distance (D) to the LMC supernova is 50 kpc, these curves are not normalized or shifted in any way, save that the arrival time of the first event for the real data is assumed to be at $t = 0$.

Figures 7.9 and 7.10 depict the signal expectations for the realistic models with different baryon masses and accretion regimes. They collectively indicate that the general theory presented here for "supernova" neutrinos is quite good. The model grid seems to tightly envelop the observations for both K II and IMB. Furthermore, these figures show graphically that the two-phase nature of the signal as described

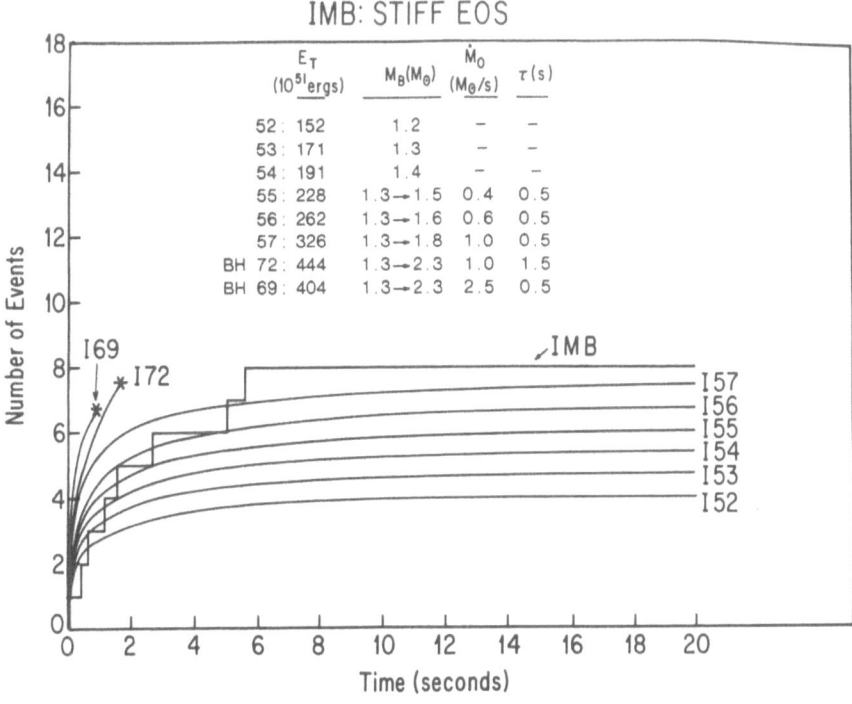

Figure 7.10. Same as Figure 7.8, but for IMB.

above is reproduced by the data. The earlier flattening of the total signal in IMB is also clearly evident. The ratio of the number of $\bar{v}_e$ events to the number of v_i–e^- events can be derived to within better than 50% by quite simple arguments, given previously. The calculation of the total number of electron scattering events in K II using eq. (7.7) yields an average value of 0.6 for the models of Table 7.1.

Both the K II and IMB data must be fit simultaneously. The two detectors must have sampled the same underlying source. The two major uncertainties in determining which model best represents the LMC remnant are our imprecise knowledge of D ($\pm 10\%$) and the inevitable $\sqrt{\mathcal{N}}$ statistical fluctuations in the sampled event number. The latter are, for 8 and 11 events, 35% and 30%, respectively. The resultant uncertainty in the fit of the data sets when taken individually is $\sim 40\%$ and when taken together is $\sim 30\%$. In addition, though the K II signal is only weakly dependent on $\langle T_{\bar{v}_e} \rangle$ ($\propto \langle T_{\bar{v}_e} \rangle^{1.35}$ at $T = 4.0$ MeV), the IMB signal, sampling as it does the high-energy tail, depends steeply on $\langle T_{\bar{v}_e} \rangle$ ($\propto \langle T_{\bar{v}_e} \rangle^4$ at $T = 4.0$ MeV). A 10% error in the calculation of $T_{\bar{v}_e}$ will result in a $\sim 50\%$ error in the predicted IMB signal. From this fact alone, we can conclude that $\langle T_{\bar{v}_e} \rangle$ is now known to better than $\sim 20\%$ and cannot be less than 3.0 MeV or greater than 5.0 MeV. From Figures 7.9 and 7.10, models 52, 53, 57, 69, and 72 seem eliminated as either too weak or too strong. Final baryon masses smaller than 1.2 $M_\odot$ or larger than 1.7 $M_\odot$ do not fit the data. This leaves models 54–56 as the best stiff equation of state fits. Their final baryon masses are in the 1.4–1.6 $M_\odot$ range, and the corresponding gravitational masses (M_G) are ~ 1.3–$1.45 \, M_\odot$. Similarly, as Burrows (1988c) shows, the best soft equation of state fits have final baryon masses of 1.3–$1.5 \, M_\odot$ and gravitational masses of 1.2–$1.37 \, M_\odot$. From these considerations, we can conclude that M_G of the LMC neutron star can be 1.3–$1.5 \, M_\odot$ with little difficulty. It would be difficult to square with the neutrino data the accretion of more than $0.3 \, M_\odot$ onto a protoneutron star whose initial mass is $\sim 1.3 \, M_\odot$ (baryon). The masses we here derive for the LMC neutron star are in the standard neutron star mass range. For these best-fit models, E_T (20 s) is $(2.0$–$2.5) \times 10^{53} \, (D/50 \text{ kpc})^2$ ergs and $E_{\bar{\nu}_e}$ (20 s) is $(3.0$–$4.0) \times 10^{52} \, (D/50 \text{ kpc})^2$ ergs. These numbers echo those in Section 7.4.1 above. Independently, the observed SN 1987A ^{56}Ni mass of $0.07 \, M_\odot$ and recent likely models for SN 1987A progenitors have led Nomoto, Shigeyama, and Hashimoto (1987) to conclude that the residue mass is indeed $\sim 1.6 \, M_\odot$ (baryon).

7.4.4. Black Holes

The $\bar{\nu}_e$ signature of the formation of black holes from protoneutron star intermediaries is also depicted in Figures 7.9 and 7.10 (stiff equation of state models 69 and 72, marked BH in Table 7.1). These models are characterized by high-accretion rates and values of $\dot{M}_0 \tau$ sufficient to push the remnants over the limit. Black-hole formation itself is marked in Figures 7.9 and 7.10 with an asterisk. The calculations were carried out until the general relativistic instability was reached, after which time the GR code could not find a solution (Burrows, 1988c). Though the precise time of instability depends on the actual accretion regime, a value of about 1 s is consistent with our current understanding of the "collapse" problem. Therefore, since the K II and IMB signals lasted many seconds, a black hole did not form immediately in SN 1987A. Interestingly, the critical mass of the hot protoneutron star is always within a few hundredths of a solar mass of the maximum mass of the corresponding cold neutron star. There seems to be little or no thermal correction to the maximum mass of such neutron stars. A hot, extended envelope, whose cooling would delay the GR instability, does not form.

The stiff models 69 and 72 must accrete $\sim 1.0 \, M_\odot$ to reach instability. In any given realistic accretion environment, such a requirement will delay criticality and allow significant heat diffusion from the interior. Further, in order to be at risk of black-hole formation, the accretion rate must be high. Hence, generically, we expect a large accretion luminosity. Therefore, as lines K69 and K72 in Figure 7.9 demonstrate, the black-hole neutrino signal in K II first climbs precipitously to large values and then suddenly terminates. Such a quick turn-off should be a distinctive and useful signature of black-hole formation. Models 69 and 72 differ only in that model 69 has a slightly higher $\dot{M}_0$ and shorter τ and is thereby forced to become unstable more quickly. Though the IMB signal, of course, cuts off as abruptly as the K II signal, it does not climb to such high values before doing so. This is demonstrated with lines I69 and I72 in Figure 7.10. Curiously, the stiff equation of state remnants become unstable before they radiate all of their binding energy. Approximately 20% is dragged into the hole in model 72 and slightly more in model 69, which reaches criticality earlier.

What has been shown here are two neutrino signatures of black-hole formation by one specific scenario. Though this is a realistic scenario, another black-hole formation scenario is possible, whose neutrino signature resembles that of canonical neutron star formation. A mass insufficient to lead to the GR instability might accumulate in the core and yield a neutrino signal like any of the non-black-hole models of Table 7.1, only to be augmented not after seconds, but after many minutes to hours, by "fall-back" from very large radii that could not be ejected in the supernova (Colgate, 1971). So long would this "rain" last that the resulting $L_{\bar{\nu}_e}$ and $T_{\bar{\nu}_e}$ would be too small to detect. If there is sufficient mass in the fall-back to push the protoneutron star over the critical mass, a black hole will form without the distinctive black-hole neutrino signature described in this section. Whether such fall-back does or can obtain is still an open question. Since ^{56}Ni, which is produced only near the core, has been identified in the SN 1987A debris, perhaps only a small amount of fall-back can be tolerated. However, if the iron core is sufficiently fat to begin with, it is reasonable to expect that it will end as a black hole within seconds. If "steep," brief neutrino signals are ever detected from massive stars, they will almost surely be announcing the birth of black holes.

7.4.5. Comments
What more could we hope to determine about stellar collapse and supernovae? The actual supernova mechanism still eludes us. The different mechanisms, prompt or delayed, are not unambiguously identified by only 19 events, though the smaller cores required by the prompt process are only barely allowed by the data. A Type II supernova within 10 kpc of the earth will provide > 250 events in K II as it is now configured. Such a number should allow us to identify the shock breakout burst of ν_e's (see Appendix).

Though the first K II event does indeed show a directionality that may be indicative of ν_e-e^- scattering, these sparse data do not allow an unambiguous determination. The predicted number of prompt neutronization burst events in K II is crudely ~ 0.05 (Burrows and Mazurek, 1983). As indicated in Section 7.3.2, this break-out burst should have lasted less than 10 ms and involved less than 3×10^{51} ergs. An electron neutrino interpretation of the first K II event, that does not invoke a statistical fluke, but that is associated with the early neutronization, requires the emission of $\sim 5 \times 10^{52}$ ergs in ν_e's over ≤ 100 ms. In the standard model, a total of only $\sim (3-5) \times 10^{52}$ ergs is thought to be available to the ν_e channel, and that is emitted over the many seconds of diffusive cooling and neutronization. Therefore, an enhancement in the ν_e luminosity during the first ~ 100 ms of perhaps more than a factor of 10 would be needed. Furthermore, in order to avoid crowding 10 or so $\bar{\nu}_e$ events into that first tenth of a second, the $\bar{\nu}_e$ luminosity must not be similarly enhanced. The only natural way to accomplish this may be by the recently proposed convective enhancement of the early emission (Arnett, 1987; Burrows, 1987g). It is difficult to obtain credible estimates of the convective enhancement of the luminosities of all the neutrino species, and more effort in simulating the convective mantle overturn is clearly required. One of the most important issues that can be addressed with future data from a galactic supernova (or collapse) is the question of convection.

Though we infer that "ν_μ's" did indeed comprise the bulk of the emission, we did not directly detect a single one. A measurement of the neutrino flavor ratios and the ν_μ temperature would go a long way towards illuminating not only collapse physics, but neutrino physics as well. The Sudbury Neutrino Observatory (SNO) detector, sensitive as it is to neutral current neutrino events, is particularly relevant in this regard (see Appendix). Hundreds of events will allow us to constrain the neutrino masses better than we were able to with the LMC supernova (Burrows, 1988d). The improved statistics of proximity far outweighs the decrease in the baseline that a multitude of events would imply.

Importantly, the end of the pause required by the long-term mechanism could be accompanied by an abrupt decrease in the accretion rate onto the protoneutron star. The resultant decrease in the luminosities and slight or moderate (see Figure 3, MW) increase in $T_{\bar{\nu}_e}$ should provide a unique signature of delayed explosion. However, to verify this, further calculation by many independent groups is still required.

Colgate and Petschek (1980) made the fascinating suggestion that the lepton-driven instability of Epstein (1979) was dynamic and would lead to the complete overturn of the entire core on dynamical times (milliseconds). However, the long duration (seconds) of the observed neutrino burst implies that this idea can no longer be entertained. One can infer, but not prove, that the entropy barrier raised by the shock in the mantle does indeed stabilize the core against dynamical overturn (Lattimer and Mazurek, 1981).

In addition, the detection of $\bar{\nu}_e$'s from a Type II supernova invalidates all models that require the complete disruption of a white dwarf-like core in a thermonuclear

LMC
•

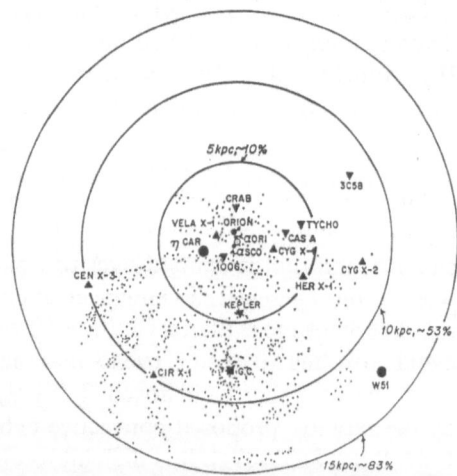

Figure 7.11. A fanciful to-scale map of the LMC and our galaxy, face on. The circles are centered on the sun and mark the 5, 10, and 15 kpc distances. Spiral structure is mimicked with dots to represent H II regions and concentrations of H I. Various binary neutron star systems, supernova remnants, and candidate nearby supernovae are indicated for interest. Note that although the LMC is quite far, in galactic terms, from us, current technology neutrino detectors were successful in catching core "collapse" neutrinos ($\bar{\nu}_e$) from it. This figure demonstrates how much more useful a collapse in our own galaxy will be.

explosion, aided or not by capture ν_e heating (see Chechetkin et al., 1980). Finally, the long-outstanding puzzle of the January 4, 1974, $\bar{\nu}_e$ events in the Homestake Cherenkov detector (Lande et al., 1974; Lande, 1979) has been clarified. If the characteristics of SN 1987A are indicative, the 24 pulses in the 20–100 MeV energy band that were observed over only $\sim$ 3 ms could not have come from a core collapse.

It is certainly amazing that out of only 19 events, many more than 19 good conclusions can be and have been drawn. The potential scientific return from *hundreds* of events from a *galactic* collapse is graphically demonstrated in Figure 7.11 and is all the more to be eagerly anticipated.

Acknowledgments
The author would like to express his gratitude to Jim Lattimer, Ted Mazurek, Gerry Brown, and to all those in the references for making this such a rich field to explore. This work was supported in part by the Alfred P. Sloan Foundation and by NSF Grant No. AST87-14176.

Appendix: Detectors

Both the K II and IMB collaborations had finished essential detector upgrades no more than 1 year before the epiphany of SN 1987A. Such serendipity cannot be relied upon in the future. It is important that future detectors maintain large buffers, accurate relative and absolute timing, and be coordinated one with the other to guarantee continuous coverage. Local neutrino bursts are sufficiently interesting and rare (10^{-1}–10^{-2} yr^{-1}) to warrant such careful cooperation. Fortunately, the experimental community is aware of these needs and is taking steps to ensure we do not miss the next one. A network with Kamiokande II (or extensions) (Koshiba, 1988), ICARUS (Cline, 1987), the LVD (Pless, 1988), and the Sudbury Neutrino Observatory (SNO) (Mak, 1988) may well exist in the next few years to provide constant surveillance of the neutrino sky.

In this appendix, the gross integrated response to collapse of the proposed Sudbury (D_2O) and ICARUS (liquid ^{40}Ar) experiments is calculated to gauge their potential for illuminating the supernova problem in its particulars. These detectors will build on the pioneering contributions to neutrino astronomy of K II, IMB, the Homestake Cherenkov experiment (Lande, 1979), the Homestake ^{37}Cl experiment (Davis, 1978), the LSD (Badino et al., 1984), and the Baksan neutrino observatory (Alekseev et al., 1987), but will offer several special features.

7.A.1. H_2O
Equation (7.7) allows us to derive a general formula for the total number of events in a neutrino "telescope" of mass M (in kilotonnes, kt):

$$\mathcal{N}_T = 2.77 \times 10^3 \left(\frac{M}{kt}\right)\left(\frac{N_i}{6.7 \times 10^{31}}\right)\left(\frac{1\ kpc}{D}\right)^2 \sum_i \left[\frac{\bar{\sigma}_i(T_i)}{10^{-41}\ cm^2}\right]\left[\frac{4\ MeV}{T_i}\right]$$

$$\times \left[\frac{E_i}{10^{52}\ ergs}\right], \tag{7.A.1}$$

where the fiducial values correspond to H_2O for convenience, T_i and E_i are the average temperature and total energy of neutrino species i, respectively, $\bar{\sigma}_i(T_i)$ is the number-averaged neutrino interaction cross section, and the sum is over all six neutrino species of the burst. A thermal spectrum is assumed ($\eta_i = 0$) and the efficiency and threshold effects are ignored.

For instance, for the $\bar{v}_e(p, n)e^+$ process in H_2O,

$$\mathcal{N}_P = 4.3 \times 10^3 \left(\frac{M}{kt}\right)\left(\frac{kpc}{D}\right)^2 \left(\frac{E_{\bar{v}_e}}{10^{52}\, ergs}\right)\left(\frac{T_{\bar{v}}}{4\, MeV}\right)^{1.08} \quad events, \quad (7.A.2)$$

where the power, 1.08, is good near $T_{\bar{v}_e} = 4.0$ MeV. If $E_{\bar{v}_e} = 4 \times 10^{52}$ ergs and $T_{\bar{v}_e} = 4$ MeV,

$$\mathcal{N}_P = 1.72 \times 10^4 \left(\frac{M}{kt}\right)\left(\frac{kpc}{D}\right)^2, \quad (7.A.3)$$

a number very close to the value of $S_{\bar{v}_e}$ derived in Section 7.4.3. The total number of e^--scattering events, $\mathcal{N}_e$, for H_2O is

$$\mathcal{N}_e \sim 1.3 \times 10^3 \left(\frac{M}{kt}\right)\left(\frac{kpc}{D}\right)^2 \left(\frac{E_T}{2.4 \times 10^{53}}\right), \quad (7.A.4)$$

roughly one-thirteenth of $\mathcal{N}_P$. Combining eqs. (7.A.3) and (7.A.4), the total number of events in K II at the distance to the galactic center ($D = 8.5$ kpc) is 509(p) + 39(e^-) = 548, where again efficiency and threshold corrections have been ignored. The number of K II v_e–e^- events per 10^{51} ergs from infall or the prompt burst at 8.5 kpc is ~ 0.5. These two numbers are useful when gauging the capabilities of other detectors.

7.A.2. Sudbury: $D_2O + H_2O$

The proposed Sudbury Neutrino Observatory (SNO) (Mak, 1988) is a Cherenkov detector similar to IMB and K II that would consist of 1 kt of D_2O surrounded by 4 m of H_2O as an anticoincidence shield. The fiducial H_2O mass for supernova detection is ~ 1.76 kt. The SNO is unique in that it would be sensitive to the v_μ's from collapse via the large cross section, neutral-current deuteron breakup reaction, $v_i + d \rightarrow n + p + v_i$. The secondary neutron is indirectly detected through the characteristic γ-ray emitted in the $^{35}Cl(n, \gamma)$ process. Chlorine-35 is present as salt (~ 2.5 t) in the D_2O. Though spectral information will be lost, the neutral current signature would be unmistakable from its Cherenkov shower. The sensitivity of the SNO to v_μ's make it an excellent probe of the neutrino species mix from supernovae.

Even if the SNO is not immediately funded, such a D_2O detector would be ideal as a long-term supernova sentry. Figure 7.12 shows the cross sections for the relevant neutrino interactions in D_2O and H_2O versus neutrino energy. The cross sections for the v–d processes were taken from Bahcall, Kubodera, and Nozawa (1988). As the figure demonstrates, the neutral current cross sections for d-processes (4) and (5) are not much smaller than those for the charge-current d-processes (6) and (7), which in turn are within a factor of 2 of the large cross section for p-process (1), $\bar{v}_e(p, n)e^+$. In contrast, the e^--scattering cross sections (2) and (3) are quite small. An excellent fit to the Bahcall, Kubodera, and Nozawa (1988) cross sections for the

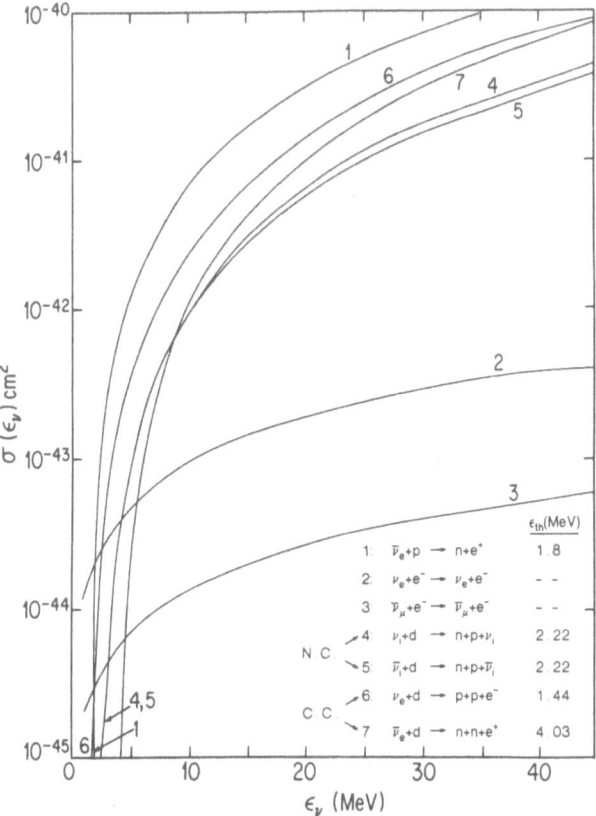

Figure 7.12. Neutrino and antineutrino cross sections (in cm²) on protons, deuterons, and electrons versus energy (in megaelectronvolts). ε_{th} (in megaelectronvolts) is the threshold of a process in the numbered list in the bottom right-hand corner of the figure. Processes (4) and (5) are neutral-current (NC) neutrino–deuteron breakup processes. Processes (6) and (7) are the corresponding charged-current processes. See the text for a discussion.

d-processes, that is good to better than 13% between 5 and 40 MeV, is

$$\sigma_i = \alpha_i (\varepsilon_\nu - \varepsilon_{th,i})^{2.3}, \tag{7.A.5}$$

where $\varepsilon_{th,i}$ is the threshold for the process i (see Figure 7.12). If the energies are in megaelectronvolts, $\alpha_i \sim 1.7 \times 10^{-44}$ cm² for the charged-current processes and $\alpha_i \sim 0.85 \times 10^{-44}$ cm² for the neutral-current processes. A good mnemonic for α_i is either σ_0 (CC) or $\sigma_0/2$ (NC). We can now derive the deuteron correlate of eq. (7.A.2)

$$\mathcal{N}_d = K[0.48 E_{\bar{\nu}_e} T_{\bar{\nu}_e}^{1.86} + 0.7 E_{\nu_e} T_{\nu_e}^{1.49} + 0.29 E_{\bar{\nu}_e} T_{\bar{\nu}_e}^{1.53}$$

$$+ 0.31 E_{\nu_e} T_{\nu_e}^{1.6} + 0.79 E_{\bar{\nu}_\mu} T_{\bar{\nu}_\mu}^{1.32} + 0.9 E_{\nu_\mu} T_{\nu_\mu}^{1.45}], \tag{7.A.6}$$

where

$$K = 2.49 \times 10^3 \left(\frac{M}{kt}\right)\left(\frac{kpc}{D}\right)^2.$$

In eq. (7.A.6), the E_i's are in units of 10^{52} ergs, $T_{\bar{\nu}_e}$ and T_{ν_e} are in units of 4 MeV, and T_{ν_μ} and $T_{\bar{\nu}_\mu}$ are units of 8 MeV. The six terms in eq. (7.A.6) refer to processes 7, 6, 5, 4, 5, 4, respectively, in Figure 7.12. Here, ν_μ refers to "ν_μ and ν_τ" and $\bar{\nu}_\mu$ refers to "$\bar{\nu}_\mu$ and $\bar{\nu}_\tau$." The moderate powers of T_i convey the moderate dependence of the signal on temperature. Note that the neutral-current processes dominate $\mathcal{N}_d$. If all the E_i's are set equal to 4, $T_{\nu_e} = T_{\bar{\nu}_e} = 4$ MeV, and $T_{\bar{\nu}_\mu} = T_{\nu_\mu} = 8$ MeV, we obtain

$$\mathcal{N}_d = 5.15 \times 10^4 \left(\frac{M}{kt}\right)\left(\frac{kpc}{D}\right)^2$$

$$= 712 \left(\frac{M}{kt}\right)\left(\frac{8.5 \text{ kpc}}{D}\right)^2, \qquad (7.A.7)$$

which is approximately three times the corresponding value for H_2O (eq. (7.A.3)). About 77% of this signal is due to neutral-current processes and 66% is due to "ν_μ's." The total galactic center ($D = 8.5$ kpc, GC) signal in the SNO with the above numbers, for the entire detector [$D_2O + H_2O$], including ν_i-e^- scattering, is 16.2 [ν_i-e^- in D_2O] + 31.7 [ν_i-e^- in H_2O] + 712 [eq. (7.A.7)] + 238 × 1.76 [eq. (7.A.3)] = 1179 events. This is better than K II by more than a factor of 2. Since in D_2O the ν_e's from both infall and the prompt burst can participate in deuteron processes (4) and (6) (NC and CC, respectively), the SNO is much more sensitive to these supernova neutrino phases than any other Cherenkov detector. Indeed, assuming for the moment that the ν_e spectrum can be represented by a $T_{\nu_e} = 4.0$ MeV spectrum, the number of GC prompt or infall ν_e's that will be caught by the SNO is ~ 4.1 ($E_{PB,IN}/10^{51}$ ergs). Therefore, the SNO is more than eight times as sensitive as K II to both infall emission and the prompt burst. If $E_{PB} \sim 3 \times 10^{51}$ ergs, $\mathcal{N}_{PB} \sim 12$, as many ν_e events as $\bar{\nu}_e$ events acquired by K II from the LMC supernova!

7.A.3. ICARUS: ^{40}Ar

Cline (1988) has proposed to construct in the Gran Sasso tunnel in Italy a drift chamber experiment containing 3.6 kt of liquid ^{40}Ar. Unlike the ^{37}Cl experiment of Davis, this would be a direct counting device. This feature would put it on a par with water Cherenkov devices that immediately record event energy, angle, and time. Unlike the SNO detector, however, ICARUS would be most sensitive to ν_e's, since the neutral current cross sections for ^{40}Ar in the 10–50 MeV range are very low. Figure 7.13 depicts the run of cross section versus energy for the ν_e–^{37}Cl(1), ν_e–^{40}Ar(2), ν_e–O(3), and $\bar{\nu}_e$–O(4) processes. The last two are from Haxton (1987) and are included to demonstrate the smallness of the neutrino–oxygen cross sections in H_2O and D_2O (see Figure 7.12). The cross sections for process (1) are from Bahcall (1978) and those for process (2) are from J.N. Bahcall as communicated by D. Joutras.

We immediately see that the ^{40}Ar cross section is almost as large as the large ^{37}Cl cross section and like the latter, it is quite energy-sensitive. A fit to the ^{40}Ar cross section, good to better than 6% between 8 and 30 MeV, is $\sigma(^{40}\text{Ar}) \sim 1.7 \times 10^{-43}$ $(\varepsilon_\nu - 5.885)^{1.8}$ cm^2, where 5.885 MeV is the absorption threshold. From this, we obtain

$$\mathcal{N}_{40} = 583 \left(\frac{M}{kt}\right)\left(\frac{kpc}{D}\right)^2 \left(\frac{T_{\nu_e}}{4 \text{ MeV}}\right)^{1.6} \left(\frac{E_{\nu_e}}{10^{52}}\right). \qquad (7.A.8)$$

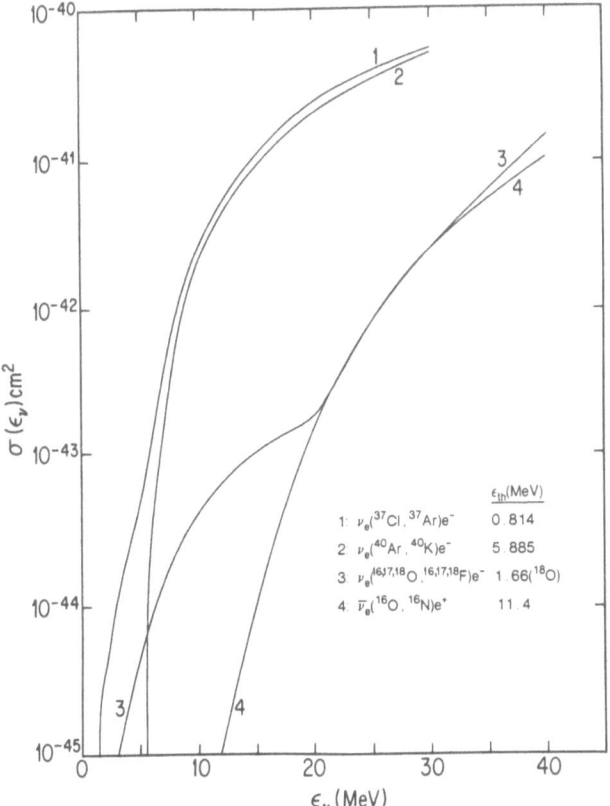

Figure 7.13. The ν_e absorption cross sections (in cm^2) on ^{37}Cl(1), ^{40}Ar(2), and ^{16,17,18}O(3), in terrestrial isotopic proportions versus neutrino energy (in megaelectronvolts). Also shown is the $\bar{\nu}_e$ absorption cross section on ^{16}O. The oxygen cross sections are from Haxton (1987), the ^{40}Ar cross sections are from calculations of J.N. Bahcall, as communicated by David Joutras, and the ^{37}Cl cross sections are from Bahcall (1978). ε_{th} (in megaelectronvolts) is the threshold for a process in the inset. The threshold for process (3) is that for ^{18}O, which dominates cross section (3) below 15.4 MeV, the threshold for ν_e absorption on ^{16}O.

If $E_{\nu_e} = 4 \times 10^{52}$ ergs, $T_{\nu_e} = 4$ MeV, $D = 8.5$ kpc, $M = 3.6$ kt, and $E_T = 2.4 \times 10^{53}$ ergs, the total number of events from the GC in ICARUS is $116(^{40}\text{Ar}) + 53$ $(\nu_i - e^-) = \underline{169}$. For comparison, the corresponding number for the Homestake ^{37}Cl experiment is 5.7. Note that e^--scattering contributes about one-third of the total signal. ICARUS's supernova detection potential is respectable, but is only about one-seventh of SNOs. However, the large cross section of the ν_e–^{40}Ar process and the large size of ICARUS (3.6 kt) conspire to make it sensitive to the infall and prompt burst phases. With the same assumptions as were employed for the SNO burst calculation above, we obtain

$$\mathcal{N}_{\text{PB,IN}} \sim 3.6\left(\frac{E_{\text{PB,IN}}}{10^{51}\ \text{ergs}}\right).$$

This number is comparable to the corresponding value for the SNO. Both are good ν_e infall and breakout detectors.

The H_2O, D_2O, and ^{40}Ar neutrino detectors complement one another so well that a constellation of such detectors would provide formidable coverage of a galactic supernova neutrino burst.

References

Alekseev, E.N., Alekseeva, L.N., Volchenko, V.I., and Krivisheina, I.V. 1987, *Pis'ma Zh. Eksp. Teor. Fiz.*, **45**, 461.
Arafune, J. and Fukugita, M. 1987, *Phys. Rev. Lett.*, **59**, 367.
Arnett, W.D. 1966, *Can. J. Phys.*, **44**, 2553.
Arnett, W.D. 1967, *Can. J. Phys.*, **45**, 1621.
Arnett, W.D. 1977, *Ap. J.*, **218**, 815.
Arnett, W.D. 1987, *Ap. J.*, **319**, 136.
Arnett, W.D. and Bowers, R.L. 1977, *Ap. J. Suppl.*, **33**, 415.
Baade, W. and Zwicky, F. 1934, *Proc. Nat. Acad. Sci.*, **20**, 259.
Badino, G. *et al.* 1984, *Nuovo Cimento*, **7C**, 573.
Bahcall, J.N. 1964, *Phys. Rev.*, **136**, B1164.
Bahcall, J.N. 1978, *Rev. Mod. Phys.*, **50**, 881.
Bahcall, J.N., Dar, A., and Piran, T. 1987, *Nature*, **326**, 135.
Bahcall, J.N., Kubodera, K., and Nozawa, S. 1988, *Phys. Rev.*, **D38**, 1030.
Bahcall, J.N., Press, W.H., and Spergel, D.N. 1987, *Nature*, **327**, 682.
Baron, E., Cooperstein, J., and Kahana, S. 1985, *Phys. Rev. Lett.*, **55**, 126.
Bethe, H.A., Brown, C.E., Applegate, J. and Lattimer, J.M. 1979, *Nucl. Phys.*, **A324**, 487.
Bethe, H.A. and Johnson, M.B. 1974, *Nucl. Phys.*, **A230**, 1.
Bethe, H.A. and Wilson J.R. 1985, *Ap. J.*, **295**, 14.
Bionta, R.M. *et al.* 1987, *Phys. Rev. Lett.*, **58**, 1494 (IMB collaboration).
Bludman, S.A. and Schinder, P.J. 1988, *Ap. J.*, **326**, 265.
Bludman, S.A. and Van Riper, K. 1978, *Ap. J.*, **224**, 631.
Bodenheimer, P. and Woosley, S.E. 1983, *Ap. J.*, **269**, 281.
Bowers, R. L. and Wilson, J.R. 1982, *Ap. J. Suppl*, **50**, 115.
Bruenn, S.W. 1975, *Ann. N.Y. Acad. Sci.*, **262**, 80.
Bruenn, S. W. 1985, *Ap. J. Suppl.*, **58**, 771.
Bruenn, S.W. 1986, *Ap. J. Suppl.*, **62**, 331.
Bruenn, S. 1987, *Phys. Rev. Lett.*, **59**, 938.
Bruenn, S.W., Arnett, W.D., and Schramm, D.N. 1977, *Ap. J.*, **213**, 213.
Burbidge, E., Burbidge, G., Fowler, W.A., and Hoyle, F. 1957, *Rev. Mod. Phys.* **29**, 547.
Burrows, A. and Lattimer, J.M. 1983, *Ap. J.*, **270**, 735.
Burrows, A. 1984, *Ap. J.*, **283**, 848.
Burrows, A. 1987a, in *Proceedings of Telemark IV: Neutrino Masses and Neutrino Astrophysics, Ashland, Wisc., March 16–18*, eds. V. Barger, F. Habzen, M. Marshak, and K. Olive (Singapore: World Scientific), p. 28.
Burrows, A. 1987b, in *Proceedings of the University of Minnesota Conference on SN 1987A, June 4–6, Minneapolis, Minn.* eds. T. Walsh and K. Olive (Minneapolis: Independence Press).
Burrows, A. 1987c, in *Proceedings of the ESO Conference on SN 1987A, July 6–8 (1987)* (Garching: European Southern Observatory).
Burrows, A. 1987d, in *Proceedings of the 4th George Mason Workshop in Astrophysics*, Oct. 12–14, 1987, eds. M. Kafatos and A.G. Michalitsianos (Cambridge: Cambridge University Press).
Burrows, A. 1987e, in *Observational Neutrino Astronomy* ed. D. Cline (Singapore: World Scientific).
Burrows, A. 1987f, *Physics Today*, **40**, 28.
Burrows, A. 1987g, *Ap. J. (Letters)*, **318**, L57.
Burrows, A. 1988a, in *Proceedings of the VIIth Rencontre de Moriond on "Neutrinos and Exotic Phenomena,"* Les Arcs, Savoie, France, Jan 21–28, ed. J. Thanh van Tran.
Burrows, A. 1988b, in *Proceedings of the INS International Symposium on "Neutrino Mass and Related Topics,"* March 16–18, Tokyo, Japan, eds. S. Kato and T. Ohshima (Singapore: World Scientific).
Burrows, A. 1988c, *Ap. J.*, **334**, 891.

Burrows, A. 1988d, *Ap. J. (Letters)* **328**, L51.

Burrows, A. and Lattimer, J.M. 1985, *Ap. J. (Letters)*, **299**, L19.

Burrows, A. and Lattimer, J.M. 1986, *Ap. J.*, **307**, 178 (BL).

Burrows, A. and Lattimer, J.M. 1987, *Ap. J. (Letters)*, **318**, L63.

Burrows, A. and Mazurek, T.J. 1982, *Ap. J.*, **259**, 330.

Burrows, A. and Mazurek, T.J. 1983, *Nature*, **301**, 315.

Burrows, A., Mazurek, T.J., and Lattimer, J.M. 1981, *Ap. J.*, **251**, 325.

Chechetkin, V.M. Gershtein, S.S., Imshennik, V.S., Ivanova, L.N., and Khlopov, M.Yu. 1980, *Astro. Space Sci.*, **67**, 61.

Chiu, H.Y. 1961, *Phys. Rev.*, **123**, 1040.

Chiu, H.Y. 1964, *Ann. Phys.*, **26**, 364.

Cline, D. (1987), in *Observational Neutrino Astronomy*, ed. D. Cline, (Singapore: World Scientific), p. 1.

Colgate, S.A. 1971, *Ap. J.*, **163**, 221.

Colgate, S.A. and Johnson, H.J. 1960, *Phys. Rev. Lett.*, **51**, 235.

Colgate, S.A. and Petschek, A.G. 1980, *Ap. J. (Letters)*, **236**, L115.

Colgate, S.A. and White, R.H. 1966, *Ap. J.*, **143**, 626 (CW).

Cooperstein, J. (1988), this volume.

Cooperstein, J., van den Horn, L.J., and Baron, E. 1986, *Ap. J.*, **309**, 653.

Cooperstein, J. and Wambach, J. 1984, *Nucl. Phys.*, **A420**, 591.

Davis, R. 1978, in Proceedings of the Informal Conference on the Status and Future of Solar Neutrino Research, Report no. BNL50879, Vol. 1, p. 1, ed. G. Friedlander (Brookhaven National Laboratory: Upton, NY).

Epstein, R.I. 1979, *M.N.R.A.S.*, **188**, 305.

Fowler, W.A. and Hoyle, F. 1964, *Ap. J. Suppl.*, **9**, 201.

Freedman, D.Z. 1974, *Phys. Rev.*, **D9**, 1389.

Freedman, D., Schramm, D. and Tubbs, D.L. 1977, Ann. Rev. Nucl. Sci., **27**, 37.

Fuller, G. 1982, *Ap. J.*, **252**, 741.

Fuller, G., Fowler, W.A. and Newman, M. 1980, *Ap. J. Suppl.*, **42**, 447.

Fuller, G., Fowler, W.A., and Newman, M. 1982, *Ap. J.*, **252**, 715.

Giovanoni, P.M., Ellison, D.C., and Bruenn, S.W. 1989, *Ap. J.*, **342**, 416.

Goodwin, B.T. 1982, *Ap. J.*, **261**, 321.

Goodwin, B.T. and Pethick C.J. 1982, *Ap. J.*, **253**, 816.

Hari Dass, N.D., Indumathi, D., Joshipura, A.S., and Murthy, M.V.N. 1987, preprint PP-IMSc/TP/-87-001.

Haxton, W. 1987, *Phys. Rev.*, **D36**, 2283.

Hirata, K. *et al.* 1987, *Phys. Rev. Lett.*, **58**, 1490 (K II collaboration).

van den Horn, L.J. and van Weert, C.G. 1981, *Ap. J. (Letters)*, **251**, L97.

Hoyle, F. and Fowler, W. 1960, *Ap. J.*, **132**, 565.

Imshennik, V.I. and Nadyezhin, D.K. 1973, *Sov. Phys.-JETP*, **36**, 821.

Imshennik, V.I. and Nadyezhin, D.K. 1979, *Astro. Space Sci.*, **62**, 309.

Kahana, S., Cooperstein, J., and Baron, E. 1987, *Phys. Lett.*, **B196**, 259.

Koshiba. M. (1988), in *Proceedings of the VIIth Recontre de Moriond, Les Arcs, France, Jan. 21–28, 1988*, ed J. Thanh van Tran.

Krauss, L.M. 1987, *Nature* 329, 689.

Krauss, L.M. 1988, Talk given at the APS meeting in Baltimore, April 18–21.

Lamb, D.Q., Lattimer, J.M. Pethick, C.J. and Ravenhall, G. 1978, *Phys. Rev. Lett.*, **41**, 1623.

Lamb, D.Q., Melia, F., and Loredo, T.J. 1987, in *Proceedings of the 4th George Mason Workshop in Astrophysics, Oct. 12–14, 1987*. eds. M. Kafatos and A.G. Michalitsianos (Cambridge: Cambridge University Press).

Lamb, D.Q., Pethick, C.J. 1976, *Ap. J. (Letters)*, **209**, L77.

Landau, L. 1932, *Phys. Z. Sowjetunion*, **1**, 285.

Lande, K. 1979, *Ann. Rev. Nucl. and Part. Sci.*, **29**, 395.

Lande, K., Bozoki, G., Frati, W., Lee, C.K., Fenyves, E., and Saavedra, O. 1974, *Nature*, **251**, 485.

Lattimer, J.M. 1987, in *Proceedings of the Minnesota conference on SN 1987A, Minneapolis, Minn., June 4–6.* eds. T. Walsh and K. Olive (Minneapolis: Independence Press).

Lattimer, J.M. and Burrows, A. 1984, in *Problems of Collapse and Numerical Relativity*, eds. D. Bancel and M. Signore (Dordrecht: Reidel), p. 147.

Lattimer, J.M. and Mazurek, T.J. 1981, *Ap. J.*, **246**, 995.

Lattimer, J.M. and Yahil, A. 1989, *Ap. J.*, **340**, 426.

Lichtenstadt, I., Ron, A., Sack, N., Wagschal, J.J., and Bludman, S.A. 1978, *Ap. J.*, **226**, 222.

Mak, H.B. 1988, in *Proceedings of the VIIth Recontre de Moriond, Les Arcs, France Jan. 21–28, 1988*.

Matthews, J. 1987, in *Proceedings of the 4th George Mason Workshop in Astrophysics, Oct. 12–14*, eds. M. Kafatos and A.G. Michalitsianos (Cambridge: Cambridge University Press).

Mayle, R. 1985, Ph.D. thesis, U.C. Berkeley (UCRL preprint no. 53713).

Mayle, R. 1987, in *Proceedings of the Minnesota Workshop on SN 1987A, June 4–6, Minneapolis, Minn*, eds. T. Walsh and K. Olive (Minneapolis: Independence Press).

Mayle, R. and Wilson, J.R. 1987, preprint.

Mayle, R., Wilson, J.R., and Schramm, D. 1987, *Ap. J.*, **318**, 288.

Mazurek, T.J. 1974, *Nature*, **252**, 287.

Mazurek, T.J. 1975, *Astro. Space Sci.*, **35**, 117.

Mazurek, T.J. 1976, *Ap. J. (Letters)*, **207**, L87.

Mazurek, T.J. 1977, *Comments Astro. Space Sci.*, **7**, 77.

Mazurek, T.J. 1982, *Ap. J. (Letters)*, **259**, 43.

Myra, E., Bludman, S.A., Hoffman, Y., Lichtenstadt, I., Sack, N., and Van Riper, K.A. 1987, *Ap. J.*, **318**, 744.

Nadyezhin, D.K. 1978, *Astro. Space Sci.*, **51**, 283.

Nadyezhin, D.K. 1983, *Astro. Space Sci.*, **2**, 75.

Nomoto, K. 1984a, in *Problems of Collapse and Numerical Relativity*, eds. D. Bancel and M. Signore (Dordrecht: Reidel), pp. 81–116.

Nomoto, K. 1984b, *Ap. J.*, **277**, 791.

Nomoto, K., Shigeyama, T., and Hashimoto, M. 1987, to appear in *Proceedings of the IAU Colloquium 108 on Atmospheric Diagnostics of Stellar Evolution: Chemical Peculiarity, Mass Loss and Explosion, Tokyo, Japan*, ed. K. Nomoto (Berlin: Springer-Verlag), p. 319.

Pless, I. 1988, in *Proceedings of the VIIth Recontre de Moriond, Les Arcs, Savoie, France, Jan. 21–28, 1988*, ed. J. Thanh van Tran.

Sato, K. 1975, *Prog. Theor. Phys.*, **54**, 1325.

Sato, K. and Suzuki, H. 1987, *Phys. Rev. Lett.*, **58**, 2722.

Sawyer, R. 1980, *Ap. J.*, 237, 187.

Sawyer, R. and Soni, A. 1979, *Ap. J.*, **230**, 859.

Schaeffer, R., Declais, Y., and Jullian, S. 1987, Saclay preprint SPhT/87-38.

Schinder, P.J. and Shapiro, S.L. 1982, *Ap. J.*, **259**, 311.

Schinder, P.J. and Shapiro, S.L. 1983, *Ap. J.*, **273**, 330.

Schramm, D.N. (1987), *Comments Nucl. Part. Phys.*, 17, 239.

Schramm, D.N. and Arnett, W.D. 1975, *Ap. J.*, **198**, 628.

Sehgal, I. 1974, *Nucl. Phys.*, **B70**, 61.

Spergel, D.N., Piran, T., Loeb, A., Goodman, J., and Bahcall, J.N. 1987, *Science*, **237**, 1471.

Suda, T. 1988, in *Proceedings of the INS International Symposium on Neutrino Mass and Related Topics, March 16–18, Tokyo, Japan*, eds. S. Kato ed T. Ohshima (Singapore: World Scientific).

Tubbs, D. 1978, *Ap. J. Suppl.*, **37**, 287.

Tubbs, D. and Schramm, D.N. 1975, *Ap. J.*, **201**, 467.

Van Riper, K.A. 1978, *Ap. J.*, **221**, 304.

Van Riper, K.A. 1988, *Ap. J.*, **326**, 235.

Van Riper, K.A. and Lattimer, J.M. 1981, *Ap. J.*, **249**, 270.

van der Velde, J. 1988, in *Proceedings of te VIIth Recontre de Moriond, Les Arcs, France, Jan. 21–28, 1988*, ed. J. Tran and Thanh ed. J. Tran van Thanh.

Vogel, P. 1984, *Phys. Rev.*, **D29**, 1918.

Wilson, J.R. 1971, *Ap. J.*, **163**, 290.

Wilson, J.R. 1974, *Phys. Rev. Lett.*, **32**, 849.

Wilson, J.R. 1985, in *Numerical Astrophysics*, eds. J. Centrella, J. LeBlanc, and R. Bowers (Boston: Jones and Bartlett), p. 422.

Woosley, S.E. and Weaver, T.A. 1986a, *Ann. Rev. Astron. Astrophys.*, **24**, 205.

Woosley, S.E. and Weaver, T.A. 1986b, in *Radiation Hydrodynamics in Stars and Compact Objects*, eds. D. Mihalas and K.-H. A. Winkler (Berlin: Springer-Verlag), p. 91.

Woosley, S.E., Wilson, J.R., and Mayle, R. 1986, *Ap. J.*, **302**, 19.

Yueh, W.P. and Buchler, J.R. 1976, *Astro. Space Sci.*, **39**, 429.

Zaringhalam, A. 1983, *Nucl. Phys.*, **A404**, 599.

8. Type I Supernovae: Carbon Deflagration and Detonation

S.E. WOOSLEY

8.1. Introduction

Type I supernovae are distinguished by the lack of prominent hydrogen lines in their spectrum at peak light. Many occur in elliptical galaxies where the rate of massive star formation is very low and, at least compared to Type II, Type I show no preference for association with spiral arms when they do occur in spiral galaxies (Maza and Van den Bergh, 1976). Taken together these facts suggest an origin for at least a major fraction of Type I supernovae in low mass objects having no hydrogen envelope. Single stars appear unlikely candidates. Above 10 $M_\odot$ a clear preference for association with spiral arms would exist. Below $\sim 8\ M_\odot$ a single star could explode only by growing a critical carbon–oxygen core mass equal to the Chandrasekhar value, but that would be impossible in a single star that did not retain its hydrogen envelope. In the narrow range from about 8 to 10 $M_\odot$ the core would evolve to iron core collapse (Miyaji et al., 1980; Woosley, Weaver, and Taam, 1980). However, solitary stars in this mass range are not expected to lose their hydrogen envelopes. If one did (presumably in a binary), the steep density gradients that exist at the edge of the degenerate core would lead to a very small amount of ^{56}Ni being synthesized (Hillebrandt, Nomoto, and Wolff, 1984; Mayle and Wilson, 1988), probably less than a few hundredths of a solar mass and far too little to provide the characteristic bright exponential tail seen in the light curves of all Type I supernovae studied at late times.

For these and other reasons a long-favored model for Type I supernovae involves a white dwarf that accretes a critical mass from a binary companion and explodes (Hoyle and Fowler, 1960; Arnett 1969; Hansen and Wheeler, 1969; Truran and Cameron, 1971; Whelan and Iben, 1971). More recently it has been found that a particular class of white dwarf models, the *carbon deflagration* models, provide especially good agreement with many observational aspects of Type Ia events (the distinction between Types Ia and Ib will be discussed in Section 8.6). These aspects include the light curve (Woosley and Weaver, 1986a; Grahm, 1987a, b; Kirshner, this volume); the spectrum near peak light (Branch et al., 1985; Branch, this volume); the spectrum at late times (Woosley, Axelrod, and Weaver, 1984; Axelrod, 1988; Branch, this volume); and the X-ray spectrum of the remnant (Hamilton, Sarazin, and Szymkowiak, 1986a, b). The other contending white dwarf models that we might think of, helium dwarf detonation (Nomoto and Sugimoto, 1977; Nomoto, 1982; Woosley, Taam, and Weaver, 1986), helium detonation on a carbon–oxygen white dwarf, or carbon *detonation* (Section 8.5), all encounter difficulties with respect to one set of observations or another. The most constraining observations are the spectrum at peak light, which shows evidence for incomplete combustion to iron, and the late-time spectrum which would exhibit unseen lines of highly ionized iron if the supernova expanded too fast (Woosley, Axelrod, and Weaver, 1984).

Still the carbon deflagration model remains essentially an empirical one. That one begins with a carbon–oxygen dwarf of about 1.38 $M_\odot$ having a central density in the range $(2-4) \times 10^9$ g cm^{-3} is undisputed. This is the ignition point where nuclear energy generation from a highly screened carbon-burning reaction becomes equal to neutrino losses by the plasma process. It also is extremely probable that nuclear burning, at least initially, proceeds all the way to nuclear statistical equilibrium. The plasma is very degenerate and will not expand appreciably until the temperature is very high. For every 0.12% of ^{12}C that is burned the temperature increases by about 10^8 K (between $T_8 = 2.5$ and 7.5). Agreement with observations is, as we shall see, best achieved if the burning front propagates at a substantial fraction of the sound speed, but not at the sound speed or faster. A slower speed does not make a supernova. Sonic or supersonic flame speeds result in almost all the star burning to iron in disagreement with spectroscopic restrictions. Best agreement is achieved if a little over one-half of the mass of the white dwarf is turned into iron before expansion gradually quenches the burning. This also leaves behind ashes of partly incinerated carbon (intermediate mass elements) as well as iron.

But these are all observational constraints. A still unsolved problem is a physical description of flame propagation that independently yields the results that the observations demand. The solution eludes us because, as we shall see, the problem is hard and intrinsically three dimensional. It is likely that residual uncertainty in the way the white dwarf ignites and burns lies at the heart of several problems still plaguing the deflagration model, namely details of its isotopic nucleosynthesis (Section 8.4.4) and the *inhomogeneity* of Type I supernovae as a class.

To dwell briefly on this latter point, while Type I supernovae are generally thought of as being a very homogeneous group, and certainly are more homogeneous a class than Type II supernovae, when they are examined closely there is substantial variation in the spectra of Type Ia supernovae (Figure 8.1; see also Branch, 1987). Though the bolometric light curves appear quite similar, they probably also vary from event to event (Barbon, Ciatti, and Rosino, 1973; Younger and Van den Bergh, 1985; Frogel *et al.*, 1987; Phillips *et al.*, 1987; Canal, Isern, and Lopez, 1988). Some differences could be masked by incomplete data, especially prior to maximum, and consequent uncertainty in the date and bolometric luminosity of the peak, but probably some of the difference is intrinsic too. From a theoretical point we might wonder, if the starting points are to be identical, namely a carbon–oxygen white dwarf of 1.38 $M_\odot$ igniting a runaway in its center, how diversity might come to exist in the ensuing explosion.

Though this paper will certainly not resolve these issues conclusively, we hope by careful consideration of some aspects of the physics of the nuclear runaway to understand how someday we might.

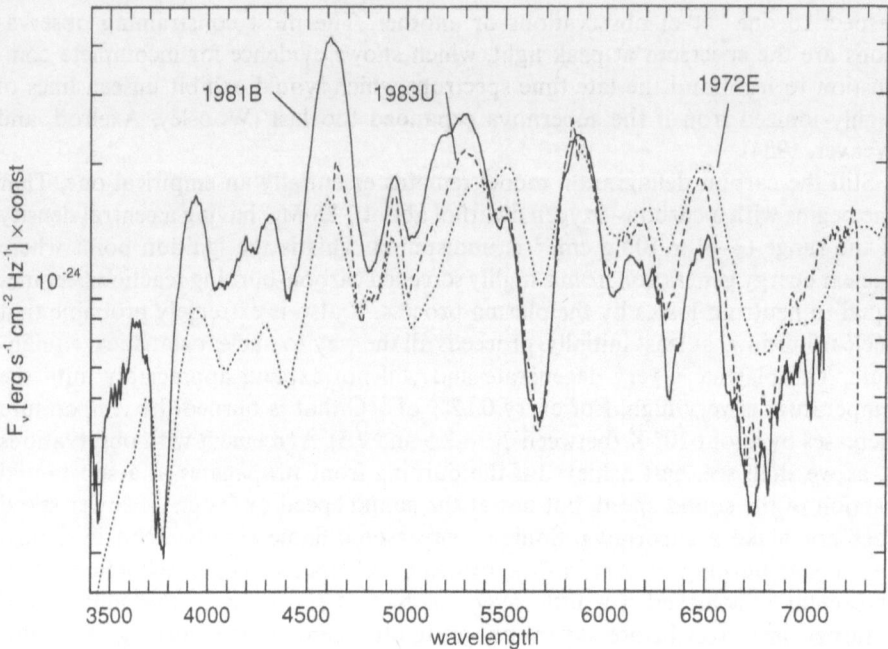

Figure 8.1. Spectra of various Type Ia supernovae taken 260 days after peak. Data are from Kirshner *et al.* (1973) for SN 1972E; Branch *et al.* (1983) for SN 1981B; and DeRobertis and Pinto (1983) for SN 1983U. Figure from Pinto (1988; unpublished Ph.D. thesis).

8.2. The Pre-Explosive Evolution

The evolution of binary systems that might produce Type I supernovae has been reviewed by Woosley and Weaver (1986a), Nomoto and Hashimoto (1987), and references therein. Considerable variation in outcome is expected for various situations—a white dwarf accreting from a helium or hydrogen-rich companion, the merger of two white dwarfs accomplished by the emission of gravitational radiation, or the formation of a common envelope. Generally speaking the kind of situation which gives rise to the kind of explosion which best fits the observations, a carbon runaway ignited at or near the center of a white dwarf, will exist provided that a carbon–oxygen dwarf having mass in the range 0.6–1.1 $M_\odot$ accretes matter at a rate in excess of about 5×10^{-8} $M_\odot$ yr^{-1} but less than a few times 10^{-6} $M_\odot$ yr^{-1} (Nomoto and Hashimoto, 1987). For lower accretion rates a potentially explosive thick layer of helium accumulates on top of the white dwarf. This shell ignites at its base, detonates, and may produce some variety of Type I supernova (Nomoto, 1982; Woosley, Taam, and Weaver, 1986; Branch and Nomoto, 1986). Depending upon the specific carbon–oxygen dwarf mass and accretion rate, a portion of the white dwarf may be left behind, or the whole star may be disrupted by detonation. In either case there are severe discrepancies, especially spectroscopic, between the models that have been calculated and observations of Type Ia supernovae. For accretion rates higher than a few times 10^{-6} $M_\odot$ yr^{-1}, the helium flashes are weak but carbon burning may ignite far off-center in a relatively weak flash that propagates into the white dwarf changing its composition into neon, oxygen, and magnesium (Nomoto and Iben, 1985; Saio and Nomoto, 1985; Woosley and Weaver, 1986b). When a white dwarf devoid of carbon or helium in its center later ignites a nuclear runaway, the result is collapse to a neutron star, not a supernova. This conclusion may be altered however if, in the case of two merging white dwarfs (Tutukov and Yungelson, 1979; Webbink, 1979; Iben and Tutukov, 1984; Paczynski, 1985; Saio and Nomoto, 1985), a common envelope is formed (Iben, 1988).

Though obviously no model can be considered complete until it is supplied with a credible pre-explosive history, the physics of the explosion process itself is sufficiently complex as to occupy the remainder of the paper. It will turn out though that major aspects of the explosion are sensitive to parameters set during the evolution of the progenitor. These include especially the composition of the white dwarf when it ignites and whether the runaway begins at the center of the star.

8.3. Some General Considerations

For a carbon–oxygen dwarf accreting matter at the specified rate ($(0.5–30) \times 10^{-7}$ $M_\odot$ yr^{-1}), ignition will occur when carbon burning begins to provide an excess of nuclear energy over that which can be carried away by plasma neutrino losses. This happens when the central density reaches about 2.8×10^9 g cm^{-3} and the temperature is about 2.1×10^8 K (dimensionless central entropy equals 0.835). The white dwarf mass at this point is about $1.38\ M_\odot$, though the exact values of these quantities depend upon Coulomb corrections to the equation of state and other corrections due to special and general relativity. The net binding energy of the white dwarf is 5.1×10^{50} ergs and its radius is 1600 km. Since the burning of a composition of carbon and oxygen (30% ^{12}C, for example) releases 7.3×10^{17} ergs g^{-1}, it is clear that *at least* $0.35\ M_\odot$ of iron will ultimately have to be produced in order to disrupt the white dwarf. To give the ejecta an average velocity of 5000 km s^{-1}, this mass is raised to $0.59\ M_\odot$. Thus given only the general validity of the carbon deflagration scenario, the iron synthesized in the explosion is rather rigidly constrained.

It is also relevant that the equation of state for this highly degenerate dwarf is characterized by Γ very nearly equal to $\frac{4}{3}$. This means that small perturbations in net energy can cause large excursions in radius. For example, though the binding energy is 5.1×10^{50} ergs, the addition of only 7.3×10^{49} ergs of energy to the star, as would be released by the burning of $0.05\ M_\odot$ of the interior to ^{56}Ni, is sufficient to cause the central density to decline (in hydrostatic equilibrium) by a factor of 3.6. If the energy is added impulsively, the white dwarf oscillates with a period of about 2 s. During its first oscillation the central density goes down by a factor of 10 and in the ensuing oscillations the core spends most of the time at relatively low densities. Because the density in the central regions is so critical in determining the amount of electron capture that occurs during the explosion and therefore the isotopic nucleosynthesis of the iron group, the possibility that a small amount of burning at a slow rate, ~ 1 s (after high temperatures, $T_9 \sim 5$, are achieved anywhere in the white dwarf), may alter the pre-explosive density structure must be seriously considered (Section 8.4.4).

Once the runaway has begun in earnest, it is virtually impossible to stop by any means other than expansion. A Chandrasekhar mass white dwarf is the biggest powder keg in the universe. As we shall find in the next section, the width of a conductive flame at the appropriate conditions is less than 10^{-3} cm in steady state, implying that, once a region of only a few grams has burned to nuclear statistical equilibrium, burning will occur faster than conduction can quench it. This number has been verified by microzoned computer calculations (Section 8.4). Interestingly, the energy released in turning 10 g of carbon into nickel is about 200 tons of high explosive. If we could somehow place a nuclear trigger in a Chandrasekhar mass white dwarf, we could make a 3×10^{28} megaton bomb. But, before anyone gets too excited, consider not only the difficulty of placing the trigger, but that the energy expended in delivering the bomb against any enemy in a reasonable time would be greater than that released by the supernova itself.

Returning to the initiation of the runaway in a real supernova, as the temperature starts to climb following ignition, the reaction rate accelerates but the pressure rises only imperceptibly. At first conduction, and then (when $T_c = 2.2 \times 10^8$ K; $\rho = 3.0 \times 10^9$ g cm^{-3}) ordinary convection is able to transport the excess energy. But as the temperature continues to rise, the rate at which nuclear energy generation is able to change the temperature eventually becomes comparable to and then faster than the time for a convective blob to execute one cycle (the Brunt–Väisälä frequency). This occurs at a temperature of $(6-7) \times 10^8$ K when both are about 10–100 s.

Beyond this point the theoretician attempting to calculate energy transport is in unknown territory (if mixing length theory may be regarded as *terra cognita!*). Simplicity is restored if a detonation wave is somehow initiated in the core, but as stated above and argued in Section 8.5 this may be unlikely. Thus we are left to consider the harsh real world of nonlinear three-dimensional instability. Quite critical in these considerations are the events that transpire as the temperature rises from 6×10^8 K to 1.0×10^9 K. At this latter temperature the burning time (~ 0.01 s) becomes much shorter than the time required for sound, at 9500 km s^{-1}, to go one pressure scale height, 450 km. At 6×10^8 K the central entropy is 0.83 and the entropy at the edge of the convective region, at 0.56 $M_\odot$, is very nearly the same. The entropy farther out in the star, at 1.0 $M_\odot$ for example, is still 0.87, only slightly greater. Thus there are no large entropy "barriers" to the extent of the pre-explosive convection and the exact placement of boundaries is very uncertain.

Energy transport by convection may also be influenced by the URCA process (e.g., Iben, 1978a, b, 1982) and, if the accretion rate is very low, by crystallization and/or phase separation in the core (Lopez et al., 1986a, b; Mochkovitch, 1983; Ichimaru, Iyetomi, and Ogata, 1988; Barrat, Hansen, and Mochkovitch, 1988). Such effects make an already hard problem even more difficult and will not be treated here.

The condition that burning occurs faster than the convective cycle time (for $T_8 \gtrsim 6$) also signals a situation in which nuclear burning may appreciably increase the entropy of a convective blob as it rises. If adiabatic expansion does not truncate the burning, a runaway in velocity may occur as the blob becomes increasingly buoyant. Provided that the blob was too large to cool by conduction, its velocity would increase until either turbulent dissipation led to fragmentation into pieces small enough to cool by conduction or else the bubble exploded. A likely result, though by no means demonstrated in any credible calculation thus far, is a runaway ignited at many discrete points throughout a substantial volume. Roughly this volume might be thought to encompass a fraction of a pressure scale height (as did the pre-explosive convection zone). For a blob to rise beyond a pressure scale height and then explode, its starting conditions would need to be precisely tuned so that burning was neither quenched by expansion nor ran away until the blob had traveled a great distance.

Following all this in proper detail will take a three-dimensional implicit hydrodynamics code capable of very fine mass resolution. Even with the recent advances in computer technology this lies quite a way down stream from here, so let's begin simple.

8.4. Carbon Deflagration

8.4.1. The One-Dimensional Problem

Consider the simplest, but physically unreasonable initial condition of a runaway igniting precisely at a point at the center of a white dwarf of given mass and composition. Then the normal flame speed (the microscopic velocity of the flame front, which may be deformed when viewed on a larger scale) can be uniquely determined either analytically or numerically by a variety of techniques (Woosley, 1986; Woosley and Weaver, 1986c, 1990). The most accurate solutions involve either:

(1) the ultrafine zoning ($\sim 10^{-5}$ cm) of the burning front in a Lagrangian hydrocode including appropriate nuclear physics and an equation of state to examine the problem (Figures 8.2 and 8.3); or
(2) use of eq. (3.10) of Zeldovich *et al.* (1985) with energy generation as a function of temperature calculated using a large nuclear reaction network.

The results of such studies can be summarized

$$v_{\text{cond}} \approx 50 \left(\frac{\rho}{2 \times 10^9 \text{ g cm}^{-3}} \right)^{0.8} \left(\frac{X_{12}}{0.5} \right) \text{km s}^{-1} \tag{8.1}$$

with ρ the density and X_{12} the carbon mass fraction. The width of the flame front is also determined by these same calculations to be $\sim 10^{-3}$ cm. These values are in approximate agreement with those obtainable by simpler means, essentially dimensional analysis (Landau and Lifshitz, 1987): $v_{\text{cond}} \sim (\sigma/C_V \tau_{\text{nuc}})^{1/2}$ with σ the conductivity, C_V the heat capacity, and τ_{nuc} the nuclear burning time scale (eq. (8.6) below), providing that an appropriate temperature, $T \sim 5 \times 10^9$ K, is chosen for when carbon begins to burn appreciably as the flame crosses. Expressions for σ, C_V, and ε_{nuc} are given by Woosley and Weaver (1990). A similar estimate of v_{cond} has been made by Buchler, Colgate, and Mazurek (1979), but beware numerical errors, in some cases, of several orders of magnitude.

Obeying the one-dimensional restriction and excluding deformation, a flame of

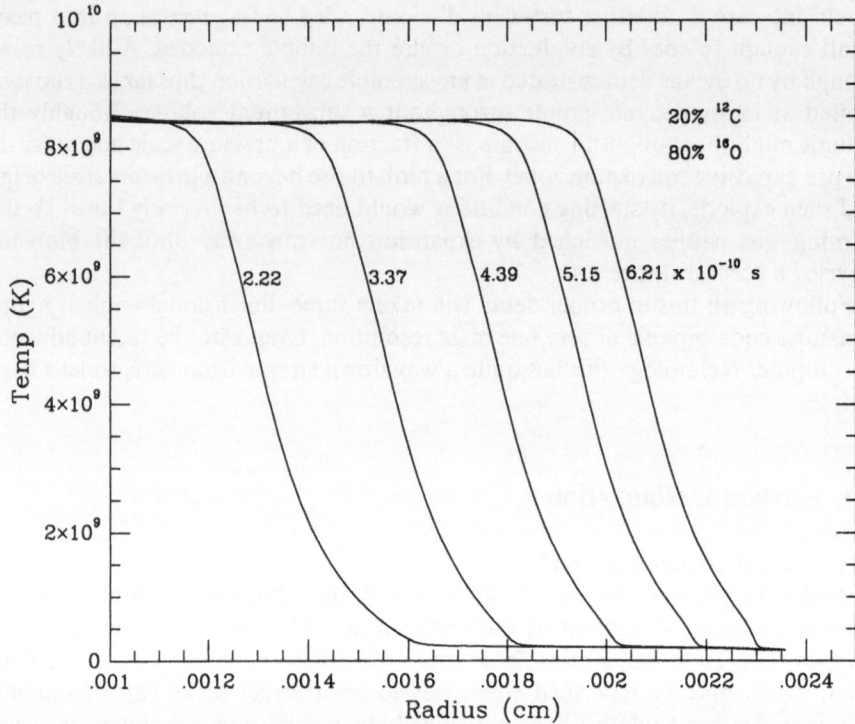

Figure 8.2. Temperature profile in a micro-zoned version of a laminar flame in degenerate carbon and oxygen at a density of 2×10^9 g cm^{-3}. The region shown comprises a sphere of 100 g that was resolved into 100 separate mass shells. The flame was artifically ignited in the inner 10 g and the subsequent propagation examined. Note the speed of the flame, about 20 km s^{-1}, and its width, about 2×10^{-4} cm. Figure taken from Woosley and Weaver (1989).

these properties initiated at a single point would slowly ($\gtrsim 50$ s) make its way through the white dwarf, decreasing in speed as regions of lower density were reached. Because the speed is so subsonic, the white dwarf would continually adjust its structure, probably by a series of flashes and oscillations until a highly extended structure developed. If it became unbound at all, the star would do so at very low velocities (see also Nomoto, Sugimoto, and Neo, 1979). In short the observed phenomenon would not at all resemble a Type Ia supernova.

8.4.2. The Multi-Dimensional Problem: Turbulence and Fractal Geometry

Of course, when considered in more than one dimension the plane flame front is not stable to deformation. Burning behind a subsonic front, in which pressure remains approximately constant, raises the temperature and lowers the density. Crossed gravity and density gradients result in Rayleigh–Taylor instability.

If the explosion is ignited precisely at a single point in the center of the white dwarf then instability does *not* immediately result for two reasons. First, there is no gravity at the center of the star and expansion leads to only small accelerations. Second, the normal speed of the conductive flame (eq. (8.1)) prohibits the growth

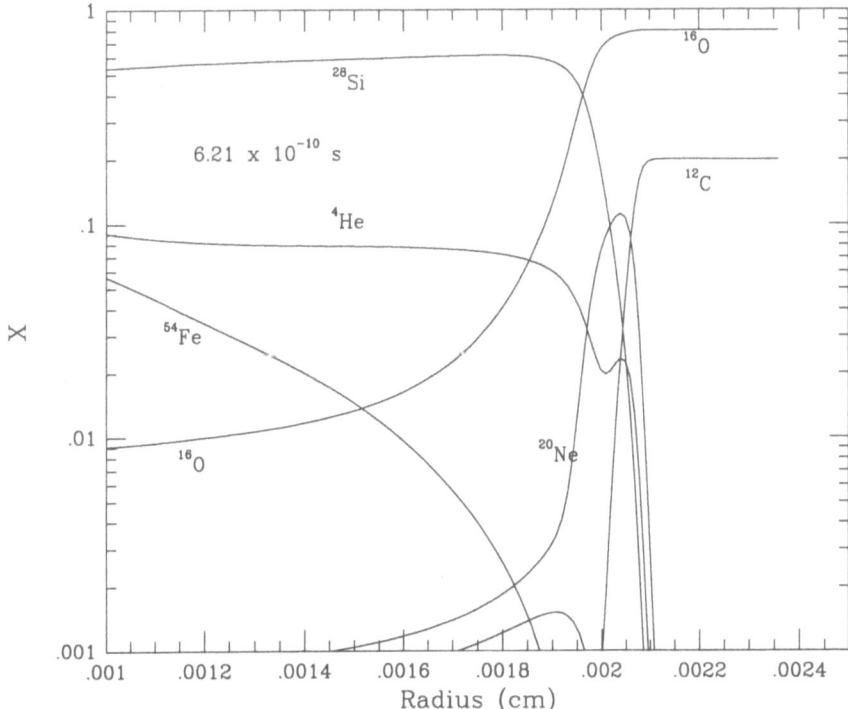

Figure 8.3. Composition across the flame boundary for conditions corresponding to Figure 8.2 at the last time point sampled. Silicon burning continues for a substantial distance behind the temperature jump. Figure from Woosley and Weaver (1989).

of small-scale deformations, and large-scale deformations cannot exist until a region of the star has been burned out that is comparable to the scale size. An incipient instability having wavelength smaller than

$$\lambda_{min} = \frac{4\pi v_{cond}^2}{g_{eff}} \tag{8.2}$$

with $g_{eff} = g(r)(\Delta\rho/\rho) = (\frac{4}{3}\pi G r \rho)(\Delta\rho/\rho)$ will not grow before the flame has already passed over it. For the time being we restrict ourselves to regions located sufficiently central that density is approximately constant. Burning at 2×10^9 g cm^{-3} gives a density reduction of typically 20%. Thus $g_{eff} \approx 10^9 r_7$ cm s^{-2} with r_7 the distance to the center of the star in units of 100 km and a typical value at 100 km for λ_{min} is a few kilometers. At larger radii g_{eff} increases and λ_{min} consequently becomes smaller.

There also exists a maximum deformation, λ_{max}, that can develop. This is given by the "event horizon," which we shall note here as r_b, the average radius of the burned out region. Henceforth we shall use λ_{max} and r_b interchangeably though there exists some constant on the order of unity relating the two. Until $\lambda_{max} > \lambda_{min}$ no deformations will develop and the flame will proceed slowly at a speed given by simple conduction (eq. (8.1)). This is true only for a very small region, ~ 20 km or

$3 \times 10^{-5} \, M_\odot$, near the center of the star. Beyond this point the flame front begins to deform and the range of allowable unstable wavelengths, λ_{min} to λ_{max}, grows rapidly, in fact as r^2, and the problem becomes increasingly complicated. Owing to the discrepant time scales associated with the growth of Rayleigh–Taylor instability for λ_{min} and λ_{max}, $(\tau_{RT} \approx (4\pi\lambda/g_{eff})^{1/2})$, instabilities of smaller wavelength may grow into a nonlinear stage, perhaps even break off and detach, while the longer wavelength deformations are still developing.

What develops is a wrinkled, quasi-simply connected sheet of flame characterized by a range of deformations from λ_{min} to λ_{max}. The larger wavelengths are responsible for the overall growth of the burned out region while the smaller wavelengths are responsible for "digesting" material at the flame boundary. An approximate steady state should exist at any particular time in which all scales of instability from λ_{min} to λ_{max} are present.

This problem is complex but not totally intractable. We are interested chiefly in an effective flame speed, v_{eff}, such that $\dot{M}_b = 4\pi r_b^2 v_{eff} \rho$. This is certainly what is needed by those of us who must ultimately map the problem into a one-dimensional computer program. It also gives the rate at which nuclear energy is being released, $q_{nuc}\dot{M}_b$ (with $q_{nuc} \approx 7 \times 10^{17}$ ergs g^{-1}), which determines the dynamic response of the star which in turn specifies nucleosynthesis and, by way of the ^{56}Ni production, the light curve and spectrum. Because the conductive flame is so thin, the rate at which mass crosses the flame can also be written $\dot{M}_b = v_{cond}\rho A$ where A is the area of the arbitrarily deformed (and perhaps not even simply connected) flame front. Obviously, $v_{eff} = (A/4\pi r_b^2)v_{cond}$, that is, if we knew the ratio of the actual area of the flame front to that of a sphere having a radius such that it encompasses an equivalent mass the desired solution would have been found.

This is a problem in fractal geometry which, being an unfamiliar field to most, warrants a brief introduction. Consider, for example, a four-sided regular solid (tetrahedron). Erect an additional tetrahedron on each face such that the edge of each addition is half the length of an original edge. Then the area of the resulting surface is increased by a factor $\frac{3}{2}$ (each face did have an area equal to four equilateral triangles; now one is covered by a new tetrahedron which displays three additional faces to the outside). This process may be continued. At each step the area is increased by a factor 1.5 and after n such steps the area has increased by $(1.5)^n$. The figure so generated has a fractal dimension of $1 + \log(3)/\log(2) = 2.585$ (Mandelbrot, 1983). A similar operation might be attempted using cubes, with each generation having an edge one-half as long as the previous one. The area multiplication factor is 2 in this case and the fractal dimension is 3. This, however, can be shown to be a degenerate case where overlap begins to occur (obvious in the third generation). If the operation were continued, space surrounding the first cube would become filled out to about twice the cube edge. A more reasonable surface could be generated by letting each edge equal one-third of the previous one in which case the area multiplies by $\frac{13}{9}$ and the dimension $D = 2 + (\log(13) - \log(9))/\log(3) = 2.335$. As the number by which we divide the length scale becomes greater and greater, the fractal dimension approaches 2, that of the surface area of the original cube. Such large changes in dimension would, however, be indicative of preferred scales to the problem, whereas ours has none.

Nor are these operations and concepts restricted to regular solids or even to simply connected surfaces. For an arbitrary surface of fractal dimension D, the area as measured at scale ε increases as $\varepsilon^{(2-D)}$. In the present problem λ_{min} sets a natural minimum scale at which we are to measure the area of the turbulent flame front (the "tile size"). Smaller surfaces are presumably smooth. The starting point is the undeformed sphere of area $4\pi r_b^2$ which in the general case is deformed to a surface having area $4\pi r_b^2 (r_b/\lambda_{min})^{(D-2)} \approx 4\pi r_b^2 (\lambda_{max}/\lambda_{min})^{(D-2)}$. Note that this equation conforms to the requisite scaling relation $(\text{area})^{1/D} \propto (\text{volume})^{1/3}$ (Mandelbrot, 1983, p. 112) provided that area and volume at each instant are measured in units of λ_{min}^2 and λ_{min}^3, respectively. The effective turbulent velocity we seek is then given by

$$v_{eff} = v_{cond} \left(\frac{\lambda_{max}}{\lambda_{min}}\right)^{(D-2)}. \tag{8.3}$$

Now it may appear that we have only substituted one unknown, the fractal dimension of the flame surface, for another, its area, but the operation is very valuable. First, we know fundamentally (Mandelbrot, 1983) that $2 < D < 3$ which gives some restrictions on v_{eff}. Since $\lambda_{min} \propto r^{-1}$ (at least for small r) and $\lambda_{max} \propto r$, eq. (8.3) tells us that the flame will accelerate rapidly as its radius increases. Actually, the situation is even better than that. The fractal dimension, D, is not likely to be very close either to 2 or to 3. Observations of turbulence in many different situations suggest $D \sim 2.5$–2.7 (Mandelbrot, 1983, see especially Plates 10 and 11 and Chapters 10 and 30). Thus the turbulent flame velocity is proportional to $(\lambda_{max}/\lambda_{min})^n$ with $n = D - 2 \sim 0.5$–0.7 and, for small r_b,

$$v_{eff} = v_{cond}\left(\frac{r_b}{\lambda_{min}}\right)^n = F\dot{X}_{12}^{(1-2n)}\rho^{(0.8-0.6n)}r_b^{2n}. \tag{8.4}$$

For $\rho = 2 \times 10^9$ g cm^{-3} and $X_{12} = 0.5$ a reasonable range in effective turbulent velocities is from $250r_7$ to $480r_7^{1.4}$ km s^{-1}, with r_7 the radius, r_b, in units of 100 km. These expressions are valid only so long as the density, the fractional decrease in density across the flame, and the fractal dimension remain constant.

For $n \sim 0.5$–0.7 the effective flame speed has a slight but significant inverse dependence upon the carbon abundance. This is initially counter-intuitive because the ordinary conductive speed (eq. (8.1)) *increases* for higher X_{12}. It occurs because for lower conductive speeds the deformed surface is measured using a smaller rule, λ_{min}, which is proportional to V_{cond}^2 (eq. (8.2)). For a given fractal dimension, the area is greater. For $D > 2.5$ this effect dominates over the fact that the speed normal to the flame surface is everywhere faster. As we shall see in the next section, the observable consequences of the explosion are very sensitive to factors of 2 in the effective flame speed. Thus qualitatively different results might be observed for supernovae whose initial carbon abundance varied by more than a factor of 4 (owing both to stellar evolution and to the settling out of oxygen to the center of the white dwarf); whose fractal dimension was 2.5 rather than 2.7; or which ignited significantly off-center rather than in the middle of the star (Iben, 1982).

Fractal dimensions near 2.5 have the interesting quality that the growth of mass behind the flame, M_b, proceeds exponentially at early times, $M_b = M_0 \exp(t/\tau)$, with $\tau = 0.13$ s for $\rho = 2 \times 10^9$, and $X_{12} = 0.5$; for $D = 2.7$ the mass grows a little faster,

$\tau = 0.069r_7^{-0.4}$ s. Thus the burned-out mass (and the nuclear energy released) increases quite slowly initially but more rapidly towards the end. This will have interesting implications for the light curve, the spectrum, and, especially, the isotopic nucleosynthesis. It also introduces aspects of *chaos* into the system. Any deviations in M_b at early times will persist in the exponential growth at late times.

Finally, we note similarities and dissimilarities between the present formulation and that expected from mixing length convection theory for the special case $D = 2.5$. The convective velocity in mixing length theory is given by (e.g., Clayton, 1968)

$$\bar{v} = \frac{1}{2}\left(\frac{GM}{\rho r^2}\Delta\nabla\rho\right)^{1/2} L \tag{8.5}$$

with L a typical distance over which acceleration of a blob, having density contrast $\Delta\rho = \frac{1}{2}L\Delta\nabla\rho$, occurs. In our terminology this is equivalent to $\bar{v} = (g_{eff}/2)^{1/2}L^{1/2}$ which may be compared to the results of eqs. (8.2) and (8.3), $v_{eff} = (g_{eff}/4\pi)^{1/2}r_b^{1/2}$. These quantities are very similar provided that the "mixing length" is identified with some fraction of r_b, the radius of the burned-out region, i.e., not the pressure scale height. The chief difference, however (aside from the fact that D is probably not exactly 2.5), is one of interpretation. In our prescription v_{eff} gives directly the rate of advancement of the burning front. In mixing length theory, $\bar{v}$ is combined with other quantities, for example, the excess of the temperature gradient over the adiabatic value, to obtain a *heat flux*. In fact, mixing length theory is not at all applicable to the present problem, though with judicious choice of parameters it may be made to mimic a correct solution.

8.4.3. Numerical Models

Four deflagration models were prepared, each starting with a carbon–oxygen white dwarf (50% each of ^{12}C and ^{16}O) that had attained a critical density in its center. Following a brief stage of stable convective carbon burning, a subsonic deflagration was naturally initiated. In contrast to the discussions to come in Section 8.5, the central zoning employed was relatively coarse, $\sim 10^{-3}$ $M_\odot$. The flame speed was parametrized to be roughly a constant, F, times the local sound speed. Further details of the models are given in Table 8.1. None of these models is particularly

Table 8.1. Carbon detonation and deflagration models.

Model	v_{flame}^a/c_s (approx.)	KE_∞ 10^{51} ergs	M_{Fe} ($M_\odot$)	$M(^{56}Ni)$ ($M_\odot$)	$Y_e^{(0)}$
DF1[b]	$\frac{1}{2}$	1.73	1.25	0.89	0.466
DF2[b]	$\frac{1}{4}$	1.04	0.85	0.51	0.468
DF3[b]	$\frac{1}{8}$	0.66	0.66	0.41	0.470
DF4[b]	$\frac{1}{16}$	0.22	0.41	0.20	0.465
DT1[c]	—	1.70	1.41	0.83	0.474
DT2[c]	—	2.46	1.41	0.91	0.471

[a] Speed as measured in the expanding frame. $c_s^0 \sim 9500$ km s^{-1}.
[b] Deflagration model.
[c] Detonation model.

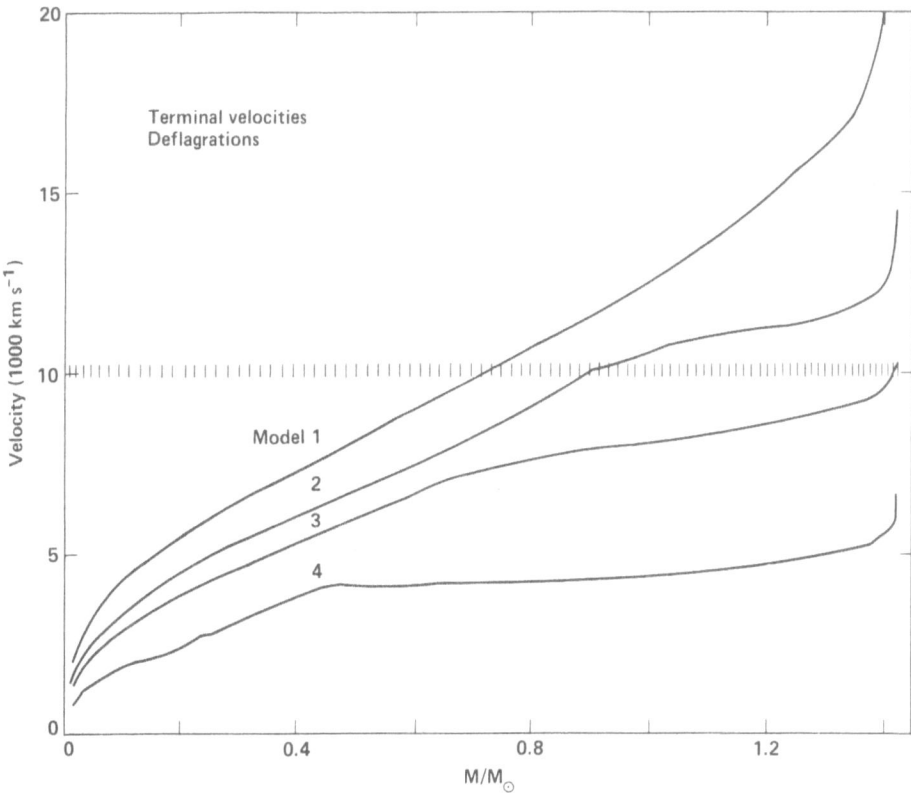

Figure 8.4. Final velocities of the white dwarfs completely disrupted by a carbon deflagration. Models 1–4 correspond to Models DF1–DF4 in Table 8.1.

realistic in light of the discussions of the previous section which strongly suggests that the flame speed should accelerate as radius increases. Such a calculation has not yet been carried out (although see Nomoto, Thielemann, and Yokoi (1984) and Woosley and Weaver (1986c) for some first attempts). The present results serve merely to indicate the qualitative effects upon the light curve and nucleosynthesis given by various values of flame speed.

As a result of the flame propagation through the white dwarf a fraction of the mass is turned into iron (some of it in the form of ^{56}Ni; some other part as ^{54,56}Fe and ^{58}Ni) and intermediate mass group elements, and adequate energy is released to unbind the white dwarf. As might be expected a faster flame produces more iron, because the flame goes farther before it is quenched by expansion, and a more energetic explosion. As we shall see this also translates into a brighter, faster light curve for those models employing a more rapid flame. Final velocities are given for the four models in Figure 8.4.

8.4.4. Nucleosynthesis

The greatest shortcoming of the carbon deflagration model (in addition to uncertainties surrounding its pre-explosive evolution, a problem generic to all Type I models) is the apparent disagreement between the calculated isotopic abundances for the iron group elements, iron and nickel in particular, and what is observed in the sun. Detailed nucleosynthesis has been calculated in this model by Nomoto, Thielemann, and Yokoi (1984), Woosley, Axelrod, and Weaver (1984), and Thielemann, Nomoto, and Yokoi (1986). Thielemann *et al.* find overproductions of $^{58}Ni/^{56}Fe$ and $^{54}Fe/^{56}Fe$ equal to 5.1 and 2.4, respectively. Providing that Type I supernovae have occurred at a rate comparable to Type II supernovae, they should greatly predominate in the synthesis of iron for Population I objects. Since it is unlikely that the elemental ratio for nickel/iron could be in error by as much as a factor of 5 or the isotopic ratio $^{54}Fe/^{56}Fe$ by a factor of 2, some mechanism must be found for curing this deficiency in the model.

One immediate improvement that could be attempted (Thielemann, Nomoto, and Yokoi, 1986) is a reduction in the metallicity of the exploding white dwarf. The number of neutrons available for incorporation into isotopes like ^{54}Fe and ^{58}Ni in nuclear statistical equilibrium is sensitive (though by no means linearly proportional) to the amount of ^{14}N that got turned into ^{18}O and ^{22}Ne during helium burning (in both the carbon–oxygen core and any mass donating star). More metallicity means more neutrons which makes a bad problem worse. Thielemann estimates that a 30% reduction in ^{54}Fe and ^{58}Ni production could be achieved by using a Population II composition for the white dwarf, though the detailed calculations have yet to be done to demonstrate it.

The main problem though is the amount of electron capture that occurs during the explosion itself. Since the peak temperature and burning of the composition to nuclear statistical equilibrium are assured by fundamental considerations, the only way out is to burn (most of) the fuel at a lower density. The carbon deflagration model (but not the carbon detonation model) offers a way to do this. As emphasized in Section 8.3, the burning of a small amount of fuel, a few hundredths of a solar mass, can cause a large expansion of the rest of the star. If the flame can then be made to move very fast after the star has already begun to expand, a larger fraction of iron will be produced at lower density and the excess neutronization problem, hopefully, cured. Preliminary calculations in this direction (Woosley and Weaver, 1986c) are encouraging.

The discussion of Section 8.4.2 offers possible justification for this procedure. We expect that the flame should be born moving very slowly and should accelerate markedly as the radius of the burned-out region increases. The possibility that ignition may occur not just at the center of the star but concurrently at many points would amplify the effect. It could even be that at late times the rate at which mass is flowing across some highly convoluted and deformed flame sheet would exceed that of a spherical flame moving at supersonic speed, even though the flame velocity was actually markedly subsonic at all points. Nucleosynthesis calculations to address this point carefully still need to be done.

8.4.5. Light Curves

The light curves of these same four models are given in Figure 8.5 and are entirely a consequence of energy released by the decay of radioactive ^{56}Ni and ^{56}Co. The light curves were calculated in collaboration with Lisa Ensman and details of the

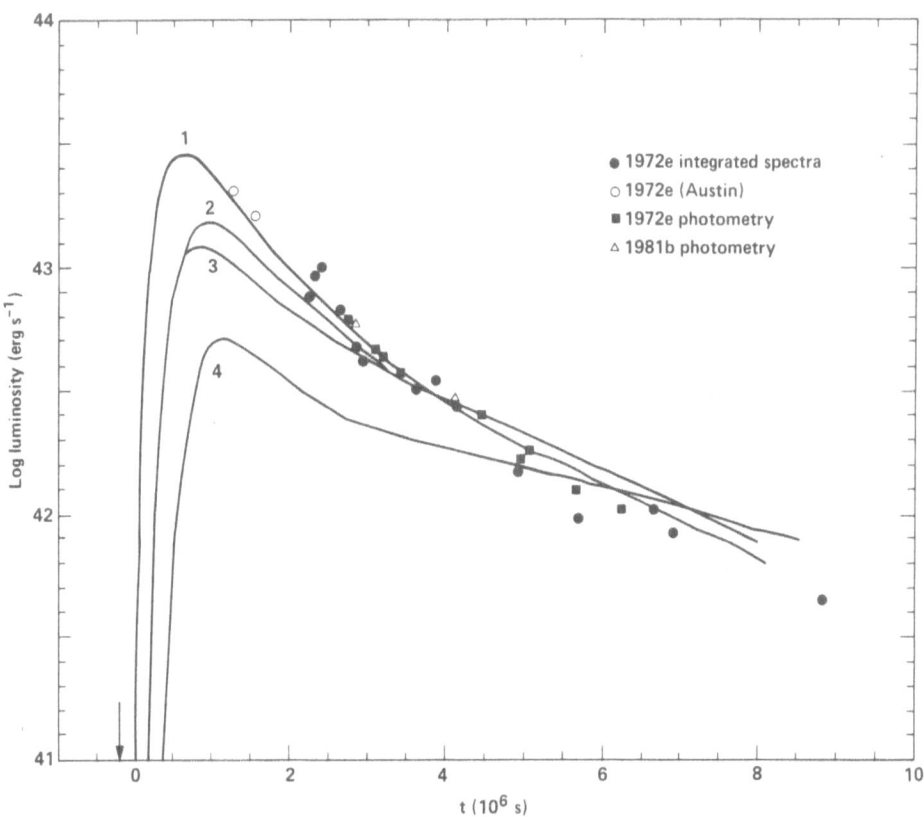

Figure 8.5. Bolometric light curves of the four deflagration models defined in Table 8.1 and Figure 8.4. Data are from a compilation by Grahm (1987b) (see also Woosley and Weaver, 1986a). The arrow just before $t = 0$ is a prediscovery limit on SN 1972E. Only models DF1 and DF2 produce enough ^{56}Ni and expand rapidly enough to be successful Type Ia models.

technique will be published elsewhere. For now we note that a complete ionization calculation was carried out to get the electron density which dominated the optical opacity. All positron kinetic energy was assumed to deposit and γ-ray deposition was calculated assuming an angle and energy-averaged opacity of 0.07 cm^2 g^{-1}. Radioactive energy was only deposited according to the local abundance of ^{56}Ni following the explosion, and in particular the important effect of electron capture in turning a portion of the iron group in the middle of the star into stable isotopes (chiefly ^{54}Fe, ^{56}Fe, and ^{58}Ni) was included. As a consequence no appreciably γ-rays were deposited in the very central regions of the star where the γ-ray optical depth was greatest. Otherwise, the light curves would have been broader than shown in Figure 8.5. Also a floor to the optical opacity, 0.03 cm^2 g^{-1} in addition to that given by electron scattering, was assumed to represent the Doppler-broadened line opacity of the expanding supernova.

Also shown in Figure 8.5 are observations of the approximate bolometric luminosity of two Type Ia supernovae, 1972E and 1981B. Only the two most energetic deflagrations are compatible with these observations. Model DF3 might just barely squeak by, providing that we liberally adjust the date and distance to the explosions, but Model DF4 is certainly too faint and too broad to be a Type Ia explosion. We note, however, that it is just these characteristics (i.e., broad and faint by roughly a factor of 4) that allegedly characterize Type Ib supernovae (Section 8.6).

8.5. Carbon Detonation

8.5.1. Spontaneous Initiation of a Detonation Wave

The considerations of Section 8.4 formally apply only to an explosion initiated at a point. For a core having a finite initial temperature gradient the location of the burning front will initially "propagate" simply because regions that are otherwise out of communication will experience nuclear runaway at comparable times (Woosley, 1986; Woosley and Weaver, 1986a, b, 1989; Blinnikov and Khokhlov, 1986, 1987). The time scale for the acceleration of nuclear burning at temperature T (henceforth the "nuclear" time scale) is given by

$$\tau_{\text{nuc}} = \left(\dot{S}_{\text{nuc}} \frac{d\dot{S}_{\text{nuc}}}{dt} \right)^{-1} = \frac{T}{j} \left(\frac{dT}{dt} \right)^{-1} \tag{8.6}$$

if the nuclear energy generation rate is approximated by $\dot{S}_{\text{nuc}} \approx D(X_{12}/0.5)^2 \, \rho_9^i T_9^j$ and the density and carbon abundance assumed constant. In the limited temperature range $(0.6–1.2) \times 10^9$ K and including the effects of electron screening, the nuclear time scale is given to better than a factor of 2 by

$$\tau_{\text{nuc}} = 0.152 \, T_9^{-20.2} \, \rho_9^{-3.05} \left(\frac{0.5}{X_{12}} \right)^2 \text{ s.} \tag{8.7}$$

The "phase velocity" of the combustion front, or how its location changes with time as a result of initial thermal conditions, is given by

$$v_{\text{phase}} = \left(\frac{d\tau_{\text{nuc}}}{dr} \right)^{-1}$$

$$= -3.26 \times 10^8 \, T_9^{21.2} \, \rho_9^{3.05} \left(\frac{X_{12}}{0.5} \right)^2 \left(\frac{dT}{dr} \right)^{-1} \text{ cm s}^{-1}. \tag{8.8}$$

Thus the initial flame speed is very sensitive both to the temperature and its gradient when the runaway starts. An important critical value for the temperature gradient is obtained by setting the phase velocity equal to the sound speed, typically about 9500 km s^{-1} ahead of the burning front

$$-\left(\frac{dT}{dr} \right)_{\text{sonic}} \lesssim 0.3 T_9^{21.2} \, \rho_9^{3.05} \left(\frac{X_{12}}{0.5} \right)^2 \text{ K cm}^{-1}. \tag{8.9}$$

A region having an actual gradient smaller than this (in absolute value) will run away on less than a sound crossing time, thus its expansion will be supersonic in the unburned medium.

An interesting and relevant case of a temperature gradient that can be evaluated analytically comes from presuming that the convection which precedes the run-away (Arnett, 1969) establishes an *adiabatic* gradient to high accuracy (Blinnikov and Khokhlov, 1986). Then the equation of hydrostatic equilibrium ($dP/dr = -GM(r)\rho/r^2$), the condition for an adiabatic temperature gradient ($dT/dr = (1 - 1/\Gamma_2)(T/P)\, dP/dr$), and the equation for relativistic degeneracy pressure ($P = 1.24 \times 10^{27}(\rho/2 \times 10^9 \text{ g cm}^{-3})^{4/3}$) can be combined, in the limit of constant density and for $\Gamma_2 \approx 1.7$, to obtain

$$T(r) = T_0\left(1 - \frac{\Gamma_2 - 1}{\Gamma_2}\frac{2\pi G\rho^2 r^2}{3P_0}\right)$$

$$= T_0\left(1 - 1.8 \times 10^{-16}\left(\frac{\rho}{2 \times 10^9 \text{ g cm}^{-3}}\right)^{2/3} r^2\right),$$

$$(8.10)$$

which is valid for a small region near the center of the star where the temperature is T_0. Thus $dT/dr \sim -2.3 \times 10^{-16} T_0 \rho_9^{2/3} r$ implying a phase velocity (eq. (8.8)) of

$$v_{\text{phase}} = 1.4 \times 10^{10} T_9^{20.2} \rho_9^{2.38}\left(\frac{X_{12}}{0.5}\right)^2 r_5 \text{ cm s}^{-1}, \qquad (8.11)$$

with r_5 the radius in kilometers. Setting this equal to the sound speed, $c_s \approx 9500$ km s^{-1} we find that the phase velocity will be supersonic in a region

$$r \lesssim 1.3 \times 10^6 \rho_9^{2.38} T_9^{20.2}\left(\frac{X_{12}}{0.5}\right)^2 \text{ cm}, \qquad (8.12)$$

which corresponds to a mass

$$M(r) \lesssim 4.8 \times 10^{-6} \rho_9^{8.15} T_9^{60.6}\left(\frac{X_{12}}{0.5}\right)^6 M_\odot. \qquad (8.13)$$

This shows the extreme sensitivity to the initial temperature. An accurate numerical calculation would need to carry zoning so as to resolve this region which, while about 0.001 $M_\odot$ at $T_9 \sim 1.0$, is only $10^{-9} M_\odot$ at $T_9 = 0.8$, and $10^{-12} M_\odot$ at $T_9 = 0.7$!

The formation of a propagating Chapman–Jouget detonation requires more than the simple existence of a region having supersonic phase velocity. This region must also be large enough that the shock does not die due to spherical dilution and the shock must be powerful enough to heat material, as it passes through, to a sufficient temperature that it will burn during the shock crossing time ($\sim \Delta r/13{,}000$ km s^{-1}). Using a numerical method for solving the shock problem (the Lagrangian method of characteristics), Blinnikov and Khokhlov (1986) conclude that the shock formed by spontaneous burning in a small central region will survive to incinerate a large fraction of the star, provided that eq. (8.10) is an accurate description of the pre-explosive temperature gradient and $T_0 \gtrsim 7 \times 10^8$ K.

8.5.2. Some Results From a Fine-Zoned Model
To study the properties of a detonation initiated by a supersonic phase velocity, Woosley and Weaver (1990) have calculated an extremely well-zoned model of a centrally ignited carbon run away. The central zone in this model was $10^{-8} M_\odot$

and the zoning was increased smoothly on a logarithmic scale so that the entire star could be carried. For certain choices of initial conditions they found the spontaneous ignition of a detonation wave which passed without decay through the entire white dwarf. As the previous discussion would suggest, the results were critically sensitive to the pre-explosive temperature distribution set up by convection. If the mixing length formalism was used to transport energy up until the time that the central temperature reached 0.85 billion K and then turned off, detonation initiated spontaneously and the entire star was incinerated (Figures 8.7 and 8.8). If, in an identical calculation, convective energy transport was halted at 0.80 billion K, detonation did not occur. Instead the conductive flame of Section 8.4 was born. The evolution to runaway, after convection was turned off, is shown for the problem which detonated in Figure 8.6. In both cases the convective velocity calculated (about 50 km s^{-1}) was a small fraction of the sound speed. Interestingly eq. (8.13)

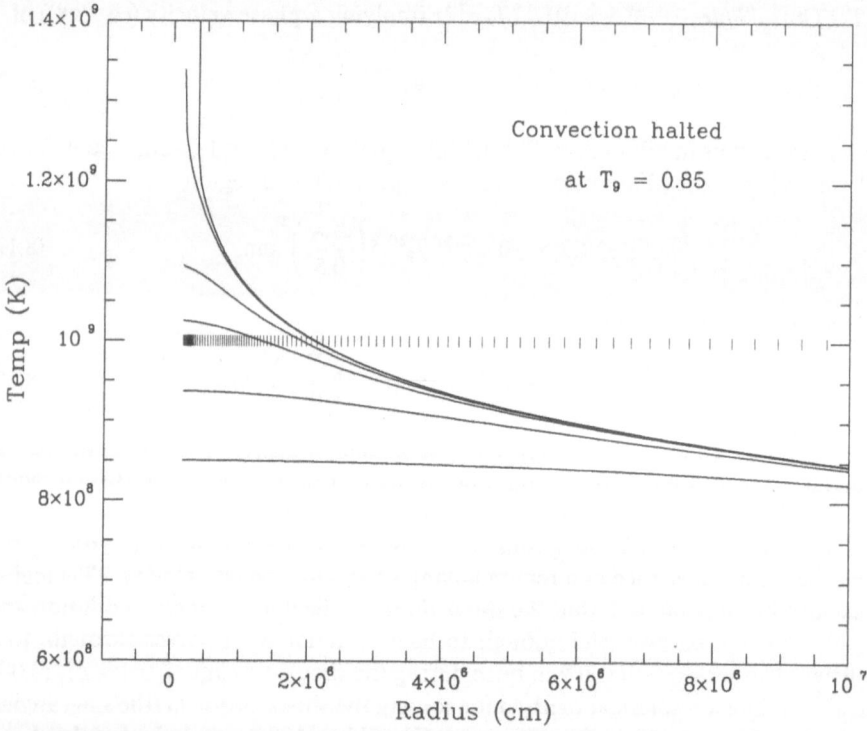

Figure 8.6. Temperature profiles set up when convection is turned off at a central temperature of 0.85 billion K. Each isolated region then continues to run away on its own time scale. When the central zone has exceeded about 1.3×10^9 K, the time step becomes very small. The outcome of the calculation is then determined by the size of the region that runs away in less than a sonic crossing time. The times of the various curves are zero (arbitrarily) for the lowest curve and 0.351, 0.4011, 0.4087, and 0.4118 s for the next four, respectively. The last curve samples the temperature 1.31×10^{-4} s after the central zone has run away. By this time a detonation front (as indicated by a substantial increase in pressure accross the burning front) has already formed. Note the small length scales and the questionable applicability of mixing length convection theory.

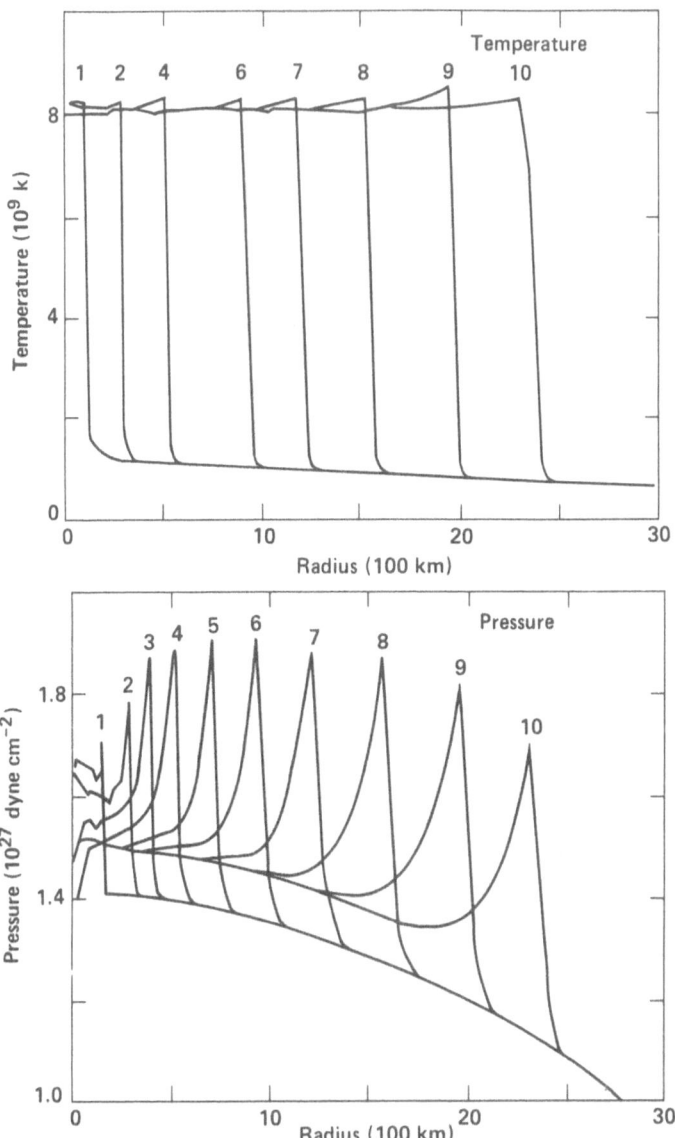

Figure 8.7. Propagation of a spontaneously ignited detonation wave in a carbon–oxygen dwarf of 1.41 $M_\odot$. Central zoning was 10^{-8} $M_\odot$. The ten time points correspond to 0.06, 0.82, 1.48, 2.37, 3.79, 5.48, 7.66, 10.3, 13.2, and 16.4 ms, respectively. The range of the figure encompasses the inner 0.125 $M_\odot$ of the exploding star, about half of a pressure scale height. Continued evolution saw the detonation wave survive until it reached the edge of the star and the conversion of the entire star into iron-group elements. Figure from Woosley and Weaver (1989).

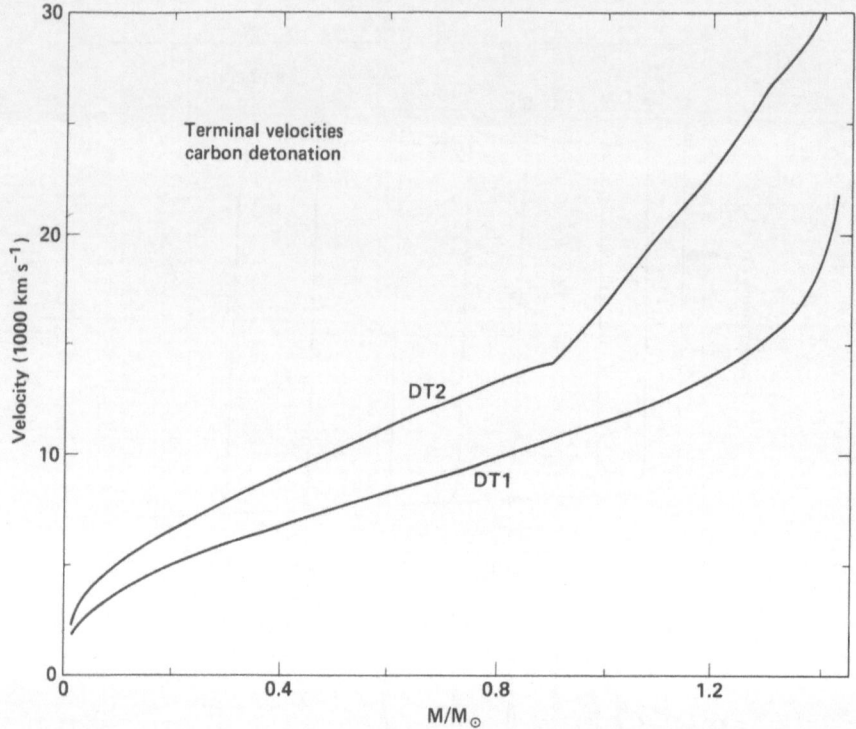

Figure 8.8. Final velocities of the detonated white dwarf models DT1 and DT2. Model DT2 developed much greater energy (Table 8.1) owing to the greater release from helium burning to iron in the outer 0.48 $M_\odot$ of the explosion. The composition–energy deposition discontinuity is also reflected as a break in the terminal velocity at 0.9 $M_\odot$.

implies that central zoning of 10^{-8} $M_\odot$ will resolve the "detonator" if the temperature is 0.85 billion K but not if it is 0.80, so a still finer-zoned model might have detonated even at the lower temperature.

Does this mean then that detonation is the correct solution to the degenerate carbon ignition problem after all and we have been misled all these years simply because calculations did not utilize adequate mass zoning? Probably not. Examine the length scale in Figure 8.6. A zone of 10^{-8} $M_\odot$ is only 1 km in radius. The pressure scale height is 450 times larger. What is the meaning of mixing length convection theory in such a microscopic context? Usually we think of convective blobs as having a dimension equal to some fraction of the scale height. What physical meaning is to be ascribed then to a temperature distribution calculated in a region that is so much smaller? Probably not much.

What really matters is how nearly isothermal we can maintain regions of mass given by eq. (8.13). As runaway proceeds are there regions that have temperatures above 0.85 billion K that are isothermal to one part in 10^5 (eq. (8.8) with $V_s = c_s$) and which contain 2×10^{-6} $M_\odot$ (at a density 3×10^9 g cm^{-3})? When the temperature at some point first exceeds 0.90 billion K, are there regions 20 km in radius around that point, regions containing $6 \times 10^{-5}\ M_\odot$, that are within 0.2% of that same temperature? If so detonation is likely; if not, it isn't.

Keeping in mind the exponential sensitivity of the reaction rate to temperature and the consequent amplification of any small variations in temperature distribution, it seems that detonation is probably not likely, but really the question cannot be answered now. So we turn, as usual, to the observations to resolve the issue.

8.5.3. Nucleosynthesis in the Carbon Detonation Model

Nucleosynthesis in the carbon detonation model of Arnett (1969) was studied by Arnett, Truran, and Woosley (1971). The conclusions of that paper were quite favorable: good agreement with solar isotopic ratios in the iron group could be obtained for the nuclear rates and models of the day. *It now appears that those calculations, frequently cited as the basis for the "good" nucleosynthetic characteristics of the carbon detonation model, were basically flawed by using much more conservative estimates for the weak interaction rates than are now credible.* Especially bad was the omission from the 1971 calculation of electron capture on free protons. The new weak rates Fuller, Fowler, and Newman (1982a, b, 1985) are also much faster than the rates of Hansen and Mazurek used in the earlier study.

For this chapter a new calculation was carried out using a simulated detonation in a zoned stellar model and employing current values of all relevant nuclear reaction rates, strong and weak (Woosley, 1986; Woosley and Hoffman, 1990). The nuclear reaction network included 100 nuclei from carbon through zinc. Because special attention was focused upon electron capture, the network was concentrated in the iron group. Specifically included were $^{44-50}$Ti, $^{47-52}$V, $^{48-54}$Cr, $^{51-56}$Mn, $^{52-58}$Fe, $^{55-60}$Co, $^{56-60}$Ni, $^{57-61}$Cu, and $^{60-61}$Zn and all strong, weak, and electromagnetic reactions coupling them. The initial configuration was a carbon–oxygen white dwarf (50% ^{12}C; 50% ^{16}O) of 1.41 $M_\odot$ that was beginning to run away in the middle (central density, 2.5×10^9 g cm^{-3}). The slightly different mass from Section 8.3 is because Coulomb corrections were not included in the equation of state for this earlier model. Detonation was simulated in the 101 zone model by depositing an energy of 7.7×10^{17} ergs g^{-1}, as would be released by the complete combustion of the initial composition to ^{56}Ni, and concurrently instantaneously turning the composition of the entire white dwarf into pure ^{56}Ni. While this procedure was adopted as a computational expedient, the aftermath of detonation wave passage should be well represented. If anything, the amount of electron capture will be slightly underestimated by assuming an instantaneous propagation of the burning front. The final neutron excess will also be underestimated by assuming a starting composition of nuclei having equal numbers of neutrons and protons. Especially in a Population I white dwarf there would already be a substantial excess of neutrons created during helium burning.

The white dwarf reacted to this energy deposition, as expected, by changing its composition into a statistical distribution of many nuclei. Within 10^{-13} s the central zone, for example, which like the rest of the star was initially pure ^{56}Ni, had photodisintegrated 99% of the ^{56}Ni into a host of other nuclei. The free α-particle abundance was 10%, and the temperature was 9.0×10^9 K. The density was essentially unchanged. As the frenzy of nuclear reactions subsided to a slower pace,

Table 8.2. Nucleosynthesis in the carbon detonation model.

^AZ	Normal		$p(e^-, v)n$ only	
	Relative[a]	Absolute[b]	Relative[a]	Absolute[b]
^{4}He	1.83E−05	3.07E−03	2.07E−05	3.47E−03
^{48}Ti	2.50E−02	3.98E−05	2.91E−02	5.79E−05
^{49}Ti	3.71E−02	4.47E−06	5.80E−02	8.76E−06
^{51}V	1.59E−01	3.81E−05	1.54E−01	4.64E−05
^{50}Cr	5.98E−01	3.04E−04	3.03E−01	1.93E−04
^{52}Cr	2.97E−01	3.05E−03	3.51E−01	4.51E−03
^{53}Cr	9.11E−01	1.09E−03	7.40E−01	1.10E−03
^{55}Mn	2.04E+00	1.93E−02	1.35E+00	1.60E−02
^{54}Fe	4.59E+00	2.42E−01	2.36E+00	1.56E−01
^{56}Fe	1.00E+00	8.59E−01	1.00E+00	1.08E+00
^{57}Fe	1.87E+00	3.91E−02	1.32E+00	3.47E−02
^{59}Co	1.19E+00	2.87E−03	9.14E−02	2.77E−04
^{58}Ni	6.84E+00	2.39E−01	2.97E+00	1.30E−01
^{60}Ni	9.60E−01	1.34E−02	—	—

[a] Relative to solar abundances, normalized at ^{56}Fe.
[b] Total produced, in $M_\odot$.

the white dwarf began to expand, eventually completely dissipating with a kinetic energy at infinity of 1.70×10^{51} ergs (see Model DT1, Table 8.1). Weak interactions as well as strong occurred during the expansion giving rise to a finite neutron excess. The final nucleosynthesis was then tabulated and compared to the sun (Table 8.2).

Since, as mentioned above, the electron capture and initial neutron excess are somewhat underestimated in this calculation, the production of ^{54}Fe and ^{58}Ni would be slightly greater than given in Table 8.2 if we had modeled an actual detonation or used a Population I white dwarf. Even so, and somewhat surprisingly, *such fundamental ratios as ^{54}Fe/^{56}Fe and ^{58}Ni/^{56}Fe are more poorly reproduced in the detonation model than in the deflagration model.* This is because in the deflagration mode a pressure wave moves ahead of the burning front and pre-expands the fuel. At the lower density less electron capture occurs. While the central zone of the detonation model may expand a little faster, this can be more than compensated for by making more ^{56}Ni farther out where the density is lower.

It is not likely that any reasonable change in weak rates will substantially alter this conclusion since, as Table 8.2 shows, setting all capture rates to *zero* except for capture on free protons and rerunning the otherwise identical calculation only halves the amount of electron capture (i.e., the excess of neutrons over half the number of baryons) that occurs. The capture rate on free protons is well determined and cannot be arbitrarily decreased.

8.5.4. Spectra of the Carbon Detonation Model
The spectrum of a carbon detonation model has not been calculated, either at peak light or at late times. However, unless the detonation wave somehow manages to decay before propagating through the entire white dwarf, the abundances of inter-mediate mass elements in the ejecta will be very small, and it will thus be very difficult

to understand the strength of silicon and calcium features in the Type Ia spectrum taken at peak light.

The late-time spectrum of the carbon detonation model would be very similar to that of Model DF1. No published calculation exists for that model either, but unpublished work by Axelrod suggests that the ionization stages of iron would shift from Fe II and Fe III, as observed in Type Ia supernovae at late time, to higher ionization stages that are not observed.

While the calculations really need to be carried out, it seems probable that the late-time spectrum and certainly the early-time spectrum of carbon deflagration (in particular Model DF2) will be superior to that of detonation.

8.5.5. Light Curve of the Carbon Detonation Model

The light curves of two carbon detonation models (DT1 and DT2 in Table 8.1) were simulated in a manner similar to the explosive nucleosynthesis calculation of the previous section. Model DT1 was calculated in an identical fashion and, because

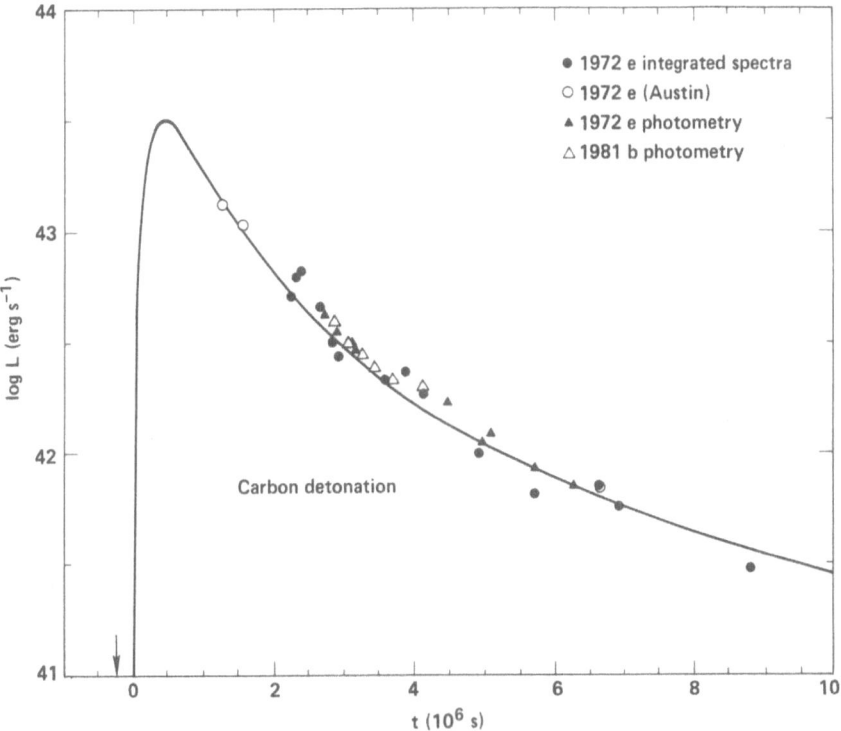

Figure 8.9. Bolometric light curve of a carbon detonation model compared to observational data (see Figure 8.5). This particular model was DT2 (Table 8.1) and the narrowness of the light curve is a consequence of both the great explosion energy and the fact that the ^{56}Ni is concentrated farther out in the star. The central regions are predominantly ^{54,56}Fe and ^{58}Ni. A pure carbon–oxygen dwarf detonation (i.e., without the outer layers initially consisting of helium) would be almost identical to Model 1 in Figure 8.5.

its parameters so closely resembled Model DF1 (Table 8.1), its light curve was virtually identical. Model DT2 included the effect of a heterogeneous composition. Since single carbon–oxygen white dwarfs of greater than 1.0 $M_\odot$ probably do not exist, a carbon–oxygen composition throughout can only be achieved in the special case of the merger of two carbon–oxygen dwarfs (Iben and Tutukov, 1984). It may be that a thick helium layer can somehow accumulate on a carbon–oxygen dwarf without prematurely detonating. If so, as the centrally ignited detonation wave reaches this helium layer, its combustion becomes even more violent and the detonation wave more healthy. The entire star would still be converted to iron group elements but a greater specific energy would be released by burning a lighter fuel. This was simulated by depositing the usual 7.3×10^{17} ergs g^{-1} in the inner 0.9 $M_\odot$ of a 1.40 $M_\odot$ dwarf (corresponding to carbon–oxygen goes to ^{56}Ni), but a greater amount, 1.51×10^{18} ergs g^{-1}, in the outer 0.5 $M_\odot$ to represent helium detonation to ^{56}Ni. The resultant explosion should be as energetic and produce as much ^{56}Ni as any credible model that uses a near Chandrasekhar mass white dwarf as its starting point. Consequently, its light curve should be the brightest possible from this class of Type I models.

The resultant light curve is shown in Figure 8.9. It agrees very well with observations of several standard Type Ia supernovae. Though the carbon detonation model may have its difficulties—spectroscopic and nucleosynthetic—its light curve is not one of them.

8.6. Type Ib Supernovae

Type Ib supernovae were first recognized as a distinct subtype in the early 1980s (Elias et al., 1985; Wheeler and Levreault, 1985), after the events SN 1983N and 1984L, which were seen to resemble the spectroscopically peculiar supernovae 1962L (Bertola, 1964) and 1964L (Bertola, Mammano, and Perinotto, 1965). Several other Type Ib supernovae have since been identified, including SN 1985F (Filippenko and Sargent, 1985), SN 1983I, 1982R, and perhaps 1954A, 1975B, 1983V, and 1986M (Elias et al., 1985; Chevalier, 1986; Gaskell et al., 1986; Grahm, 1986, and references therein; Wheeler, Harkness, and Capellaro, 1987). Most of the characteristics of Type Ib supernovae are inferred from a rather sparse data set. Few have been discovered and none has been observed over a long span of time at all wavelengths. Porter and Filippenko (1987) and Ensman and Woosley (1988) give a good summary of observed Type Ib properties.

Except for the fact that, on average, their light curves are roughly 1.5–2 mag fainter in the blue, SN Ib are very similar photometrically to SN Ia. Chief differences are spectroscopic—Type Ib lacks the 6347 and 6371 Å blend of Si II lines near peak light and displays, at times later than 100 days or so, a very strong emission line (6300 Å) of [O I]. Thus far Type Ib supernovae have all been discovered in spiral galaxies, always in or near H II regions. They also seem to have unusual properties in their infrared spectra (Elias et al., 1985) and two Type Ib supernovae are the only Type I supernovae ever to be observed as strong radio sources (Sramek, Panagia, and Weiler, 1984; Panagia, Sramek, and Weiler, 1986; Sramek and Weiler, this volume). There is some diversity in the light curves of SN Ib supernovae; SN 1985F for example, seems to have declined much more slowly after peak than 1964L (Ensman and Woosley, 1988), but the lack of bolometric data for both makes this an unreliable premise.

Many models have been proposed to explain Type Ib supernovae (see Woosley and Weaver 1986a for a review). The favored model of the day is that Type Ib supernovae are explosions of massive stars that, owing to mass exchange in a binary system or mass loss by stellar wind, have lost their hydrogen envelope. This would account for the lack of hydrogen lines, the association with star-forming regions, the oxygen emission spectrum at late times (Axelrod, 1988; Fransson, 1988), and possibly the strong radio emission associated with the supernova blast wave interacting with its own pre-explosive ejecta. Because the amount of ^{56}Ni synthesized in the explosion of a massive star is much smaller (SN 1987A, a supernova with a 6 $M_\odot$ helium core, ejected only 0.075 $M_\odot$ of ^{56}Ni, about six times less than Model DF2), it is expected that such explosions would have fainter light curves than Type Ia. Why the Si II feature should be suppressed in the spectrum at peak light remains unclear; the explosion of a massive star should make as much silicon as a carbon deflagration. But this aspect of the spectrum has not yet been accurately modeled. Filippenko (1988) has also found, in the unusual case of SN 1987K, what might be regarded as the "missing link"—a supernova that initially has a spectrum like that of a Type II, presumably the explosion of a single massive star, but at late times lacked hydrogen lines and looked like a Ib.

The greatest concern regarding the massive helium core model for SN Ib supernovae, as has been recently emphasized by Ensman and Woosley (1988), is that, for any credible explosion energy, the ejecta expand too slowly to give the observed rapid decline in luminosity after peak observed for all Type Ib supernovae except perhaps 1985F (Figure 8.10). Though bolometric data for Ib supernovae are unfortunately quite scarce, Ensman's study appears to rule out helium cores of greater than 6 $M_\odot$ as Type Ib progenitors. Further, since helium cores lighter than about 4 $M_\odot$ produce very little ^{56}Ni, it seems that the progenitors of Ib supernovae would have to lie in the narrow mass range 4–6 $M_\odot$. Even so, the decline after peak is too slow unless rather severe clumping is allowed in the ejecta in order to facilitate γ-ray escape.

For this reason, as well as the possible existence in nature of more than one kind of Type Ib progenitor, it remains important to consider alternative models. Woosley (1986) suggested that carbon deflagration might provide *both* Type Ia and Ib supernovae. If the flame speed is turned down to about 10% or less that of sound (as in Model DF4, for example), still about 20 times the normal conductive flame speed (Section 8.4.1; Figure 8.2), the observational properties of the supernova become markedly different from a supernova of Type Ia. The light curve becomes fainter and broader (Figure 8.5). Moreover less silicon is produced, which *could* account for the suppression of the Si II feature at peak, and the spectrum calculated at late times (Axelrod and Pinto, unpublished) becomes dominated by the emission of [O I] rather than Fe II and Fe III. In short the model takes on properties very much like that of a Type Ib.

Important issues, such as why the explosions would occur in spiral galaxies, near

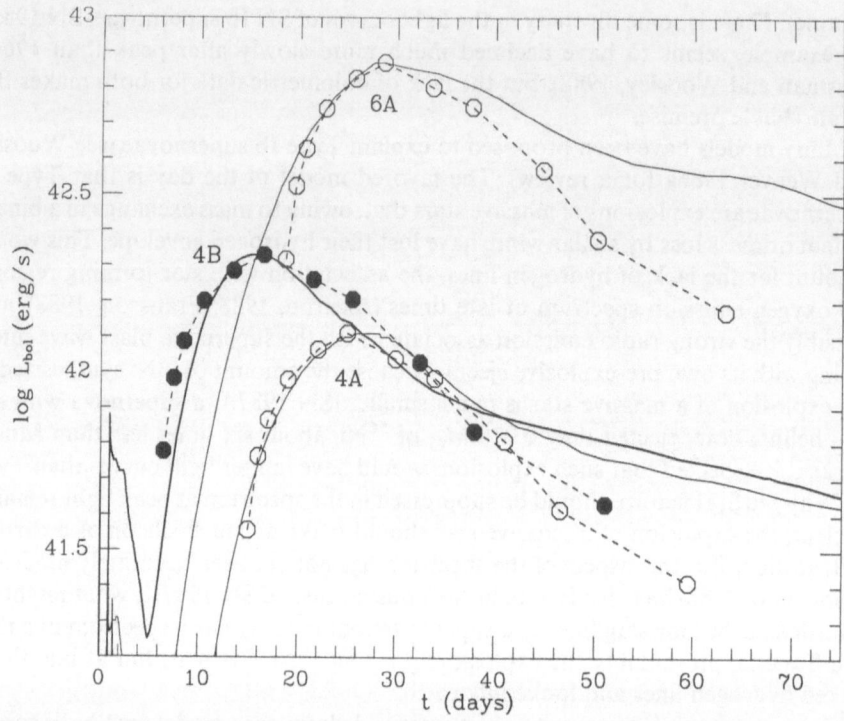

Figure 8.10. The bolometric light curve of Type Ib SN 1983N compared to models calculated by Ensman and Woosley (1988). The model progenitors were massive stellar cores that, when stripped of their hydrogen envelopes, had masses of 4 and 6 $M_\odot$. Models 4A, 4B, and 6A had kinetic energies at infinity equal to 0.5, 3.2, and 1.4 × 10^{51} ergs, respectively. Observational data from Blair and Panagia (1987) has been normalized so as to fit the time and flux at peak emission in the models. After about 40 days the observed luminosity declines more rapidly than any of the models.

H II regions, and exhibit radio emission, are not immediately addressed by this sort of model (though see Branch and Nomoto, 1986), but it is interesting that carbon deflagration is capable of providing a continuum of models whose properties span the space from the brightest Type Ia supernovae observed down to Type Ib supernovae. The essential parameter is the effective flame speed. Ideally, we would like to relate variables upon which the flame speed depends, notably the composition and thermal structure of the white dwarf, to environmental conditions.

Branch and Nomoto (1986) have proposed their own variety of white dwarf model for Type Ib—that they are helium detonation on an accreting carbon–oxygen core. Such a model explains the faintness of the light curve and the lack of silicon at peak. It also gives very high velocity helium lines that have been observed. Unfortunately, it seems very unlikely that this model can give a late-time spectrum dominated by oxygen emission. What little oxygen is ejected comes out at low velocity and is underneath the ^{56}Ni in a relatively small volume of space that would intercept few γ-rays. Axelrod (in Woosley, Axelrod, and Weaver, 1984) found that the late spectrum of helium detonations was dominated by the emission lines of highly ionized iron (Fe V and Fe VI), resembling in no way the observations of SN Ib. Perhaps added detail—clumping, three-dimensional modeling of the detonation and oxygen ejection, better treatment of radiation transport—will alter these results. For now the model has difficulties, but then so do all the others.

8.7. Type I$\frac{1}{2}$ Supernovae

While observationally not a Type I supernova, because its spectrum would contain hydrogen lines, a carbon–oxygen dwarf could equally well explode inside of an asymptotic giant branch star that had failed to lose all of its hydrogen envelope before the core mass reached 1.38 $M_\odot$. The explosion mechanism would be carbon deflagration or detonation, just as we have discussed for Type Ia supernovae, but the light curve as well as spectrum would be greatly modified by the massive but tenuous hydrogen envelope that tamps the explosion. Iben and Renzini (1983) have called such a supernova—Type I by mechanism, Type II by observation—a Type I$\frac{1}{2}$ supernova.

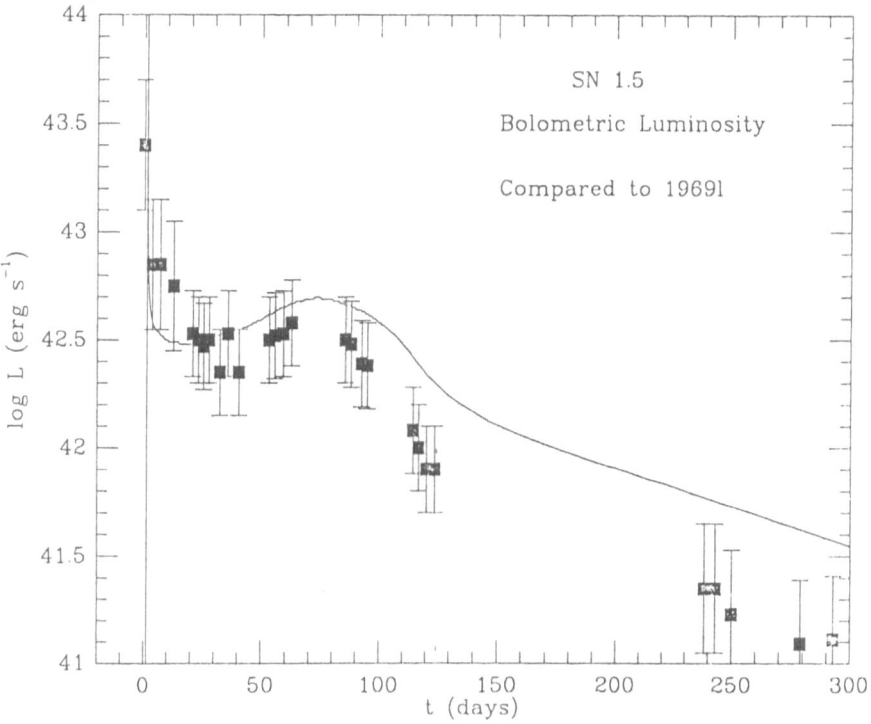

Figure 8.11. The light curve of a Model Type I$\frac{1}{2}$ supernova would be virtually indistinguishable from that of an ordinary Type II save for an exceptionally bright ratio of tail to plateau luminosity. Figure from Woosley and Iben (1988).

Recently, Woosley and Iben (1988) have evaluated the probable light curve for such a model. Iben (1977) had evolved a 7 $M_\odot$ star, without mass loss, to the point where the degenerate carbon–oxygen core had a mass of 1.36 $M_\odot$. The degenerate core was surrounded by a thin, thermally pulsing helium shell that was slowly increasing its mass. The rest of the star's mass was contained in a low-density hydrogen-rich envelope which constituted a red supergiant having luminosity 5.8×10^4 $L_\odot$. The core of this star was replaced by model DF2 (Table 8.1) which was in the process of exploding by carbon deflagration. As a result of the explosion the star and core were totally disrupted with a final kinetic energy of 1.0×10^{51} ergs. Core velocities were greatly tamped by the hydrodynamical interaction with the envelope and ended up moving slower than 2000 km s^{-1} throughout. The hydrogen envelope had higher velocities ranging to above 10,000 km s^{-1} in the outer layers. Energy deposition from the decay of radioactive ^{56}Ni and ^{56}Co was followed as in Section 8.4.5, as was radiation transport and electronic recombination in the envelope. The resultant light curve (Figure 8.11) was virtually indistinguishable from that of an ordinary Type II supernova, SN 1969L, except that the ratio of the bolometric luminosity on the tail to that on the plateau was about three times larger than observations. This reflects the roughly three times greater amount of ^{56}Ni created in Model DF2 compared to that which Weaver and Woosley (1980) calculated for SN 1969L.

Based purely upon the observational data, there is no reason to exclude a large fraction of Type II supernovae from being powered by this sort of mechanism. One key difference, however, is that there would be no neutron star or black hole remnant and no neutrino burst.

8.8. Conclusions

Besides the obvious, that we are still far from a complete understanding of Type I supernovae (be they Ia, Ib, Ip, or what), several general results come from our present deliberations.

First, over all, the carbon deflagration model fares quite well as a standard model for Type Ia supernovae. The same model explains several distinct sets of data—the light curve and the early and late spectrum in particular. Nucleosynthesis presents a problem, but the nature of the deflagration allows a possible solution once the physics of propagation are better modeled. It is interesting that the brightest optical events in the present universe may be a consequence of Rayleigh–Taylor instability. Were the flame front to propagate undeformed through the star, the outcome would be a marginal unbinding of the white dwarf and a very subluminous display. It is only because the flame becomes highly wrinkled, or in the language of this chapter, because the fractal dimension of the surface is considerably greater than 2, that an effective flame speed close to the speed of sound (the white dwarf's own natural expansion speed) becomes practical.

Most past models of carbon deflagration have been calculated using whatever prescription existed in the code for (mixing-length) convection (for an exception see Müller and Arnett, 1986). We find for special circumstances (notably D, the fractal dimension, $= 2.5$) that mixing length convection can be used to represent the actual physics of the flame propagation, so long as the length scale is adjusted to be the radius of the burned-out region. However, there are notable differences in the general case. First, the fractal surface may not be $D = 2.5$. Second, we have found that the rate of propagation will be sensitive to the composition in a complex fashion. Mixing length theory would give no dependence. Surprisingly, the effective flame speed does not necessarily increase with carbon abundance in the same manner as the normal conductive speed (eqs. (8.1) and (8.3)). This is because slowing the conductive speed allows the growth of smaller scale instabilities and thus a more convoluted flame sheet (i.e., D is greater *and* the tile size is smaller). Just which effect predominates is not easily determined. Another initially counter-intuitive result is that the effective flame speed *could* eventually become supersonic without there existing any shock anywhere in the star. This is because v_{eff} measured the rate at which mass is crossing a deformed surface, a surface that may not even be simply connected. *There is no real limit to v_{eff}*, though in practice a value somewhat less than the speed of sound is probably preferred to satisfy observational constraints. In ordinary convection a propagation faster than the sound speed would never have been allowed.

The visualization of the flame as a highly convoluted surface also helps to understand why the spectrum may indicate "mixing" in the ejecta between shells of differing composition (Branch *et al.*, 1985). Actually, what is observed is that elements are spread out in velocity, or, to quote Branch "the most important role of the mixing is to bring cobalt and calcium to higher velocities than in the unmixed model." If burning is ignited at many different radii or if the flame front develops significant protuberances (see also Müller and Arnett, 1986), this is easy to understand without invoking the kind of microscopic compositional mixing in space that is implied by the word "convection." It has never made much sense to invoke "convection" in material that is expanding supersonically.

Surprising to some will be the result that there is no fundamental reason why carbon cannot detonate leading to a complete incineration of the white dwarf (Section 8.5). On the other hand, there is no compelling reason that it has to either. Stringent restrictions on the temperature gradient may not be satisfied. Observationally, the carbon detonation model produces a fine light curve, but the nucleosynthesis and spectrum are not good. Unlike in the deflagration model, there may be no way to cure the excess of ^{54}Fe and ^{58}Ni that characterizes the detonation model.

Also surprising may be the result that carbon deflagration can produce a credible *Type Ib* supernova. The light curve and spectrum, at least the late-time spectrum, seem to be well reproduced. It is a problem though to understand why, if they are white dwarf deflagrations, Type Ib supernovae would be preferentially associated with star forming regions and sometimes radio sources.

Indeed for a credible range of parameters, carbon deflagration produces a continuum of events having luminosities ranging from that of Type Ib to the brightest Type Ia. For the models that are capable of reproducing the shapes of observed Type Ia light curves (Figure 8.5) the extrema of the peak luminosities are in the range $2.5 \pm 1 \times 10^{43}$ ergs s^{-1} (Figures 8.5 and 8.9). This is similar to but slightly brighter on the lower end than the range found by Arnett, Branch, and Wheeler (1985) who used analytic arguments. Let those who will, use this to get the Hubble constant, but take care that the supernova peak is obtained (with at least one pre-maximum point), that the *bolometric* luminosity is determined, and that a spectrum is taken to make sure that the supernova is Type Ia. We note in closing that the luminosity range cited corresponds to a bolometric magnitude range of -20.2 to -19.2 and, since according to van den Bergh (1988) this magnitude should equal $-18.45 \pm 0.23 + 5 \log H_0/(100 \text{ km s}^{-1} \text{ Mpc}^{-1})$, the present work implies a range for H_0 of 40–80 km s^{-1} Mpc^{-1}. Though the error bars are a little bit larger, this is consistent with the value 56 ± 15 km s^{-1} Mpc^{-1} found by Branch (1979). If we recognize that a portion, or even all, of the *observational* spread in measured bolometric peak magnitudes (the 0.23 in Van den Bergh's equation) may be due to intrinsic differences in the supernova luminosity itself, then combining observational and theoretical error bars double counts the same effect. Neglecting the observational spread and keeping only the theoretical one gives $H_0 = 58 \pm 13$ km s^{-1} Mpc^{-1}, virtually identical to Branch's (1979) estimate.

Acknowledgments

Most of the computer calculations presented here were carried out using a code jointly developed by the author and Tom Weaver at Lawrence Livermore National Laboratory. The author is grateful to Tom Weaver not only for use of the code, but for many helpful physical insights into the nature of Type I supernovae. He is also grateful to Lisa Ensman and Phil Pinto for permission to use unpublished information and to Drs. Blinnikov and Khokhlov for helpful correspondence and preprints regarding the manner in which a detonation wave ignites in degenerate carbon. The idea of spontaneous ignition of a detonation wave, developed in Section 8.5.1, was originally suggested to the author in 1985 by Jim Wilson of the Lawrence Livermore National Laboratory. This research has been supported by the National Science Foundation (AST 84-18185; AST 88-13649); the NASA Theory Program (NAGW 1273), and, at Livermore, by the Department of Energy (W-7405-ENG-48).

References

Arnett, W.D. 1969, *Ap. and Spac. Sci.*, **5**, 180.
Arnett, W.D., Branch, D., and Wheeler, J.C. 1985, *Nature*, **314**, 337.
Arnett, W.D., Truran, J.W., and Woosley, S.E. 1971, *Ap. J.*, **165**, 87.
Axelrod, T.S. 1988, in *Atmospheric Diagnostics of Stellar Evolution: Chemical Peculiarity, Mass Loss, and Explosion*, ed. K. Nomoto (Berlin: Springer-Verlag), p. 375.
Barbon, R., Ciatti, F., and Rosino, L. 1973, *Astron. and Ap.*, **25**, 241.
Barrat, J.L., Hansen, J.P., and Mochkovitch, R. 1988, *Astron. and Ap.*, **199**, L15.
Bertola, F. 1964, *Ann. d'Ap.*, **27**, 319.
Bertola, F., Mammano, A., and Perinotto, M. 1965, in *Contr. Asiago Obs.*, No. 174, p. 51.
Blair, W.P. and Panagia, N. 1987 in *Exploring the Universe with the IUE Satellite*, eds. Y. Kondo, W. Wamstecker, A. Boggess, M. Grewing, C. DeJager, A.L. Lane, J.L. Linsky, and R. Wilson (Dordrecht: Reidel), p. 549.
Blinnikov, S.I. and Khokhlov, A.M. 1986, *Pis'ma Astron. Zh.*, **12**, 318; *Sov. Astron. Lettr.*, **12**, 131.
Blinnikov, S.I. and Khokhlov, A.M. 1987, *Pis'ma Astron. Zh.*, **13**, 867; *Sov. Astron. Lettr.*, **13**, 364.
Branch, D. 1979, *M.N.R.A.S.*, **186**, 609.
Branch, D. 1987, *Ap. J. (Letters)*, **316**, L81.

Branch, D., Doggett, J.B., Nomoto, K., and Thielemenn, F.-K. 1985, *Ap. J.*, **294**, 619.

Branch, D., Lacy, C.H., MacCall, M.L., Sutherland, P.G., Uomoto, A., Wheeler, J.C., and Wills, B.J. 1983, *Ap. J.*, **270**, 123.

Branch, D. and Nomoto, K. 1986, *Astron. and Ap.*, **164**, L13.

Buchler, J.R., Colgate, S.A., and Mazurek, T.J. 1979, *J. Phys.*, **C2**, 159.

Canal, R., Isern, J., and Lopez, R. 1988, preprint submitted to *Ap. J.* (*Letters*).

Chevalier, R.A. 1986, in *Highlights Astr.*, Vol. 7, ed. J.P. Swings (Dordrecht: Reidel), p. 599.

Clayton, D.D. 1968, *Principles of Stellar Evolution and Nucleosynthesis* (New York: McGraw-Hill), p. 257.

DeRobertis, M.M. and Pinto, P.A. 1983, *Ap. J.* (*Letters*), **293**, L77.

Elias, J.H., Mathews, K., Neugebauer, G., and Persson, S.E. 1985, *Ap. J.*, **296**, 379.

Ensman, L.M. and Woosley, S.E. 1988, *Ap. J.*, **333**, 754.

Filippenko, A.V. 1988, *Astron. J.*, **96**, 1941.

Filippenko, A.V. and Sargent, W. 1985, *Nature*, **316**, 407.

Fransson, C. 1988, in *Atmospheric Diagnostics of Stellar Evolution: Chemical Peculiarity, Mass Loss, and Explosion*, ed. K. Nomoto (Berlin: Springer-Verlag), p. 383.

Frogel, J.A., Brook, G., Kawara, K., Laney. D., Phillips, M.M., Terndrup, D., Vrba, F., and Whitford, A.E. 1987, *Ap. J.* (*Letters*), **315**, L129.

Fuller, G.M., Fowler, W.A., and Newman, M.J. 1982a, *Ap. J. Suppl.*, **48**, 279.

Fuller, G.M., Fowler, W.A., and Newman, M.J. 1982b, *Ap. J.*, **252**, 715.

Fuller, G.M., Fowler, W.A., and Newman, M.J. 1985, *Ap. J.*, **293**, 1.

Gaskell, C.M., Cappellaro, E., Dinerstein, H.L., Garnett, D., Harkness, R., and Wheeler, J.C. 1986, *Ap. J.* (*Letters*), **306**, L77.

Grahm, J.R. 1986, *M.N.R.A.S.*, **220**, 27P.

Grahm, J.R. 1987a, *Ap. J.* (*Letters*), **318**, L47.

Grahm, J.R. 1987b, *Ap. J.*, **315**, 588.

Hamilton, A.J.S., Sarazin, C.L., and Szymkowiak, A.E. 1986a, *Ap. J.*, **300**, 698.

Hamilton, A.J.S., Sarazin, C.L., and Szymkowiak, A.E. 1986b, *Ap. J.*, **300**, 713.

Hansen, C.J. and Wheeler, J.C. 1969, *Ap. and Spac. Sci.*, **3**, 464.

Hillebrandt, W., Nomoto, K., and Wolff, R.G. 1984, *Astron. and Ap.*, **133**, 175.

Hoyle, F. and Fowler, W.A. 1960, *Ap. J.*, **132**, 565.

Iben, I. Jr. 1977, *Ap. J.*, **217**, 788.

Iben, I. Jr. 1978a, *Ap. J.*, **219**, 213.

Iben, I. Jr. 1978b, *Ap. J.*, **226**, 996.

Iben, I. Jr. 1982, *Ap. J.*, **253**, 248.

Iben, I. Jr. 1988, *Ap. J.*, **324**, 355.

Iben, I. Jr. and Renzini, A. 1983, *Ann. Rev. Astron. and Ap.*, **21**, 271.

Iben, I. Jr. and Tutukov, A.V. 1984, *Ap. J. Suppl.*, **54**, 535.

Ichimaru, S., Iyetomi, H., and Ogata, S. 1988, *Ap. J.*, **334**, L17.

Krishner, R.P., Oke, J.B., Penston, M.V., and Searle, L. 1973, *Ap. J.*, **185**, 303.

Landau, L.D. and Lifshitz, E.M. 1987, *Fluid Mechanics*, 2nd ed. (Oxford: Pergamon Press), pp. 484ff.

Lopez, R., Isern, J., Canal, R., and Labay, J. 1986a, *Astron. and Ap.*, **155**, 1.

Lopez, R., Isern, J., Labay, J., and Canal, R. 1986b, *Rev. Mex. Astron.*, **13**, 41.

Mandelbrot, B.B. 1983, *The Fractal Geometry of Nature* (New York: W.H. Freeman).

Mayle, R. and Wilson, J.R. 1988, *Ap. J.*, **334**, 909.

Maza, J. and van den Bergh, S. 1976, *Ap. J.*, **204**, 519.

Miyaji, S., Nomoto, K., Yokoi, K., and Sugimoto, D. 1980, *Publ. Astron. Soc. Japan*, **32**, 303.

Mochkovitch, R. 1983, *Astron. and Ap.*, **122**, 212.

Müller, E. and Arnett, W.D. 1986, *Ap. J.*, **307**, 619.

Nomoto, K. 1982, *Ap. J.*, **257**, 780.

Nomoto, K. and Hashimoto, M. 1987, *Ap. and Spac. Sci.*, **131**, 395.

Nomoto, K. and Iben, I. Jr. 1985, *Ap. J.*, **297**, 531.

Nomoto, K. and Sugimoto, D. 1977, *Publ. Astron. Soc. Japan*, **29**, 165.

Nomoto, K., Sugimoto, D., and Neo, S. 1979, *Ap. and Spac. Sci.*, **39**, L37.

Nomoto, K., Thielemann, F.K., and Yokoi, K. 1984, *Ap. J.*, **286**, 644.

Paczynski, B. 1985 in *Cataclysmic Variables and Low Mass X-Ray Binaries*, eds. D.Q. Lamb and J. Patterson (Dordrecht: Reidel), p. 1.

Panagia, N., Sramek, R.A., and Weiler, K.W. 1986, *Ap. J. (Letters)*, **300**, L55.

Porter, A. and Filippenko, A. 1987, *Astron. J.*, **93**, 1372.

Phillips, M.M., *et al.* 1987, *Pub. Astron. Soc. Pacific*, **99**, 592.

Saio, H. and Nomoto, K. 1985, *Astron. and Ap.*, **150**, L21.

Sramek, R.A., Panagia, N., and Weiler, K.W. 1984, *Ap. J. (Letters)*, **285**, L59.

Thielemann, F.-K., Nomoto, K., and Yokoi, K. 1986, *Astron. and Ap.*, **158**, 17.

Truran, J.W. and Cameron, A.G.W. 1971, *Ap. and Spac. Sci.*, **14**, 179.

Tutukov, A.V. and Yungelson, L.R. 1979, *Acta Astr.*, **23**, 665.

Van den Bergh, S. 1988, preprint of paper presented at *Symposium on the Extragalactic Distance Scale*, to be published in Vol. 2 of the Astron. Soc. Pacific Symposium Series.

Weaver, T.A. and Woosley, S.E. 1980, *Ann N.Y. Acad. Sci.*, **336**, 335, Figure 10.

Webbink, R.F. 1979 in *White Dwarfs and Variable Degenerate Stars*, eds. H.M. Van Horn and V. Weidemann (New York: University of Rochester Press), p. 426.

Wheeler, J.C., Harkness, R.P., and Cappellaro, E. 1987, in *Proceedings of the 13th Texas Symposium on Relativistic Astrophysics*, ed. M.P. Ulmer (Singapore: World Scientific), p. 402.

Wheeler, J.C. and Levreault, R. 1985, *Ap. J. (Letters)*, **294**, L17.

Whelan, J.C. and Iben, I. Jr. 1971, *Ap. J.*, **186**, 1007.

Woosley, S.E. 1986, in *Nucleosynthesis and Chemical Evolution*, 16th Advanced Course, Swiss Society of Astrophysics and Astronomy, eds. B. Hauck, A. Maeder, and G. Meynet (Geneva: Geneva Observatory), p. 1.

Woosley, S.E., Axelrod, T.S., and Weaver, T.A. 1984, in *Stellar Nucleosynthesis*, eds. C. Chiosi and A. Renzini (Dordrecht: Reidel), p. 263.

Woosley, S.E. and Hoffman, R. 1990, *Ap. J. Suppl.*, in preparation.

Woosley, S.E. and Iben, I. Jr. 1988, *Bull. Amer. Astron. Soc.*, **19**, 1036.

Woosley, S.E. and Weaver, T.A. 1986a, *Ann. Rev. Astron. and Ap.*, **24**, 205.

Woosley, S.E. and Weaver, T.A. 1986b, in *Nucleosynthesis and Its Implication for Particle Physics*, eds. J. Audouze and T. van Thuan (Dordrecht: Reidel), p. 145.

Woosley, S.E. and Weaver, T.A. 1986c, in *Radiation Hydrodynamics in Stars and Compact Objects*, Proc. IAU Symposium 89, eds. D. Mihalas and K.-H. Winkler (Berlin: Springer-Verlag), p. 91.

Woosley, S.E. and Weaver, T.A. 1990, *Ap. J.*, in preparation.

Woosley, S.E., Taam, R.E., and Weaver, T.A. 1986, *Ap. J.*, **301**, 601.

Woosley, S.E., Weaver, T.A., and Taam, R.E. 1980, in *Type I Supernovae*, ed. J.C. Wheeler (Austin: University of Texas), p. 96.

Younger, P.F. and Van den Bergh, S. 1985, *Astron. Ap. Suppl.*, **61**, 365.

Zeldovich, Ya. B., Barenblatt, G.I., Librovich, V.B., and Makhviladze, G.M. 1985, *The Mathematical Theory of Combustions and Explosions* (New York: Consultant's Bureau, a division of Plenum Publishing Corporation), p. 264ff.

9. Supernovae: The Direct Mechanism and the Equation of State

JERRY COOPERSTEIN[1] and EDWARD A. BARON

9.1. Introduction

For the last 25 years or so there have been two basic theoretical models for Type II supernovae. Both have worn several costumes and have been defrocked repeatedly, yet both remain. The first model is called the "direct" or "prompt" mechanism, and is the simpler of the two. Here stellar collapse is followed by a hydrodynamical rebound of the inner core, unleashed when the stiffness of nuclear matter puts a halt to further implosion. The bounce of the core drives a shock wave which proceeds rapidly outwards and eventually ejects the overlying layers of the star, producing the visible supernova explosion. The direct method has been the focus of our work and is the subject of this chapter.

The problem that has frustrated investigators time and time again is that when the best presupernova conditions are combined with the best physics and numerical techniques in the numerical simulations, the shock wave may be not strong enough to do its work. It falters, enters an accretion mode, and cannot produce a supernova explosion unless a new agency enters the picture. We will show why shocks fail and indicate how they may be made to succeed.

At any rate, if the shock wave does fail, either all or part of the time, there is still hope that the second mode of producing supernovae, the so-called "delayed" mechanism, will save the day. There have been several distinct varieties over the years, but in all pictures neutrinos play a crucial role. In the most recent incarnation, that invented by Jim Wilson and explored by him together with Ron Mayle (Bethe and Wilson, 1985; Mayle, 1985; Wilson, 1985), it is the absorption of a small fraction of the departing neutrinos on nucleons which heats the matter and breathes new life into the shock. This chapter will not attempt to evaluate this scenario. It is the subject of Ron Mayle's chapter in this volume.

Observed supernova energies are of the order of a foe (1 foe = 10^{51} ergs). A vigorous explosion is one with more than 1 foe, up to perhaps 3 foe. Some supernovae, such as the one which created the Crab Nebula, have inferred energies of only ~ 0.1 foe. Kinetic energy dominates the total integrated light curve, which generally is less than <0.1 foe. Yet the total energy available is the binding energy of the produced neutron star, 100–300 foes. This energy escapes as neutrinos.

The process is driven by an extremely precarious balance of opposing forces. The gravitational energy and internal energy of a collapsing star are almost equal in

---

[1] Present address: Department of Physics, Randall Laboratory, University of Michigan, Ann Arbor, MI 48109-1120, U.S.A.

magnitude and opposite in sign, and are each individually about one hundred times the net energy released in the early part of the explosion. These numbers mean that we are forced to calculate an inefficient process in which at most only a few percent of the energy released couples to the expanding shell. They also warn us that small errors (as compared to the 100-foe scale) can produce large mistakes, and that an inefficient neutrino process could produce a supernova.

The thrust of our work has been the numerical simulation of Type II supernovae. A historical review of more than two decades of simulations would present a roller coaster ride through successes and failures. Promising scenarios have been dashed on the rocks of new observations either of supernovae or of fundamental processes which play prominent roles in the phenomenon, such as the weak interaction physics of neutrinos, or the strong interaction physics of the equation of state. Unfortunately, there has also been much time spent pursuing the false trails left by computer errors, although these have often produced extremely useful pedagogical lessons.

Supernova calculations are difficult because of the inherent complexity of the problem; after all, we must deal with the full and dirty system that nature has presented us because we cannot control the explosions experimentally. It is easy to get lost in the details of only one individual piece of the picture; the art of supernova engineering lies in the coupling of the various ingredients.

However, nature solves the problem of supernovae, and surely astrophysicists are making progress in deciphering the clues. Just as surely, the fortuitous appearance of Sn 1987A has given us a good boost, both because of the richness of the new data, and the heightened sense of importance and urgency it has brought to our research.

Several succeeding stages must be considered. These are: the supernova progenitor's evolution up to dynamic instability; the catastrophic gravitational implosion to supranuclear density; the bounce of the innermost stellar core and formation of the shock wave at the sonic point; the blast's progression outwards to the surface; and the birth of the neutron star. Theoretical knowledge of each of these phases retains considerable uncertainties despite intensive study by many groups. However, we believe that the engineering is sufficiently mature to map out the propagation of these uncertainties through succeeding stages.

The organization of this chapter is as follows. First, in Section 9.2, we discuss the collapse of the star. In Section 9.2.1 we examine the inner core of the progenitor star at the completion of silicon burning at the center. In Section 9.2.2 we analyze the instability which leads to catastrophic gravitational collapse by seeing how the adiabatic index, γ, drops below the critical value for stability, $\gamma = \frac{4}{3}$. In Section 9.2.3 we use the theory of self-similar collapse as developed by Yahil (1983) to show how the star separates into an inner and outer core, with a homologous flow in the central subsonic region.

The next two sections concern the equation of state of dense supernova matter, an essential and decisive ingredient in the theory. Section 9.3 is about material below nuclear saturation density, and Section 9.4 is about material above saturation density. Both regimes are to be found in supernovae. Several years ago the tendency was to downplay the influence of the high-density equation of state (as long as nuclear matter was quite stiff, we did not think it mattered precisely how stiff). Today, there has developed the opposite tendency to emphasize only the high-density equation of state—a case of things getting out of balance. We will argue, in fact, that it is important to put these two disparate regimes on an equal footing and to use the same physics for each.

In Section 9.5 we give results of numerical simulations. Section 9.5.1 describes the infall phase. Section 9.5.2 considers the dynamics of the bounce phase of the explosion, explains the importance of general relativity, and looks at how the equation of state controls the efficiency of energy transfer to the shock wave. Section 9.5.3 discusses the further propagation of the shock wave through the iron core, the effects of neutrino losses upon it, and whether it succeeds or fails.

Section 9.6 contains a brief discussion of the possible effects of rotation on the supernova, and Section 9.7 gives our final discussion.

9.2. Gravitational Collapse

The infall of the star is the best understood part of the supernova problem. While there remain important uncertainties in the input physics, the collapse calculations are mature and the various simulators are in essential agreement. Furthermore, the analytical theories we will relate work quite well. There is still room for improvement in the equation of state, the calculation of electron capture, and the testing of new interactions or processes; but the simulations are mature enough that gauging the consequences of variations in the input physics is generally a straightforward experiment. The largest problem is the question of the initial model, with which we will begin.

9.2.1. The Presupernova Star

A blue supergiant gave birth to SN 1987A, rather than the canonical red supergiant star visualized as the main site of Type II supernovae. While there were pre-existing calculations indicating that giant stars could go unstable while shining blue (Brunish and Truran, 1982), the massive stars that evolved with the most sophistication were all red. Fortunately (for the purposes of this chapter), the outer structure has very little to do with the viability of the direct mechanism, which is solely the responsibility of the inner core of the star. The outer layers do affect the observable signature, but not the basic mechanism, except in how the presupernova evolution helps determine the precise size and profile of the inner core of the star.

The onionshell structure of this central region of the progenitor star is schematically represented in Figure 9.1. The centermost "iron" core of elements in nuclear statistical equilibrium (NSE) lies underneath successive burning shells of silicon, oxygen–neon, carbon, helium, and hydrogen. Calculation of the presupernova structure of massive stars has been done by two groups, one in California and one in Japan (see Woosley and Weaver, 1986, 1988; and Nomoto and Hashimoto, 1988, for reviews). Many of these calculations follow only the helium core and the total mass of hydrogen over this is inferred. The initial composition must be assumed in all cases. In particular, it has been indicated that a low metallicity, mass

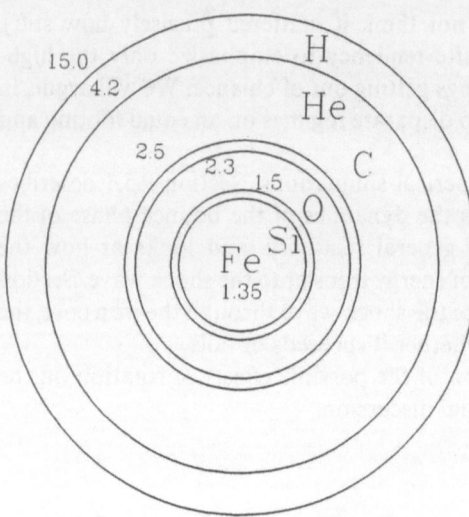

Figure 9.1. Supernova progenitor structure. Elemental shells of the presupernova 15 $M_\odot$ star of Woosley and Weaver (1984) schematically represented. Next to each shell's dominant element is listed the enclosed mass of the shell in $M_\odot$, so the mass in the shell is obtained by subtraction of the enclosed mass of the next inner shell. Except for the outermost hydrogen shell, radii of the shells are proportional to the mass enclosed.

loss, and various assumptions about mixing can lead to a blue supernova such as 1987A (Brunish and Truran, 1982; Hillebrandt *et al.*, 1987; Woosley, Pinto, and Ensman, 1988; Saio Kato and Nomoto, 1988; Saio Nomoto and Kato, 1988; Truran and Weiss, 1988). The details of the ensuing stellar evolution and thus the structure of the star sketched in Figure 9.1 depend both on the initial conditions assumed and on the physics used to evolve the stars. Many difficult questions are involved, such as mass loss, convection, semiconvection, nuclear reaction rates, etc. We cannot adequately discuss these matters as they are enormous topics in themselves. However, we will comment upon them in Section 9.7.

In what follows we are most concerned with the innermost iron core. The critical parameters which the presupernova evolution supplies are the total mass of the iron core, its thermal structure, and its composition (particularly Y_e, the number of protons per nucleon.) The outer shells are important for nucleosynthesis and observational consequences such as the total energy of the explosion and its light curve, but do not control whether or not the explosion occurs or what its early characteristics are.

In Figure 9.2 we show the structure of three different recent initial model calculations. Solid lines give the results for the Nomoto and Hashimoto (1988) 13 $M_\odot$ (3.3 $M_\odot$ helium core) star, short-dashed lines the Woosley and Weaver (1988) 15 $M_\odot$ star, and the long-dashed lines the Weaver and Woosley (1988) 18 $M_\odot$ star. The effects of the initial conditions depicted here will be discussed in Section 9.7.

Why is the mass of the iron core so important? As we shall see in Section 9.5.2, the shock wave does not begin to heat the matter at the center of the star, but is launched from about 10 km, at a Lagrangean mass point we denote as M_{form}, which is typically found to satisfy $0.5\ M_\odot < M_{form} < 0.9\ M_\odot$. It must make its way out to the edge of the iron core, which we denote as M_{core}, and which presupernova simulations find satisfies $1.15\ M_\odot < M_{core} < 2.0\ M_\odot$. (Results for numerical calcula-

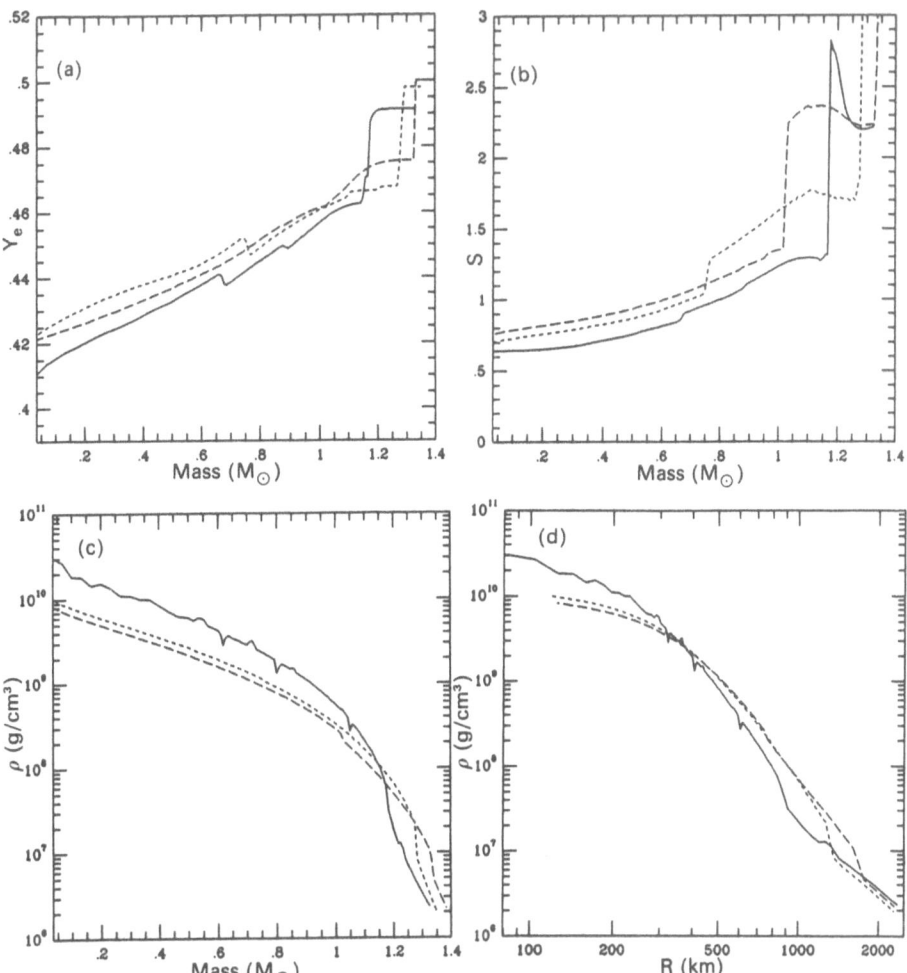

Figure 9.2. Initial models. A comparison of the structure of three initial models. The solid line refers to the 13 $M_\odot$ (3.3 $M_\odot$ helium core) model of Nomoto and Hashimoto (1988), the short-dashed line is the 15 $M_\odot$ model of Woosley and Weaver (1988), and the long-dashed line is the 18 $M_\odot$ model of Weaver an Woosley (1988). In (a) we display the profile of the electron fraction Y_e; in (b) the profile of the entropy S in units of k_b per nucleon; and we display the density as a function of mass in (c) and of radius in (d). Note the initial central density differs for the different models.

tions are given in Section 9.5.) Most of the energy of the shock wave is spent on smashing complex nuclei such as iron into their nucleonic constituents. These are bound by about 10 MeV per nucleon (10^{19} ergs g^{-1}, or 20 foes per $M_\odot$.)

It is difficult to calculate the initial shock energy. There are many definitions and it is difficult to ascertain the useful one. It is less ambiguous to analyze the directly calculated quantities in the computer outputs, such as the density, velocity, entropy, and composition. However, if the shock begins with

$$E_{\text{shock, initial}} > \frac{M_{\text{core}} - M_{\text{form}}}{M_\odot} \times 20 \text{ foes}, \tag{9.1}$$

success would seem indicated (Burrows and Lattimer, 1983; Cooperstein and Brown, 1985). This equation, while useful, is imprecise because the shock's movement is influenced strongly by the density profile of the initial model and other initial characteristics. Failed shock waves look pretty uniform, but small adjustments in the input conditions and physics can lead to large changes once the threshold to success has been crossed.

Equation (9.1) indicates that if we could reduce the iron core mass, M_{core}, by just $0.1\ M_\odot$, we would reduce the energy expenditure of the shock by 2 foes, an amount equal to the total final energy of a vigorous direct explosion. Therefore, anything which reduces M_{core} in the presupernova evolution is warmly welcomed by the direct mechanism. Fortunately, such a decrease has been obtained by the theoretical simulations. This reduction has come from a number of places, especially inclusion of Coulomb lattice effects on the equation of state (Nomoto and Hashimoto, 1988; Woosley and Weaver, 1988), inclusion of changed nuclear reactions and electron capture rates (particularly the $^{12}C(\alpha, \gamma)^{16}O$ rate, Weaver and Woosley, 1988), and fiddling with the process of semiconvection (Woosley and Weaver, 1988).

It turns out that successful shocks need dissociate less matter than indicated in eq. (9.1). They come to lower temperatures more quickly because they expand faster, and thus need not dissociate the entire iron core (Cooperstein, Bethe, and Brown, 1984).

Besides its mass and density profile, the other critical parameters of the iron core are its entropy, S, which we will always take in units of k_B per nucleon, and Y_e, the number of protons per nucleon (Bethe et al., 1979). These are shown in Figure 9.2. Higher initial S and Y_e each increase M_{core}. Furthermore, a high initial S will lead to a reduction in M_{form} due to additional electron captures during the infall. A high initial Y_e also tends to have the same effect, because the entropy released by the early captures bootstraps through captures by protons boiled off of nuclei by the heating. (Normally we expect escaping neutrinos to cool the matter as the skin is cooled by perspiration, but when Y_e is high the nucleus tends to be left in an excited state because the protons are less bound than the neutrons (Cooperstein and Wambach, 1984).) Thus the progress of the presupernova evolution towards lower entropy has been quite helpful, with S dropping from $S \sim 1$ to $S \sim 0.7$ in the most recent calculations.

9.2.2. Why the Star Goes Unstable

The unstable stars that are the end product of the presupernova evolution already have material falling in at a velocity of about 1000 km s^{-1} and central densities of about ten billion grams per cubic centimeter. How do the instabilities which precipitate this collapse arise and how will they further develop and eventually be halted?

The iron core is incapable of generating stellar energy as it is the last ashes of nuclear burning. Neutrinos are already carrying off much more energy than photons. Because they leave the central region and undergo no important further interaction with the overlying material, the inner and outer cores are dynamically decoupled. As with the conclusion of all previous burning stages the star contracts raising the temperature, while searching for the ignition point of a new fuel, but this time the search is in vain.

The iron core of the progenitor perched on the brink of collapse approximates a hot relativistic white dwarf, its pressure dominated by $\gamma = \frac{4}{3}$ electrons.[2] A relativistic star admits no length scale; the gravitational and internal energies vary inversely with scale size under a homologous transformation and have opposite sign. Thus their sum, the total energy, is independent of scale, and furthermore, it identically vanishes. The entire star can communicate, regardless of length scale, to preserve the homologous structure. There being no length scale, the star remains in its equilibrium configuration upon expansion or contraction.

Real objects such as the iron core do not exhibit perfect $\gamma = \frac{4}{3}$ behavior in their equations of state. For instance, in the white dwarf case the less relativistic outer layers provide stability. In our case we shall see that the nuclear influence on the equation of state can upset the balance, because while the nuclear contribution to the pressure is small, the relativistic electron gas essentially cancels itself out as far as stability is concerned. In addition, weak interaction processes can and do drive instabilities because they are not generally in equilibrium in supernova material, except at the highest densities. The escape of neutrinos from the star and their diffusion within the star are nonadiabatic processes, and the coming to equilibrium of the weak interactions increases the entropy. Because the material is so close to the relativistic $\gamma = \frac{4}{3}$ situation, slight pushes can lead to runaway behavior as soon as $\gamma < \frac{4}{3}$. We will see that the difference, $\gamma - \frac{4}{3}$, is the most important quantity during the gravitational collapse.

Two processes work hand in hand to propel collapse; photodisintegration and electron capture. A third influence is the negative Coulomb pressure. We will consider each in terms of how they affect γ.

Now there exist various kinds of γ's that enter into supernova collapse, the most basic being the adiabatic one,

$$\Gamma_S \equiv \frac{d \ln P}{d \ln \rho}\bigg|_{S, Y_e}, \tag{9.2}$$

the derivative taken with S and Y_e held constant. (If neutrinos are present, and in equilibrium with the matter, $Y_1 = Y_e + Y_{ve}$ should be held constant.) This Γ_S is purely a function of the equation of state. For the relativistic gas $\Gamma_S = \frac{4}{3}$.

A second critical adiabatic index is that calculated along the collapse trajectory of a given Lagrangean mass element, i.e.,

$$\Gamma_M = \frac{d \ln P}{d \ln \rho}\bigg|_M = \Gamma_S + \frac{\partial \ln P}{\partial S}\bigg|_{\rho, Y_e} \frac{\delta S}{\delta \ln \rho}\bigg|_M + \frac{\partial \ln P}{\partial Y_e}\bigg|_{\rho, S} \frac{\delta Y_e}{\delta \ln \rho}\bigg|_M \tag{9.3}$$

giving the value of Γ_M over a small change $\delta \ln \rho$.

---

[2] We will use a number of different adiabatic indices, γ. When we use the symbol "γ" we are speaking quite loosely, and will be more specific in our notation when matters require it.

A third critical adiabatic index is that which gives the structure at a given time, i.e.,

$$\Gamma_R = \frac{d\ln P}{d\ln\rho}\bigg|_R = \Gamma_S + \frac{\partial\ln P}{\partial S}\bigg|_{\rho,Y_e}\frac{\partial S}{\partial\ln\rho}\bigg|_R + \frac{\partial\ln P}{\partial Y_e}\bigg|_{\rho,S}\frac{\partial Y_e}{\partial\ln\rho}\bigg|_R \qquad (9.4)$$

which includes the effects of radial gradients in entropy and composition. Γ_M and Γ_R are related in the same sense as Lagrangean and Eulerian coordinates. All three indices, Γ_S, Γ_M, and Γ_R, can drop below $\gamma = \frac{4}{3}$ during supernova collapse.

Photodisintegration involves nuclear statistical equilibrium reactions of the type

$$(A, Z) + \bar{\gamma} \rightarrow \left(\frac{Z}{2}\right)\alpha + (A - 2Z)n \qquad (9.5)$$

or in the case of iron,

$$Fe^{56} \rightarrow 13\alpha + 4n. \qquad (9.6)$$

These reactions take place as the center contracts adiabatically attempting to ignite a new fuel, raising the temperature with only a small increase in pressure. Such reactions are endothermic, costing ~ 124 MeV, for example, for the breakup of iron into α-particles and neutrons. This energy must come from somewhere, and where it comes from is the electrons. In this process 56 nucleons marching in step go on 17 different paths and thus the nucleonic entropy must increase. Since the total entropy cannot change in an adiabatic compression, entropy must be transferred from the relativistic electrons to the nucleons. In this process no electrons are destroyed, but thermal energy is transferred to the nucleons. Now, at fixed Y_e,

$$\Gamma_S = \frac{P_e}{P_e + P_N}\left[\Gamma_e + \frac{\partial\ln P_e}{\partial S_e}\frac{\partial S_e}{\partial\ln\rho}\bigg|_S\right] + \frac{P_N}{P_e + P_N}\left[\Gamma_N + \frac{\partial\ln P_N}{\partial S_N}\frac{\partial S_N}{\partial\ln\rho}\bigg|_S\right], \qquad (9.7)$$

where

$$\Gamma_e = \frac{\partial\ln P_e}{\partial\ln\rho}\bigg|_{S_e} = \frac{4}{3}; \qquad \Gamma_N = \frac{\partial\ln P_N}{\partial\ln\rho}\bigg|_{S_N}. \qquad (9.8)$$

The pressure per nucleon in relativistic electrons is given by

$$\frac{P_e}{\rho} \cong \frac{Y_e\varepsilon_F}{4}\left(1 + \frac{2}{3}\left(\frac{S_e}{\pi Y_e}\right)^2\right). \qquad (9.9)$$

where the electron Fermi energy and the entropy per nucleon in electrons are given by

$$\varepsilon_F = (3\pi^2 Y_e\rho)^{1/3}; \qquad S_e = \frac{Y_e\pi^2 T}{\varepsilon_F}. \qquad (9.10)$$

Putting in numbers we find

$$\frac{P_e}{\rho} = 0.87\left(\frac{Y_e}{0.42}\right)^{4/3}\rho_{10}^{1/3}\left(1 + \frac{2}{3}\left(\frac{S_e}{\pi Y_e}\right)^2\right)\text{MeV}, \qquad (9.11)$$

$$S_e = 0.50\rho_{10}^{-1/3}\left(\frac{Y_e}{0.42}\right)^{2/3}T. \qquad (9.12)$$

The pressure due to nucleons has two components and is given by

$$\frac{P_N}{\rho} = \sum_{A,Z} \frac{\phi_{A,Z}}{A} T + \frac{P_{coul}}{\rho} = YT + \frac{P_{coul}}{\rho}, \tag{9.13}$$

where $\phi_{A,Z}$ is the mass fraction of the (A, Z) nucleus normalized so that $1 = \sum \phi_{A,Z}$, Y is the number of independent particles per nucleon (the reciprocal of the mean atomic weight), and P_{coul} is the Coulomb lattice pressure and is negative. (We will discuss this later in Section 9.3.2.) It is given roughly by

$$\frac{P_{coul}}{\rho} \cong -0.036 \rho_{10}^{1/3} \tag{9.14}$$

(for $Y_e \sim 0.42$ and nucleons clustered mostly in one large nuclear species). Thus,

$$\frac{P_{coul}}{P_e} \simeq -\frac{0.036}{0.86} \simeq -0.04. \tag{9.15}$$

Note P_{coul} and P_e have the same density dependence, or $\gamma = \frac{4}{3}$, and so the Coulomb pressure does not decrease the adiabatic index although it does decrease the pressure and hence the effective Chandrasekhar mass, which varies as $P^{3/2}$. The nuclear thermal pressure, YT, depends on Y. For example, consider a mixture of 10% free nucleons and iron. Then

$$Y = 0.1 \cdot \left(\frac{1}{1}\right) + 0.9 \cdot \left(\frac{1}{56}\right) = 0.116, \tag{9.16}$$

and for pure iron, $Y = 1/56 = 0.018$. The two cases would give

$$\frac{YT}{P_e/\rho} = \frac{YT}{Y_e \varepsilon_F/4} \cong \frac{YT}{0.87} \sim (0.02 - 0.13)T. \tag{9.17}$$

Since $T < 1$ MeV for the early stages of collapse, the 2% correction is more appropriate. Figure 9.3(a) shows the quantity Y along relevant adiabats from $S = 0.75$ to $S = 2.5$. The ratio of the nucleonic to the electron pressure is shown in Figure 9.3(b) and we see that the nuclear contribution to the pressure is much smaller than the electron one, and can even be negative if all the nucleons are in large nuclei. The electron pressure dominates. The adiabatic index Γ_S and the temperature are given in Figure 9.3(c) and (d).

Thus we can neglect the second term in eq. (9.7), the one due to nucleons, in estimating the drop in Γ_S due to photodisintegration and we find

$$\Gamma_S \simeq \frac{4}{3} \frac{1}{1 + P_N/P_e} \left[1 + \frac{(S_e/\pi Y_e)^2}{1 + \frac{2}{3}(S_e/\pi Y_e)^2} \frac{1}{S_e} \frac{\partial S_e}{\partial \ln \rho} \right]. \tag{9.18}$$

Looking at Figure 9.4 we see that the electronic share of the entropy decreases with increasing density and thus the second term is indeed negative and Γ_S is reduced as is seen if Figure 9.3(c). Photodisintegration, which actually *cools* the electrons as matter heats up, promotes instability. Note that if we simply mix a $\gamma = \frac{5}{3}$ component into a predominantly $\gamma = \frac{4}{3}$ gas without dissociation reactions things would go the

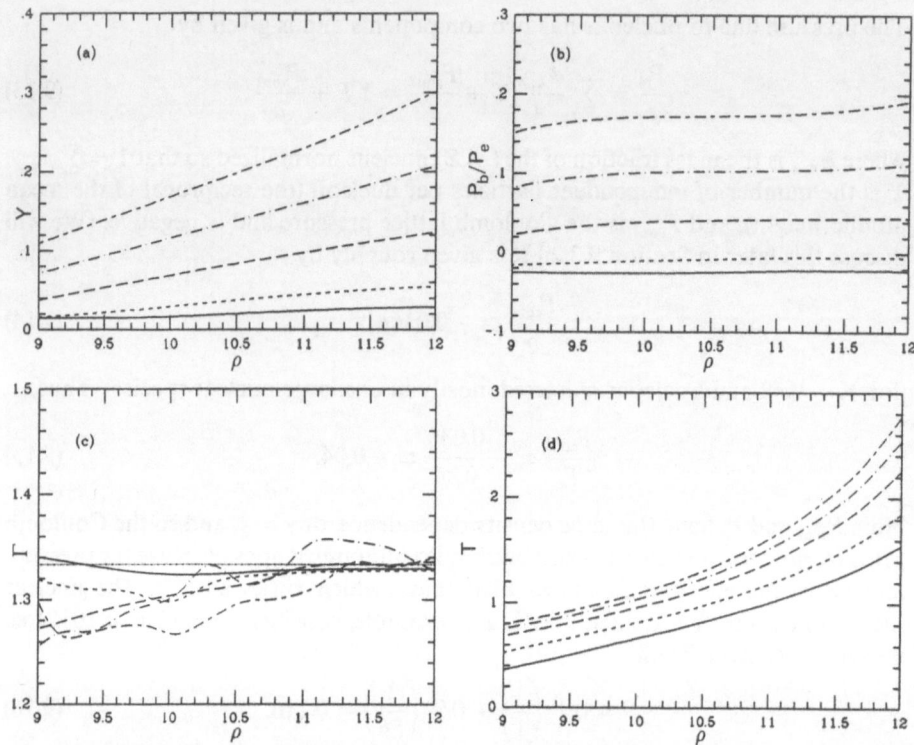

Figure 9.3. Thermodynamic quantities during the early collapse. Results for the adiabats $S = 0.75$ (solid line), $S = 1.0$ (short-dashed line), $S = 1.5$ (long-dashed line), $S = 2.0$ (dashed–dotted line), and $S = 2.5$ (short-dashed–long-dashed line), for the case $Y_e = 0.42$ and no neutrinos. The abscissa labels are $\log_{10}(\rho)$. (a) Y, the reciprocal mean molecular weight of nucleons $\langle 1/A \rangle$, averaged by mass fraction; (b) P_N/P_e where P_N includes all nuclear terms (lattice, translational, and excited states; (c) Γ_s, the adiabatic index, with a solid line showing $\Gamma_s = \frac{4}{3}$; (d) T, the temperature, in MeV. The graphs of Γ_s are quite noisy, but notice that they fall well below $\Gamma_s = \frac{4}{3}$.

opposite way; since the temperature would rise faster along a nonrelativistic adiabat ($T \sim \rho^{2/3}$ instead of $T \sim \rho^{1/3}$) entropy would be transferred to the electrons as the density increased.

The net effect in supernova collapse is that the correction factor in eq. (9.9), $\frac{2}{3}(S_e/\pi Y_e)^2$, can drop from about 0.15 to about 0.05, which is a 10% pressure deficit.

Electron capture induces a drop in the second adiabatic index, Γ_M, as we now show. If we ignore the pressure due to nucleons and the change in entropy due to neutrino diffusion and escape, we can write

$$\Gamma_M \sim \frac{4}{3}\left(1 + \frac{\delta \ln Y_e}{\delta \ln \rho}\right). \tag{9.19}$$

(Actually this change in Γ ($\frac{4}{3}(\delta \ln Y_e)/(\delta \ln \rho)$) should be added to the change due to photodisintegration.) During the early collapse the process $e^- + p \rightarrow n + \nu_e$ drops Y_e from 0.45 to 0.40 as ρ_{10} increases from 1 to 100 (we are being approximate here,

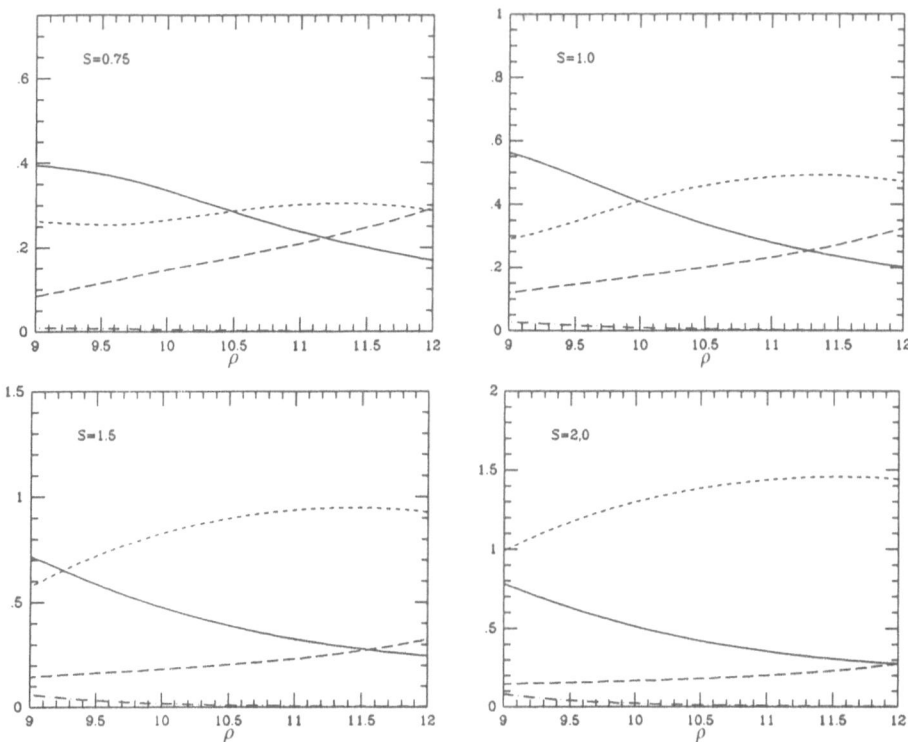

Figure 9.4. Transfer of entropy from electrons to nuclei. Contributions to the entropy. Displayed are S_e, the electron entropy (solid line), S_d the entropy in baryonic translational motion (short-dashed line), S^*, the entropy stored in nuclear excited states (long-dashed line), and S_{RAD}, the entropy in photons (dashed–dotted line), for the adiabats $S = 0.75$, 1.0, 1.5, and 2.0 during the collapse of the star, for the case $Y_e = 0.42$ and no neutrinos. The abscissa labels are $\log_{10}(\rho)$. Note that as the density increases along the adiabats, entropy is transferred from the electrons to the nucleons.

so see Section 9.5.1), and thus we find

$$\Gamma_M \sim \frac{4}{3}\left(1 + \frac{\ln(0.40/0.45)}{\ln(100)}\right) \simeq \frac{4}{3}(1 - 0.0256) = 1.30. \qquad (9.20)$$

So electron capture decreases γ to less than $\frac{4}{3}$. In fact, this is an underestimate because even once neutrinos become trapped (at a density $\rho_{trap} \sim 10^{12}$ g/cm^{-3}), there will be further electron captures until β-equilibrium is reached. These will reduce the effective Y_e from $Y_e \sim 0.40$ to $Y_{e,\,eff} \sim 0.37$ (see eq. (9.92)) over less than a magnitude of density. Thus even after the net lepton number stops decreasing the effective adiabatic index continues to remain low. Actually, in the relation $P = K\rho^\gamma$, it is K which drops through these agencies, but we can approximate this by a decrease in γ. Since the Chandrasekhar mass satisfies $M_{Ch} \propto Y_e^2$, it is reduced by a factor $\sim(0.37/0.45)^2$, which gives about a one-third reduction in the Chandrasekhar mass, about 0.45 $M_\odot$. This controls the size of the region that can collapse homologously and is a critical factor in the following Section 9.2.3 on self-similar collapse.

The radial structure of the star reduces the third adiabatic index, Γ_R. Looking at eq. (9.4) we see that while the derivatives of the pressure with respect to S and Y_e must be positive for thermodynamic stability of the equation of state, the variations with respect to ρ of S and Y_e are negative. Looking at Figure 9.2, we see that while the radial gradient of ρ is surely negative, the radial gradients of S and Y_e are positive. This is because in the presupernova evolution electron captures proceed most rapidly in the center, reducing Y_e the most there, and furthermore reduce S because the escaping neutrinos cool the material. Additionally, there is a sharp increase in S at the edge of the iron core upon going into the silicon shell.

Thus while it is true that thermal pressure increases the effective Chandrasekhar mass of the iron core, the other effects (transfer of entropy to the nucleonic matter from the electrons reducing Γ_S, electron capture reducing Γ_M, and S and Y_e gradients reducing Γ_R) reduce the effective Chandrasekhar mass. The result is that the mass of the iron core when it goes unstable is essentially the same as the zero temperature Chandrasekhar mass.

In summary, we note that $\Gamma_M < \Gamma_S$ and $\Gamma_R < \Gamma_S$, and so no matter which Γ we take as the relevant one, the iron core is in deep trouble. Since once collapse gets under way Γ_S and Γ_M are bound to decrease even further, we expect that the implosion will accelerate and even run away. The general features of the infall stage can be understood quite well analytically, and that is the subject of the next section.

9.2.3. Self-Similar Homologous Collapse

The main features of the collapse phase can be well understood from the beautiful analytical work of Yahil (1983) who considered the case $1.2 \leq \gamma \leq \frac{4}{3}$, and was motivated by the earlier and also quite elegant work of Goldreich and Weber (1980).

Yahil notes there is only a small change in the entropy during the collapse, and suggests that the polytropic equation of state

$$P = K\rho^\gamma \tag{9.21}$$

should suffice to describe at least the early stages of the problem. We assume that $\gamma = \Gamma_M = \Gamma_S$ and remains constant throughout the epoch to be described. There are only two dimensional constants in the problem, K and G, and only one dimensionless combination of r and t linear in r can be made with them:

$$X = K^{-1/2}G^{(\gamma-1)/2}r(-t)^{\gamma-2}. \tag{9.22}$$

The minus sign occurs with t because $t = 0$ is taken as the moment of catastrophe, or infinite central density. If any other parameters in the problem besides K and G, such as the initial density and velocity, have only a transient effect, all the hydrodynamical variables (density, velocity, energy, etc.) should be functions of only the self-similar variable X. Yahil says "thus, the solutions are self-similar; i.e., they have the same spatial structure at all times." Goldreich and Weber did consider, in the more limited case of $\gamma = \frac{4}{3}$, whether perturbations to the ordered flow would amplify faster than conditions changed due to collapse, and found their solutions to be stable. In the more general case Yahil's conjecture is well borne out by many numerical simulations.

Yahil continues to construct the dimensionless hydrodynamical variables:

$$\rho = G^{-1}(-t)^{-2}D(X), \tag{9.23}$$

$$v = K^{1/2}G^{(1-\gamma)/2}(-t)^{1-\gamma}V(X), \tag{9.24}$$

$$m = K^{3/2}G^{(1-3\gamma)/2}(-t)^{4-3\gamma}M(X), \tag{9.25}$$

$$\varepsilon = K^{5/2}G^{(3-5\gamma)/2}(-t)^{6-5\gamma}E(X), \tag{9.26}$$

where ρ, v, m, ε (D, V, M, E) are the dimensional (dimensionless) density, velocity, enclosed mass, and enclosed energy. These last two are defined by

$$M(X) = \int_0^X dX\, 4\pi X^2 D(X), \tag{9.27}$$

$$E(X) = \int_0^X dX\, 4\pi X^2 D(X)\left[\frac{V^2(X)}{2} + \frac{D(X)^{\gamma-1}}{(\gamma-1)} - \frac{M(X)}{X}\right]. \tag{9.28}$$

The dimensionless velocity V is then broken into two pieces

$$V = (\gamma - 2)X + U. \tag{9.29}$$

This is an important step. If $V = (\gamma - 2)X$ then a fluid element remains stationary in that frame; i.e., always has the same X coordinate. (Because $X \sim r(-t)^{\gamma-2}$.) Thus, the "wind velocity" U is measured with respect to a comoving, homologous frame. After putting the acceleration equation and the equation of continuity into dimensionless form, the solution for the wind velocity becomes trivial and Yahil obtains

$$4\pi X^2 DU = (4 - 3\gamma)M. \tag{9.30}$$

Note that in the case $\gamma = \frac{4}{3}$, U vanishes everywhere. This means the entire core collapses homologously. But for $\gamma < \frac{4}{3}$, $U > 0$; the interpretation is that the outer core is unable to keep up with the inner core. It is this feature incorporated into the Yahil theory that represents a great improvement over the earlier Goldreich–Weber theory. In the latter case, a limiting homologous core mass, equal to 1.45 M_{ch}, was found corresponding to a boundary condition of the outer edge being in free fall. Yahil's theory extends to larger core masses as we indeed find in supernova theory. (The initial iron core Chandrasekhar mass is $M_{\text{Ch}} \sim 1.45\, M_\odot$, but due to electron capture it drops to $\sim 0.9\, M_\odot$, and thus there truly is an inner and outer core.)

Yahil obtains the asymptotic behavior of the hydrodynamical variables as $X \to \infty$ ($r \to \infty$ or $-t \to 0$) by noting that in order to avoid singularities, their t-dependence must be canceled by the t-dependence implicit in the X-dependence of the dimensionless variables. This results for $X \gg 1$ ($r \gg 0$) in

$$D(X) \sim X^{-2/(2-\gamma)}, \tag{9.31}$$

$$M(X) \sim X^{(4-3\gamma)/(2-\gamma)}, \tag{9.32}$$

$$V(X) \sim X^{(1-\gamma)/(2-\gamma)}, \tag{9.33}$$

$$E(X) \sim X^{(6-5\gamma)/(2-\gamma)}. \tag{9.34}$$

(We can also cast these relations in dimensional form, i.e., $\rho(r) \sim r^{-2/(2-\gamma)}$ as $r \to \infty$.)

Table 9.1. Asymptotic behavior of
self-similar solution.

	$\gamma = \frac{4}{3}$	$\gamma = 1.3$
ρ	$r^{-1/3}$	$r^{-2.86}$
v	$r^{1/2}$	$r^{-0.43}$
$v_{ff} = 2Gm/r$	r^{-1}	$r^{-0.86}$
v/v_{ff}	$r^{3/2}$	$r^{0.43}$
ρv^2	$r^{2/3}$	$r^{-3.72}$

At the time of catastrophe this gives the asymptotic behavior of the entire infalling matter. However, by this point the self-similar solution breaks down at the center because the stiffening of the equation of state above nuclear density introduces new dimensional parameters into the theory. (Note $\gamma \leq 1.2$ is excluded because the energy diverges in this case.)

Let us consider the asymptotic behavior of several important quantities at large r. We display this for $\gamma = \frac{4}{3}$ and $\gamma = 1.3$ in Table 9.1. As γ is reduced the density in the outer core drops less slowly with radius because the inner core falls away more rapidly. We shall see that this will be harmful to the shock.

Further progress requires the constants in front of the asymptotic relationships and so the remaining dimensionless differential equation must be integrated. (The solution for U has eliminated one of them.) Yahil does this with some care by noting that for $X \ll 1$, $U = (\frac{4}{3} - \gamma)X$. Since the dimensionless sound speed is given by

$$A = \gamma^{1/2} D^{(\gamma-1)/2}, \tag{9.35}$$

at small X the wind velocity is subsonic. For $X \gg 1$ we have the asymptotic relation, $A \sim X^{(1-\gamma)(2-\gamma)}$, and since for $X \gg 1$, $U \to \infty$ because $V \to 0$, the wind becomes supersonic. Since U and A have the same asymptotic behavior, $U/A \to$ constant as $X \to \infty$. The correct solution must obey a regularity condition at the critical point, or the sonic point, where the wind becomes supersonic. Other solutions either arrive at the critical point with infinite derivatives, or U reaches a subsonic maximum and then falls to zero at infinity, which is unphysical. (Yahil's paper explains the mathematical considerations.)

Figure 9.5 shows the solutions of Yahil's equations for the seven cases $\gamma = 1.20$, $1.22, \ldots, 1.32$. The case $\gamma = 1.30$, a good representation for supernova collapse, is denoted with a dashed line. The stars indicate the position of the maximum velocity surface. We note the following: $-V/V_{ff}$ reaches an asymptotic value in the outer core, which drops with γ. Likewise, the Mach number $|V/A|$ reaches an asymptotic value which also decreases with γ. This has important consequences. It means that if γ drops appreciably the outer core would be much less able to keep up with the collapsing inner core, which runs away.

The position of the maximum infall velocity surface moves outward somewhat as γ drops, but the density distribution (shown by the combination $DX^{2/(2-\gamma)}$) remains higher for low γ. Note that the physical mass, $m \sim (-t)^{4-3\gamma} M(X)$, so in the case of γ appreciably less than $\frac{4}{3}$ the maximum velocity surface moves in quite a bit with time. Since this is where the shock wave will form (actually at the sonic point

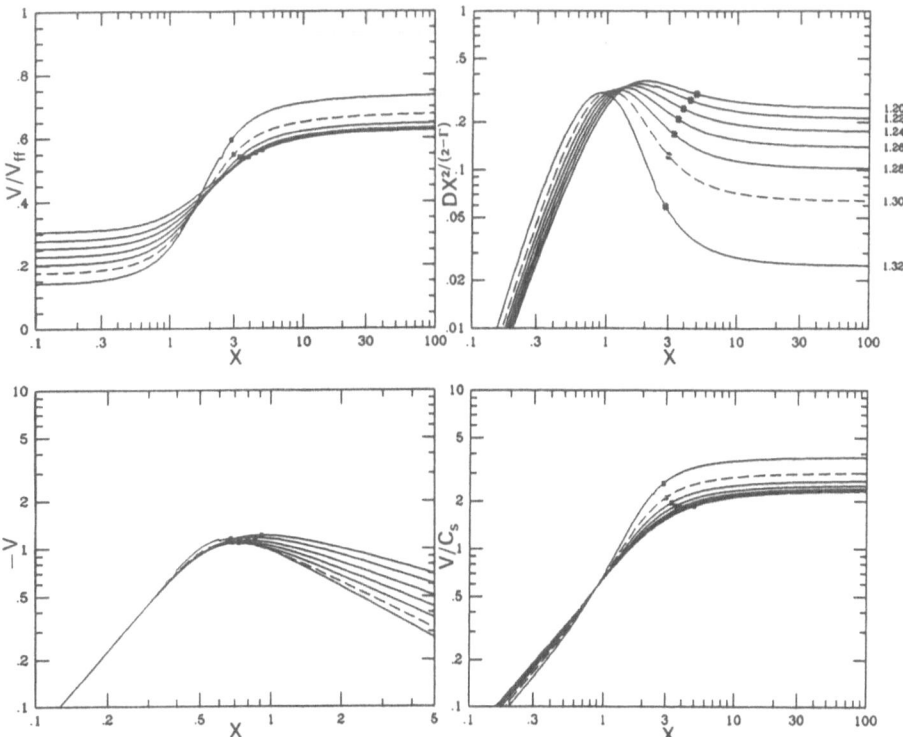

Figure 9.5. Yahil's self-similar collapse. For $\gamma = 1.20, 1.22, \ldots, 1.32$, with $\gamma = 1.30$ the dashed line, we give the Yahil (1983) solutions for V/V_{ff}, $DX^{2/(2-\gamma)}$, $-V$, and V/C_s, as a function of X. The star gives the position of the maximum of infall velocity. γ values are denoted alongside the top right figure.

which also moves inward), then this is harmful. For all these reasons Yahil claimed "the difference $\frac{4}{3} - \gamma$ may be the most important parameter in determining success or failure."

An important quantity in the Yahil analysis is the ratio of the infall velocity to the freefall velocity at the center

$$\Omega_0^{-2} = \left(\frac{v}{v_{ff}}\right)^2\bigg|_{r=0} = (6\pi D_0)^{-1}. \tag{9.36}$$

This is shown in Figure 9.5 and is $(0.175)^2 = 0.0303$ for $\gamma = 1.3$, $(0.226)^2 = 0.052$ for $\gamma = 1.26$ and will figure prominently in Section 9.5.2. There we will consider what happens when the central self-similarity solution breaks down, as the core bounces when the stiffness of nuclear matter produces a rapid increase in the adiabatic index.

9.3. Subsaturation Density Equation of State

The calculation of the equation of state for matter with $\rho < \rho_0$ is technically the most difficult part of numerically modeling supernova collapse, and the part where computers spend much of their time—in our calculations more than 75% when we do simple neutrino transport, and about 50% even when we do detailed neutrino diffusion. Therefore, both physical accuracy and ease of calculation are urgent priorities. Fortunately, these objectives are not orthogonal. Indeed, despite the propensity to equate detail with quality, the simpler treatments do in fact give the best physical results.

9.3.1 Nuclear Statistical Equilibrium

We begin by considering the nature of the material at the onset of gravitational collapse. The matter is actually rather cold; even though the temperature is quite high, the entropy is quite low, $S \lesssim 1$ in units of Boltzmann's constant per nucleon (Bethe et al., 1979). Now, e^s is essentially the number of ways each nucleon can arrange itself, and so we see things are highly ordered. This means that nucleons will cluster into large nuclei, since by doing so the translational degree of freedom is shared amongst all the participants in each nucleus.

Thus, at the beginning of the collapse the matter consists of a nuclear statistical equilibrium (NSE) mixture of various nuclei. (It is easy to show that time scales are sufficiently long and the density high enough for the strong and electromagnetic forces to equilibrate.) The abundance of each element in the mixture is then implicitly given by the relation

$$\mu_{A,x} = A(\mu_n - x\hat{\mu}); \qquad x \equiv Z/A, \qquad \hat{\mu} \equiv \mu_n - \mu_p, \qquad (9.37)$$

where the μ's are the chemical potentials. Mass and charge conservation can be expressed as

$$\sum \phi_{A,Z} = 1; \qquad \sum \phi_{A,Z} x_{A,Z} = Y_e, \qquad (9.38)$$

where $\phi_{A,Z}$ is the mass fraction of the (A, Z) nucleus and $Y_e = Y_{e^-} - Y_{e^+}$ is the total number of protons per nucleon. If we can express the chemical potentials of the ingredients in terms of the concentrations and other thermodynamic variables such as the temperature and density, we can then proceed to calculate equation of state quantities, such as the pressure, simply by summing them up weighted by mass fractions.

Unfortunately, supernova material is not very easy to describe, because the properties of the constituents of the NSE mixture are not precisely known. There are several reasons for this. For one thing, the matter is quite neutron rich; initially $Y_e \sim 0.4$–0.5, but it soon descends to $Y_e \sim 0.33$. (Remember that the lepton number, Y_l, will not drop below about $Y_l = 0.37$, but some of the leptons will be neutrinos, so Y_e is considerably lower than Y_l.) Such nuclei are not terrestrially stable and their properties can only be inferred through extrapolation. Furthermore, the temperature in supernovae soon climbs above 1 MeV. This is high enough to break pairing and wash out shell effects. While it does make things smoother, it means we need to know the partition function for each poorly understood nucleus. Finally, supernova matter is a charge neutral system, and the uniform background of relativistic electrons produces some modifications in the picture. Most important, it reduces the repulsive Coulomb energy.

The most popular and best approach has been to replace the entire ensemble of nuclei with a representative system of four components (Epstein and Arnett, 1975). These are free neutrons and protons, α-particles, and one representative heavy nucleus, whose properties are chosen so as to model the true mixture. The accuracy and sensibility of this approximation depends completely on how well the mean nucleus represents the full ensemble in its properties. The most successful approach has been to consider this nucleus as a compressible liquid drop, as we will discuss in Section 9.3.2, with its energy functional drawn from the semiempirical mass formula including compressional terms and extended for neutron excess. The partition function must match the properties of known nuclei but must also account for the decrease in the density of states as the nuclei become more massive and neutron rich.

While some have preferred using networks and have objected to the above approach (see, e.g., El Eid and Hillebrandt, 1980), we believe it preferable to consolidate ignorance in a few adjustable parameters of one representative nucleus, tuned by empirical information and theoretical studies, rather than spread it out over up to 250 poorly understood objects. Moreover, it has been shown that the thermodynamic shifts introduced by truncating the ensemble at four elements are quite small, especially when compared with the other uncertainties in the problem (Burrows and Lattimer, 1984; Cooperstein and Wambach, 1984). However, we should note that the NSE distribution is actually quite broad in mass number, and that while equation of state shifts may be small, effects on other arenas, such as electron capture, could be more appreciable, and corrections must be introduced to account for the spread in the ensemble.[3]

9.3.2. The Compressible Liquid Drop Model

We now consider the equation of state for $\rho < \rho_0$, following Cooperstein (1985) and Bethe et al. (1983), except that we will give K_0 a Y_e-dependence as in Baron (1985), and modifications will be made to the density of excited states.

The compressible liquid drop model was first used to describe neutron star matter by Baym, Bethe, and Pethick (1971). An improved treatment of the surface was given by Ravenhall, Bennett, and Pethick (1972), and finite temperatures and higher proton fractions were considered by Lattimer and Ravenhall (1978). While a number of groups have contributed to this approach, it has been most fully developed by the Illinois group (whose work we will generically denote as LLPR) in a long series of papers (Lamb et al., 1978, 1981, 1983, 1985; Pethick, Ravenhall, and Lattimer, 1983, 1984; Ravenhall, Pethick, and Lattimer, 1983). Excellent reviews of this work are given in Lattimer (1981) and Lamb et al. (1985).

An important feature of this work is that the properties of both the liquid and the vapor phase, as well as all the constituent properties (bulk, surface, Coulomb energies, etc.), are calculated from the same microscopic Hamiltonian. In addition the finite temperature qualities of these contributions are also computed. Use of the same effective nuclear Hamiltonian for all phases permits calculation up to the critical point of the liquid–vapor phase transition.

The approach we will use begins with the fits of Bethe et al. (1979), to the early

---

[3] However, as far as electron capture through thermally unblocked Gamow–Teller transitions (Fuller, 1982; Fuller, Fowler, and Newman, 1982) are concerned, these corrections were shown to be small by Cooperstein and Wambach (1984) because there is enough thermal unblocking in the mean nucleus to provide sufficient channels for captures to proceed.

work of the Illinois group. While the important features of our equation of state are the same as that of LLPR, our approach has a very important advantage. It takes the relevant nuclear parameters, such as the bulk incompressibility and the saturation density (and their dependences on neutron excess), the effective mass, the bulk symmetry energy coefficient, etc., as input quantities rather than as the output from a specific set of force parameters. This permits a convenient alteration of the physical inputs to gauge the effects upon supernova collapse. While the LLPR treatment rests upon stronger microscopic grounds, things depend on the particular interaction chosen, and upon details of the engineering of many of the pieces. We have thus chosen to turn matters inside out by basing our parametrization on the final phenomenological quantities of relevance.

A further advantage of our treatment is that it permits the equation of state to be used efficiently directly in line in the hydrodynamical programs, rather than requiring the use of thermodynamic tables and interpolation. (The problems with tables are: changing one parameter requires a whole new table, and tables are excruciating to construct for a full grid; special steps must be taken to ensure thermodynamic self-consistency so that the interpolating procedure must be of a high order; many subsidiary quantities are also needed, such as the chemical composition and opacities, which may vary exponentially with the tabulated thermodynamic quantities, and so unless these are tabulated too the calculation can be inaccurate; the grid must be quite large to cover the supernova phase space and thus consumes machine resources. In many cases, table interpolation actually consumes more computer time than direct inline computation.) One final advantage of our treatment is that our equation of state has now been used by several groups in supernova simulations (besides ourselves, by Bruenn (1988a, b) and Myra and Bludman (1989)) and this has made possible good comparisons of other parts of the problem, such as the influence of neutrino transport methods.

Other approaches besides the compressible drop model have been used to describe warm dense matter, such as the Thomas–Fermi and Hartree–Fock approximations. The article by Lamb et al. (1985) contrasts these approaches and explains their advantages and shortcomings and contains a complete set of references to this work.

We begin with the $T = 0$ case, and for the energy per nucleon of nucleons we take

$$W = E_{\text{bulk}} + E_{\text{size}}, \qquad (9.39)$$

where E_{bulk} is the bulk nuclear matter term, and E_{size} includes surface energies, Coulomb energies, etc., i.e., the things involving finite size. For the moment we ignore translational terms and neutron drip.[4]

9.3.2.1. Bulk Energy

We assume the nucleons are clustered at a density near ρ_0, even when $\rho \ll \rho_0$. Then we can write

$$E_{\text{bulk}} = E_{nm} + W_s(1 - 2x)^2 + \frac{K_0(x)}{18}(1 - \theta)^2, \qquad (9.40)$$

---

[4] Lamb et al. (1985) have shown that for $Y_l \sim 0.40$, neutrons do not drip at $T = 0$ in supernova conditions, unlike the situation in β-equilibrium neutron stars.

where $E_{nm} \simeq -16$ MeV is the energy of infinite symmetric nuclear matter, and $x = Z/A$ ($= Y_e$ at $T = 0$). The density compression factor θ is given by

$$\theta = \frac{\rho_0}{\rho_s(x)},\tag{9.41}$$

where $\rho_s(x)$ is the saturation density at the given neutron excess, and ρ_0 is the actual density of the nuclear matter. At the symmetric nuclear matter saturation density, $\rho_s(1/2) = 0.16$ fm^{-3}, the bulk energy is minimized, and thus E_{bulk} is assumed to vary quadratically both in $(1 - 2x)$, the neutron excess, and $(1 - \theta)$, the density compression factor. (For $\theta \gg 1$ this is inadequate, and so our expression for high-density uniform matter will be more complicated.)

Both ρ_s and the nuclear bulk incompressibility, K_0, given by

$$K_0(x) = 9\frac{\partial P}{\partial \rho}\bigg|_{\rho=\rho_s(x)} = 9\rho^2\frac{\partial^2 E}{\partial \rho^2}\bigg|_{\rho=\rho_s(x)},\tag{9.42}$$

have x-dependences. (K is related to the bulk modulus. It decreases as the matter becomes softer or more compressible. Factors of 3 and 9 enter from nuclear physics conventions) Each drops with increasing neutron excess, as x decreases towards neutron matter, the drop in K_0 coming mostly from ρ_s, upon which it depends quadratically (see Kolehmainen et al., 1985). We need not put an explicit density dependence in the bulk symmetry coefficient, W_s, as that is understood to be included in K_0 and ρ_s. For the size energy term we will use below, $W_s = 31.5$ MeV is the best value to use, as will be discussed in Section 9.3.5.

With the Skyrme force Skm* parameters we obtain

$$\rho_s(x) = \frac{\rho_s(\tfrac{1}{2})}{1 + 0.75(1 - 2x)^2},\tag{9.43}$$

$$K_0(x) = \frac{K_0(\tfrac{1}{2})}{1 + 2.0(1 - 2x)^2},\tag{9.44}$$

where $\rho_s(1/2) \simeq 0.16$ fm^{-3} ($=2.7\cdot10^{14}$ g cm^{-3}). From the data on the Pb Giant Monopole Resonance breathing mode, Blaizot, Gogny, and Grammaticos (1976) and Blaizot (1980) deduced a value of $K_0(\tfrac{1}{2}) \sim 210 \pm 30$ MeV (see Section 9.3.3 for a discussion of getting K_0 from experiment). Working at the bottom of the above error bars in K_0, Baron, Cooperstein, and Kahana (1985a, b, BCK), and Baron et al. (1987) have chosen to use $K_0(\tfrac{1}{2}) = 180$. This choice leads to $K_0(\tfrac{1}{3}) = 140$ MeV, and $\rho_s(\tfrac{1}{3}) = 0.145$ fm^{-3} for matter with twice as many neutrons as protons. The quadratic expressions for K_0 and ρ_s should not be used for very small x.

9.3.2.2. Size Energy
The size energy, E_{size}, must be a function of the fraction of the fraction of the total volume occupied by the nuclear material. This is the "packing fraction,"

$$u \equiv \frac{\rho}{\rho_0} = \frac{\rho}{\rho_s\theta},\tag{9.45}$$

where ρ is the total baryon density. In general, E_{size} will be a complicated function of the geometrical arrangement which minimizes the energy, but we assume that it is a smooth function of basic mean quantities, like the ratio of surface area to volume. We have in mind a description of low but finite temperatures where small energy differences between different geometrical distributions will be unimportant.

In the case of a nucleus we write

$$E_{\text{size}} = \frac{\omega_{\text{surf}} A^{2/3} + \omega_{\text{coul}} A^{5/3}}{A}, \tag{9.46}$$

where we have not included curvature terms, which are poorly known, and which tend to be canceled by translational terms we also neglect (see Cooperstein and Wambach, 1984). It is difficult to separate curvature from surface contributions, and while curvature does have some effects on the nuclei to bubbles transition (Ravenhall, Pethick, and Lattimer, 1983), it need not be included here.

Since $A \sim R^{1/3}$, we find for the A dependence of the surface and Coulomb energies, $4\pi R^2 \sim A^{2/3}$ and $Z^2/R \sim A^2/A^{1/3} \sim A^{5/3}$. We can find A via

$$\left. \frac{\partial E_{\text{size}}}{\partial A} \right|_{u,x} = 0 \rightarrow A = \frac{1}{2} \frac{\omega_{\text{surf}}}{\omega_{\text{coul}}}. \tag{9.47}$$

Equivalently, this minimization could been seen as giving the best radius for the nucleus or the best ratio of surface area to volume. The result is a virial theorem (Baym, Bethe, and Pethick, 1971) which says

$$\text{Coulomb energy} = \tfrac{1}{2} \text{ surface energy.} \tag{9.48}$$

As A increases, the surface energy per nucleon goes down because a lower proportion of the nucleons are on the surface, while the Coulomb energy still goes up. The surface energy can be written as

$$E_{\text{surf}} = \omega_{\text{surf}} A^{-1/3} = \frac{1}{A} 4\pi R^2 \gamma_{\text{surf}}(x). \tag{9.49}$$

where γ_{surf} is the surface energy per unit area. We will assume that the neutron excess $(1 - 2x)$ is constant through the space inhabited by nucleons. In fact, things are more complicated, and there is a "neutron skin" (Lamb et al., 1985) but we will neglect this. An approximation to the surface energy per unit area, adapted from the fit of Bethe et al. (1979) to the calculation of Lattimer and Ravenhall (1978), is given by

$$\gamma_{\text{s}}(x) = 17.8x^2(1 - x)^2 \text{ MeV fm}^{-2}, \tag{9.50}$$

which gives us

$$\omega_{\text{surf}} = 290x^2(1 - x)^2 \left(\frac{0.16}{\rho}\right)^{2/3} \text{ MeV.} \tag{9.51}$$

For $x = \tfrac{1}{2}$, $\gamma_{\text{s}} = 1.1$ MeV fm^{-2}.

The Coulomb energy is calculated by consideration of an electrically neutral Wigner–Seitz cell with a nucleus containing Z protons and $A-Z$ neutrons at the center as displayed in Figure 9.6(a). We assume the electrons inhabit the total volume uniformly because screening is not a big correction, as has been shown by

Figure 9.6. Wigner–Seitz cells. (a) Nuclei; (b) bubbles. The hatched regions are inhabited by the dense nuclear phase.

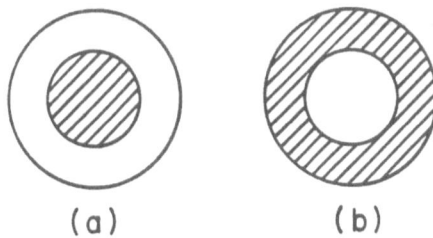

(a)   (b)

Lamb *et al.* (1985). Then the electrostatic energy per nucleon is given by

$$E_{coul} = \frac{1}{A}\frac{3}{5}\frac{Z^2 e^2}{R}g(u),$$ (9.52)

where

$$g(u) = 1 - \tfrac{3}{2}u^{1/3} + \frac{u}{2}.$$ (9.53)

(Actually, more generally, $E_{coul} = (1/AR)\tfrac{3}{5}(Z'e/(1-u))^2 g(u)$, where $Z'e$ is the actual charge contained within the nuclear volume. If all the protons are inside the nucleus then $Z' = Z(1 - u)$, the reduction factor coming because a fraction u of the electrons are inside the nuclear volume. If there are protons outside the nucleus, either free or in α-particles, then things must be corrected to take this into account.) This gives

$$E_{coul} = 0.75x^2\left(\frac{\rho_0}{0.16}\right)^{1/3}g(u)A^{2/3} \text{ MeV}.$$ (9.54)

The virial theorem now becomes

$$A = \frac{1}{2}\frac{\omega_{surf}}{\omega_{coul}} = 193(1 - x)^2\left(\frac{0.16}{\rho_0}\right)\frac{1}{g(u)},$$ (9.55)

which would give, for an isolated iron nucleus $(x = 26/56 = 0.464, u = 0)$, $A = 55.3$. However, under supernova conditions it is important that u is greater than 0, and A increases because the Coulomb energy is less than in the case of an isolated nucleus.

We now have for E_{size},

$$E_{size}(x, \theta, u) = \beta(x)\theta^{-1/3}G_{nuc}(u),$$ (9.56)

where

$$\beta(x) = 75.4x^2(1 - x)^{4/3}(1 + 0.75(1 - 2x)^2))^{1/3} \text{ MeV},$$ (9.57)

$$G_{nuc}(u) = g^{1/3}(u) = \left(1 - \tfrac{3}{2}u^{1/3} + \frac{u}{2}\right)^{1/3}.$$ (9.58)

This is very convenient because the x, θ, and u dependences are separated.

9.3.2.3. Bubbles

Baym, Bethe, and Pethick (1971), Lamb *et al.* (1978), and Pethick, Ravenhall, and Lattimer (1984) noted the appearance of different geometries. Suppose, as in Figure 9.6(b), we invert the geometry, putting the matter outside bubbles. If the matter were to turn "inside out" when $u = \frac{1}{2}$, then the amount of surface area would be the same, and it is clear that for $u > \frac{1}{2}$ energy would be better minimized by shrinking a bubble rather than filling up the Wigner–Seitz cell at its outer boundary. It is possible to calculate E_{size} in this bubble phase by reinterpreting the nucleus case. All we have to do is make the substitution that

$$G_{bub}(u) = \left(\frac{1-u}{u}\right) G_{nuc}(1-u) = \left(\frac{1-u}{u}\right) g^{1/3}(1-u). \tag{9.59}$$

(See Bethe *et al.* (1983) and Cooperstein (1985); this substitution is possible because the θ-dependence is the same in either phase.) In either the nucleus or bubble phases the Coulomb energy wants the denser nuclear material to fill the entire cell because a uniform charge distribution has vanishing electrostatic energy. But the surface tension in the nuclear case fights this, by compressing the nucleus. In the bubble case the surface tension tries to collapse the bubble to eliminate the surface; thus the bubble may collapse unless the compressional force proportional to K_0, which resists this expansion of the nuclear matter phase, is large enough, as we shall see in Section 9.3.2.5. There is also an asymmetry in the curvature energy, which changes sign in the transition (Ravenhall, Pethick, and Lattimer, 1983), but the curvature coefficient is very poorly known and we will ignore this refinement because the transition is quite undramatic anyway, as we will show shortly, due to the presence of the electrons.

9.3.2.4. *Other Geometries*

It is possible to have a whole series of other geometries besides nuclei and bubbles. These are displayed by order of their symmetries in Figure 9.7. Such geometries were calculated by Ravenhall, Pethick, and Wilson (1983, RPW) and also by Williams and Koonin (1985). Of course the real situation may have a very complicated geometry but comparing these symmetries will give us the scale of the geometric possibilities. For the geometrical function RPW obtain

$$G_{RPW}(u) = \left[\frac{5}{9}\frac{d^2}{d+2}\left(\frac{1-(d/2)u^{1-2/d}}{d-2}+\frac{u}{2}\right)\right]^{1/3}, \tag{9.60}$$

where d is the dimension of the symmetry. (The bubble phases are obtained by $G_{bub}(u) = [(1-u)/u]\,G_{nuc}(1-u)$.) At a given density matter takes the geometry with minimal energy, but the differences between succeeding phases are small (keV). In

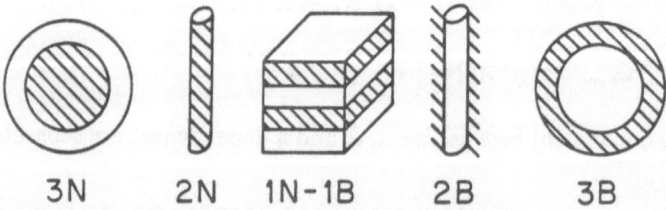

$$3N \qquad 2N \qquad 1N-1B \qquad 2B \qquad 3B$$

Figure 9.7. Nuclear pasta. Possible geometric configurations suggested by Ravenhall, Pethick, and Wilson (1983). The configurations are: $3N$ (nuclei); $2N$ (nuclear spaghetti); $1N-1B$ (lasagne and anti-lasagne, which are the same); $2B$ (antispaghetti); and $3B$ (bubbles).

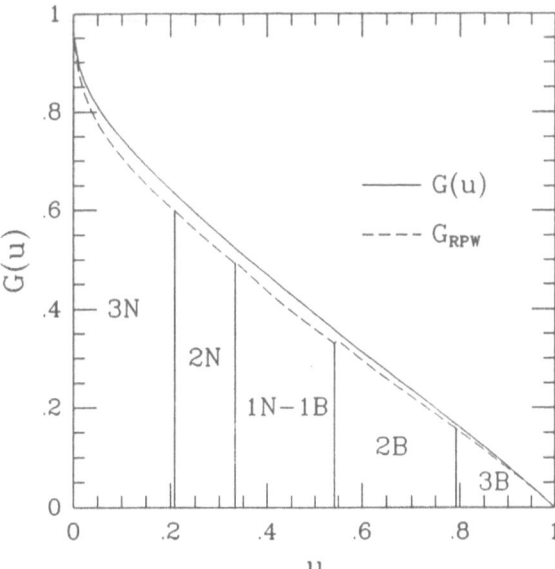

Figure 9.8. Geometrical function for subsaturation configurations. The geometric function, $G(u)$. The solid line gives $G(u)$ of Cooperstein (1985), and the dashed line the solution for the energetically favored phases of Ravenhall, Pethick, and Wilson (1983), $G_{RPW}(u)$. The regions labeled are: $3N$ (nuclei), $2N$ (nuclear spaghetti), $1N$–$1B$ (lasagne and antilasagne, which are the same), $2B$ (antispaghetti), and $3B$ (bubbles).

the supernova problem we have $T \sim$ several MeV at the relevant densities and we should expect a mixture of the different phases. Thus the true situation is better described as a "casserole" of various kinds of nuclear pasta. In addition, the RPW geometries were calculated for $Y_e = 0.5$; for nonsymmetric matter some of the geometries will disappear as the surface tension depends strongly on neutron excess.

Our solution was to adopt the following interpolating formula (Cooperstein, 1985):

$$G(u) = (1 - u)G_{nuc}(u) + uG_{bub}(u)$$
$$= (1 - u)[g^{1/3}(u) + g^{1/3}(1 - u)], \tag{9.61}$$

which is graphed in Figure 9.8, along with $G_{RPW}(u)$. The simple formula passes smoothly along all the geometries. The physical content is that E_{size} should depend only on the ratio of the the total surface area to volume of the system, and this should be a smooth function of density (actually u, the packing fraction).

Use of the smooth $G(u)$ function passes over the weakly first-order nucleus–bubbles phase transition. At $u = \frac{1}{2}$, $E_{bub} = E_{nuc}$, so we do not expect a large discontinuity in density. It takes a considerable amount of algebra (or a small amount of numerics!) to show that the discontinuity is indeed small near the transition, as can be seen in the graphs of Lamb et al. (1985). This is largely because the leptonic pressure (electrons and neutrinos) dominates the nuclear pressure. In fact, without the presence of the electrons, there would be no transition, as the total pressure would be negative for both nucleonic phases, and each would be unstable.

9.3.2.5. *Equilibrium Constraints and* Γ_N

Regardless of the geometry we have

$$E = E(x, \theta, u); \qquad \rho = \rho_0 u = \rho_s(x)\theta u, \tag{9.62}$$

the geometry giving only the specific form of $G(u)$, and thus at fixed x

$$d \ln \rho = d \ln \theta + d \ln u, \tag{9.63}$$

$$\frac{P}{\rho} = \frac{dE}{d \ln \rho} = \frac{\partial E}{\partial \ln \theta}\bigg|_u + \left[\frac{\partial E}{\partial \ln u}\bigg|_\theta - \frac{\partial E}{\partial \ln \theta}\bigg|_u \right] \frac{d \ln u}{d \ln \rho}. \tag{9.64}$$

The equilibrium u (and hence θ as well) at given ρ and x is obtained by observing that at fixed ρ, δE must vanish. This yields for the pressure per particle

$$\frac{P}{\rho} = \frac{\partial E}{\partial \ln \theta}\bigg|_{x,u} = \frac{\partial E}{\partial \ln u}\bigg|_{x,\theta}. \tag{9.65}$$

The above equilibrium constraint can be written as

$$\theta^{4/3}(1 - \theta) + \varepsilon_{\text{size}}\left(\frac{G}{3} + G'\right) = 0, \tag{9.66}$$

where the prime denotes differentiation with respect to $\ln u$, and its solution is given in Figure 9.9. Given x, this equation determines θ versus u; and then $\rho = \rho_s(x)\theta u$ is obtained. The solution depends only on

$$\varepsilon_{\text{size}} = \frac{\text{size energy of isolated nucleus}}{\text{bulk incompressibility}} = \frac{9\beta(x)}{K_0(x)}, \tag{9.67}$$

which is the basic parameter in the problem. In the case of iron, we would have $\varepsilon_{\text{size}} \simeq 9 \times 7/K_0 \sim 0.25$.

The nuclear adiabatic index is given by

$$\Gamma_N = \frac{\partial \ln P_N}{\partial \ln \rho}\bigg|_x = \frac{2}{3} + \frac{G''/G' + \frac{1}{3}}{1 + (d \ln \theta)/(d \ln u)}, \tag{9.68}$$

$$\frac{d \ln \theta}{d \ln u} = \tfrac{3}{7}\varepsilon_{\text{size}} \frac{G'/3 + G''}{\theta^{4/3}(\theta - \frac{4}{7})}. \tag{9.69}$$

When $(d \ln \theta)/(d \ln u) = -1$, $\Gamma_N \to \infty$ and this signifies a phase transition. Since θ decreases as u increases this means there is a maximum density at which the either the bubbles will collapse or the nuclei will explode, and that values for lower θ will represent energy maxima rather than minima. This is essentially the spinoidal transition since both Γ_N and the sound speed vanish (at a nearby density). This catastrophe occurs when

$$\theta^{4/3}(\theta - \tfrac{4}{7}) + \tfrac{3}{7}\varepsilon_{\text{size}}\left(\frac{G'}{3} + G''\right) = 0. \tag{9.70}$$

Together with eq. (9.66) this locates the collapse point, given $\varepsilon_{\text{size}}$, and the solution is displayed by the starred point in Figure 9.9. We see that except for rather small $\varepsilon_{\text{size}}$ (stiff equations of state) the nonuniform system becomes unstable before $u = 1$. However, in reality such an instability is unlikely to occur in supernova matter for two reasons. The first is that neutrons will drip out of the dense phase and form a vapor and fill up the space outside the nuclei (or inside the bubbles) with a vapor whose pressure resists expansion of the denser phase. Second, at finite temperature

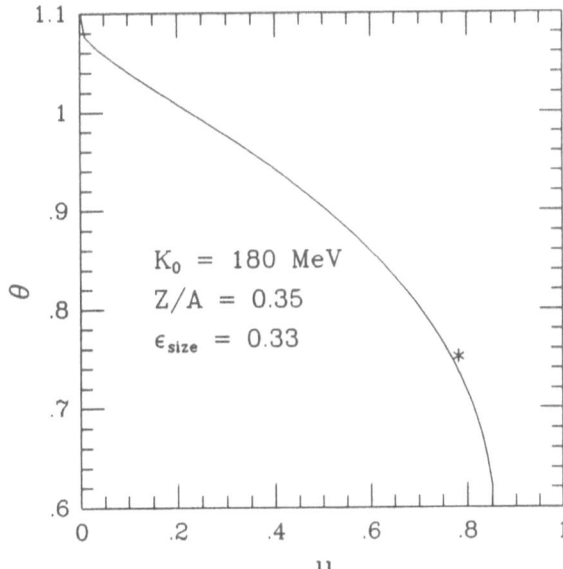

Figure 9.9. Solution of the equilibrium constraints. Displayed is $\theta = \rho_0/\rho_s$, the nuclear density compression factor, versus $u = \rho/\rho_0$, the nuclear packing fraction, at $T = 0$ for the case $Y_e = Z/A = 0.35$, which gives $\varepsilon_{\text{size}} = 0.33$. The starred point shows where the equilibrium solution fails and the matter collapses to a uniform phase.

besides an enhancement of the drip fraction, the partition function favors excited states associated with the surface, because these increase the entropy and lower the free energy. In doing so, they make it more favorable for bubbles to open up and prevent nuclei from exploding than at $T = 0$. These factors prevent this instability from taking place in supernova matter.

Some special cases can be easily derived and are listed in Table 9.2. For $u \ll 1$ with $\varepsilon_{\text{size}} \sim 0.25$ for symmetric matter $\theta \sim 1.08$. Thus the nucleus is compressed by $\sim 8\%$ over saturation due to the surface tension winning out over the Coulomb force.

For u of order unity, $G(u) \sim 1 - u$, and $u = 0.25$, we find $\theta = 1$ independent of $\varepsilon_{\text{size}}$; i.e., at a packing fraction of one-fourth, the nuclear matter phase is at saturation density.

For $u \sim 1$ the bubble disappears when $\theta < 1$ ($\theta \sim 0.75, 0.8$ for $\varepsilon_{\text{size}} \sim 0.25, 0.20$). Thus we go to uniform nuclear matter at $\rho \sim 0.8\rho_s(x)$ while the nuclear pressure is still negative, a condition known as "cavitation."

The processes which reduce γ during the infall of the star below $\gamma = \frac{4}{3}$, photodissociation and electron captures, are no longer important in the later stages of collapse ($\rho > 10^{12}$ g cm^{-3}, $u > 0.01$.) In fact, photodissociation begins to go the

Table 9.2. Limits on θ, the nuclear compression.

u	$G/3 + G'$	θ
0	$\frac{1}{3}$	$\sim 1 + \frac{1}{3}\varepsilon_{\text{size}} \simeq 1.08$
$\frac{1}{4}$	~ 0	~ 1
1	-1	$\sim 1 - \varepsilon_{\text{size}} \simeq 0.75$

other way as α-particles get squeezed back into nuclei (Bethe et al., 1979). Electron captures are unimportant because the neutrinos are now trapped on the infall timescale, and neutrino absorption and emission balance. But now the importance of Γ_N increases and this serves to keep γ pretty much constant during the infall, at slightly less than $\frac{4}{3}$, until the matter goes above nuclear saturation density.

For the mixture of relativistic leptons and nuclear material the full adiabatic index is given by

$$\Gamma = \frac{P_e}{P_e + P_N} \Gamma_e + \frac{P_N}{P_e + P_N} \Gamma_N = \frac{4}{3} + \frac{P_N}{P_e + P_N}\left(\Gamma_N - \frac{4}{3}\right). \qquad (9.71)$$

Although $\Gamma_N > \frac{4}{3}$, $P_N < 0$, so the nuclear material decreases Γ. This effect becomes quite large because both Γ_N and the magnitude of the negative nuclear pressure increase as we pass through the bubble phase. Thus, in the last stages of collapse the negative nuclear pressure acts as a vacuum cleaner and tends to accelerate the collapse. However, at finite temperatures the reduction of Γ is not as large as the positive thermal pressure and the reduction of the effective mass gives additional compensatory contributions.

9.3.3. Determinations of K_0 from the Monopole Resonance

Blaizot and coworkers (Blaizot, Gogny, and Grammaticos, 1976; Blaizot, 1980) used measurements of the monopole resonance in heavy nuclei to determine the incompressibility K_0. This has been the most sophisticated and reliable approach to date. K_0 is not, however, a directly measurable quantity. To determine the strength and position of the monopole resonance, which is clearly a collective effect, we must perform a random phase approximation (RPA) calculation and go back and fit the data. Blaizot's approach was to perform RPA calculations with a Skyrme force and then readjust the parameters of the Skyrme force until the data were fit. This procedure yields what is commonly called the empirical value for the incompressibility $K_0 = 210 \pm 30$ MeV, where the determination is for isospin symmetric matter ($Y_e = 0.5$). This is a difficult procedure, as surface and bulk effects are hard to separate, and the difference in boundary conditions between finite and infinite systems is not unambiguous.

Another approach to the determination has been through the use of the Landau forward-scattering sum rule (Brown and Osnes, 1985; Brown, 1988b). They pointed out that with a value for the effective mass of $m^*/m \sim 0.9$ that the Landau coefficient F_0 must be negative, which leads to a rather low value of the incompressibility $K_0 \sim 120$ MeV. Pines, Quader, and Wambach (1988) using a value of $m^*/m = 0.8$ also find a value for K_0 of 120 MeV. Brown (1988b) has argued that fitting the position of the monopole resonance in ^{208}Pb does not give the value of K_0 appropriate to nuclear matter, which is what is important for the equation of state. Brown (1988b) argues that the position of the giant monopole resonance in lead is insensitive to the effective mass, because the velocity of the interacting nucleons is low. Correcting for this, and for the finite size of the nucleus he finds that the empirical value should be lowered by a factor of about $\frac{2}{3} \times 180 = 120$ MeV.

Recently, a Dutch group (Sharma et al., 1988) has claimed to have re-fit the monopole resonance and finds a high value, $K_0 \sim 300$ MeV. The reason for the difference between their results and those of Blaizot is that they have not performed the appropriate RPA calculation, using instead scaling laws and semiempirical mass formulas. In fact, Blaizot (1980) warns of the limitations of such macroscopic calculations. In any case, the value of Blaizot seems to be a reasonable value and we should keep in mind that lower values are quite possible.

The monopole resonance measures density variations of at most a few percent about saturation density, and supernova matter goes to densities of about four times greater. Hence K_0 is not the sole parameter of relevance and we discuss this further in Section 9.4.

9.3.4. Finite Temperatures

At finite temperatures, the picture changes considerably. The equation of state behaves more smoothly due to the additional flexibility afforded by the increase in the available degrees of freedom. However, two important new degrees of freedom increase the complexity and make the actual computations a formidable numerical task. First, we have the population of excited intrinsic nuclear states. Second, there is the appearance of a nuclear statistical equilibrium mixture of species, with free protons and neutrons dripping out of the condensed phase, and a variety of nuclear species now existing, including light fragments such as α-particles.

We work with the free energy, $F = E - TS$. The entropy is obtained as

$$S = -\frac{\partial F}{\partial T}\bigg|_{Y_e, \rho} . \tag{9.72}$$

Before including "drip" and translational terms, we add an excited state contribution to the nuclear free energy B

$$B = F = E_{\text{bulk}} + E_{\text{size}} - \frac{a}{A}\frac{m^*}{m}T^2, \tag{9.73}$$

$$\frac{a}{A} = \frac{\pi^2}{4E_f} = \left(\frac{\rho_0}{0.16}\right)^{-2/3}(14.9)^{-1}\text{ MeV}, \tag{9.74}$$

$$\frac{m^*}{m} = m_s + \frac{(m_0 - m_s)}{(1 + T/T^*)^2}\frac{E_{\text{size}}}{E_{\text{size}}[\text{Fe}^{56}]}. \tag{9.75}$$

E_f is the nucleon Fermi energy, and $E_{\text{size}}[\text{Fe}^{56}] = 6.88$ MeV. We have assumed the additional density of states due to the surface is proportional to the size energy and can be effectively parametrized in the effective mass (Bethe et al., 1983; Cooperstein, 1985) In our expression for the effective mass, $m_0 = 2$ is m^*/m in the iron peak, and $m_s = 0.7$ is m^*/m for uniform matter. The cutoff factor $(1 + T/T^*)^2$ in the denominator of the surface effective mass term (not present in the earlier work of Cooperstein, 1985), prevents nuclei from persisting at very high temperatures and makes sure the entropy, excitation energy, and pressure due to this term remain positive at high temperature. We employ $T^* = 12$ MeV in our calculations.

Using eqs. (9.72)–(9.75), we obtain for the entropy per nucleus in excited states

$$S^* = 2\frac{a}{A}T\left(\frac{m^*}{m} + \frac{T}{2}\frac{\partial m^*/m}{\partial T}\right) = 2\frac{a}{A}T\left(m_s + \frac{m^*/m - m_s}{1 + T/T^*}\right), \tag{9.76}$$

and for the excitation energy

$$E^* = F^* + TS^* = \frac{a}{A}T^2\left(m_s + \left(\frac{m^*}{m} - m_s\right)\frac{1 - T/T^*}{1 + T/T^*}\right). \tag{9.77}$$

For T well below T^* the negative pressure due to the finite size terms becomes (Cooperstein, 1985)

$$\frac{P}{\rho_{size}} \simeq \beta\theta^{-1/3}G' \cdot \left[1 - \left(\frac{T}{9}\right)^2\right]. \tag{9.78}$$

This is an important modification; the thermal pressure arising from the reduction of the effective mass with increasing density raises γ before the star collapses all the way to $\rho = \rho_s$.

If we include drip and the translational contribution of the heavy nucleus as well, then we now have a four-component mixture: n, ρ, α, and a heavy nucleus, H, with (Z, A). The mass and charge conservation equations (eqs. (9.38)), become

$$\phi_n + \phi_p + \phi_\alpha + \phi_H = 1, \tag{9.79}$$

$$\phi_p + \phi_\alpha\tfrac{1}{2} + \phi_H x = Y_e, \tag{9.80}$$

where the ϕ's are the mass fractions. If we ignore the finite volume of the α-particles, shown to be an unimportant correction by Lamb et al. (1985), then the equation defining the packing fraction (Eq. (9.45)), becomes

$$u = \phi_H\frac{\rho}{\rho_0} = \phi_H\frac{\rho}{\rho_s\theta}. \tag{9.81}$$

Four additional constraints must be satisfied in addition to mass and charge. These are

$$\mu_n - \tfrac{1}{2}\hat\mu = \frac{\mu_\alpha}{4}, \tag{9.82}$$

$$\mu_n - x\hat\mu = B + \frac{\mu_H^{tr}}{A} + \frac{\partial B}{\partial \log\theta}, \tag{9.83}$$

$$-\hat\mu = \frac{\partial B}{\partial x} + \left(\frac{\partial B}{\partial \log(\rho_s/0.16)} - \frac{\partial B}{\partial \log\theta}\right)\frac{\partial \log(\rho_s/0.16)}{\partial x}, \tag{9.84}$$

$$\frac{YT}{1 - u}\frac{\rho}{\rho_0} = \frac{\partial B}{\partial \log\theta} - \frac{\partial B}{\partial \log u}, \tag{9.85}$$

where the heavy fragment translational chemical potential is given by

$$\mu_H^{tr} = \mu_H - AB = \left(\frac{T}{A}\right)\left(\frac{\phi_H\rho}{(1 - u)A}\left(\frac{2\pi\hbar^2}{m_n AT}\right)^{3/2}\right), \tag{9.86}$$

and $Y = \phi_n + \phi_p + \tfrac{1}{4}\phi_\alpha + (1/A)\phi_H$. The last constraint is the pressure balance at the surface, and

$$\frac{P}{\rho} = \frac{YT}{1 - u} + \phi_H\frac{\partial B}{\partial \log u} = YT + \phi_H\frac{\partial B}{\partial \log\theta} \tag{9.87}$$

is the nuclear contribution to the pressure.

Figure 9.10 gives the behavior of several of the important thermodynamic quantities at the lower densities met early in the collapse, from densities between 10^9 g cm^{-3} and 10^{12} g cm^{-3}. The calculations took $W_s = 31.5$ MeV, $K_0(\frac{1}{2}) = 180$ MeV, and assumed $Y_e = 0.42$ with no neutrinos and were carried out along the adiabats $S = 0.75$, 1.0, 1.5, 2.0, and 2.5. Figure 9.10(a), (b), (c), (d) shows the mass fractions ϕ_n, ϕ_p, ϕ_α, and ϕ_H. Figure 9.10(e) shows Z/A and Figure 9.10(f) gives $\hat\mu$.

Figures 9.11 and 9.12 give the behavior of several of the important thermodynamic quantities in the last stages of collapse, from densities between 10^{12} g cm^{-3} and 10^{15} g cm^{-3}, or $\sim 4\rho_s$. The calculations assumed $Y_e = 0.33$, $Y_\nu = 0.07$. Figure 9.11(a), (b), (c), (d) gives ϕ_H, Z/A, θ, and m^*/m and Figure 9.12(a), (b), (c), (d) gives Γ, P/ρ, and P_N/ρ where P_N/ρ is the full nucleonic contribution to the pressure including both the heavy phase's size pressure and the drip and translational pressure.

Important new behavior follows upon the inclusion of the drip. Most of the nucleons which drip into the vapor are neutrons. Thus $Z/A > Y_e$, and as the entropy

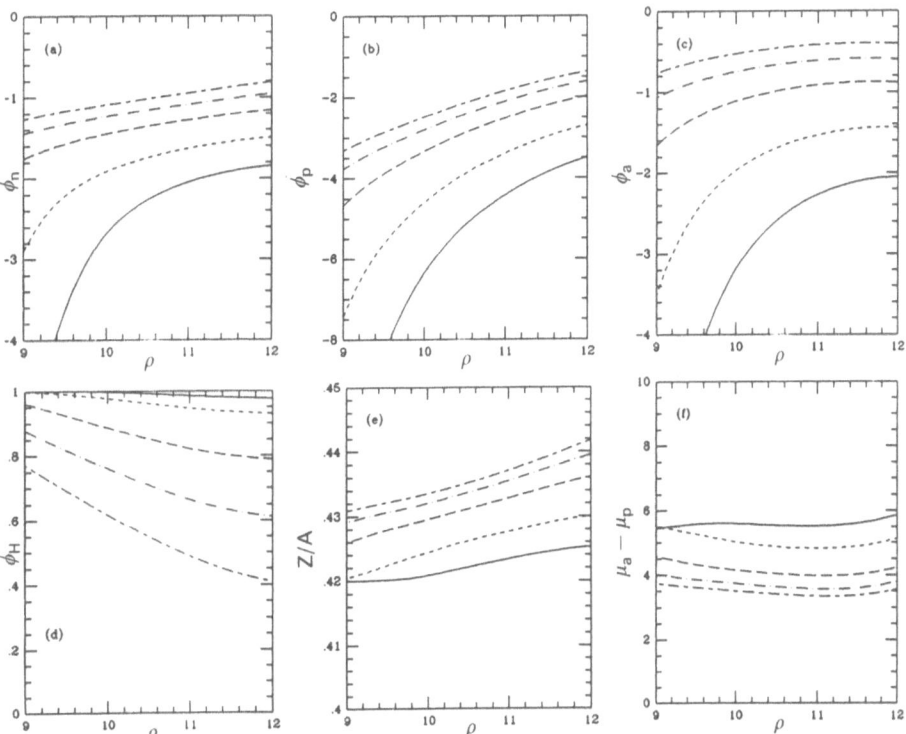

Figure 9.10. Thermodynamic quantities during the early collapse. Results for the adiabats $S = 0.75$ (solid line), $S = 1.0$ (short-dashed line), $S = 1.5$ (long-dashed line), $S = 2.0$ (dashed–dotted line), and $S = 2.5$ (short-dashed–long-dashed line), for the case $Y_e = 0.42$ and no neutrinos. The log of ρ is again given, as in several later figures. (a) φ_n, the free neutron mass fraction; (b) φ_p, the free proton mass fraction; (c) φ_α, the α-particle mass fraction; (d) φ_H, the heavy nucleus mass fraction; (e) Z/A, the nuclear charge ratio; and (f) $\hat\mu = \mu_n - \mu_p$. $\varphi_{n,p,\alpha}$ are given in units of log 10.

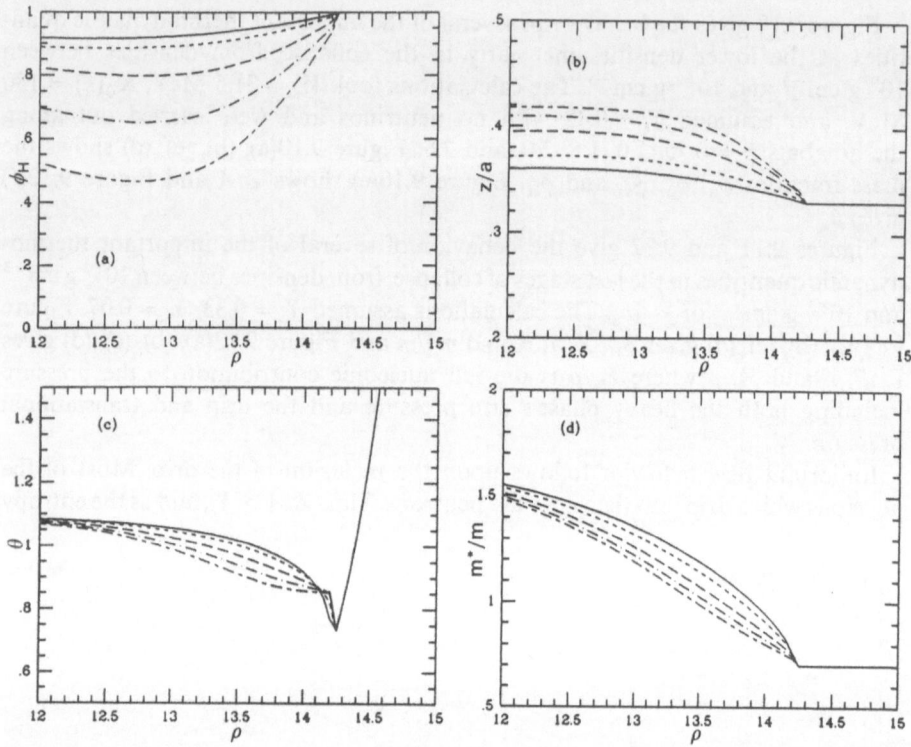

Figure 9.11. Thermodynamic quantities during the final collapse. Results for the adiabats $S = 0.75$ (solid line), $S = 1.0$ (short-dashed line), $S = 1.5$ (long-dashed line), $S = 2.0$ (dashed–dotted line), and $S = 2.5$ (short-dashed–long-dashed line), for the case $Y_e = 0.33$ and $Y_v = 0.07$. (a) φ_H, the heavy phase mass fraction; (b) Z/A the heavy phase proton fraction; (c) θ the heavy phase density compression factor; and (d) m^*/m, the nucleon effective mass.

increases the dense phase becomes increasingly proton rich. This has a major impact because it increases the surface tension, increases the bulk incompressibility, etc. While m^*/m decreases with temperature at fixed density, θ also tends to decrease, with the result that the quantity S^*/T is almost constant until T^*. The fraction, by mass of the matter in the dense phase, ϕ_H decreases with temperature, but goes down with density as the nucleons are squeezed back into the dense phase as the available volume decreases for them. We see that the heavy phase dominates until at least $S \sim 3$, as was pointed out by Lamb et al. (1978).

It turns out that the thermal effects and the proton enrichment of the dense phase tend to be larger than many $T = 0$ phenomena.

9.3.5. Bulk and Size Symmetry Energies and W_s

In Figure 9.13, taken from Bruenn (1989a), we display the free proton fraction X_p in the 15 $M_\odot$ initial model of Woosley and Weaver (1988) for various choices of the bulk symmetry coefficient W_s. Small changes in W_s lead to rather large changes in the proton fraction, because the value of W_s controls the value of $\hat\mu = \mu_n - \mu_p$ and

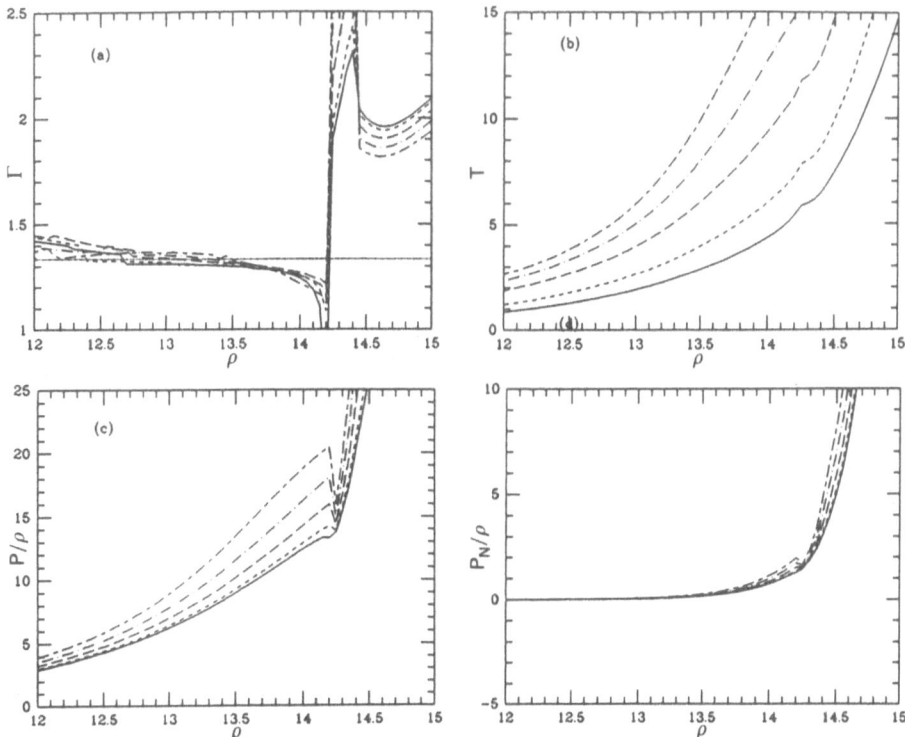

Figure 9.12. Thermodynamic quantities during the final collapse. Results for the adiabats $S = 0.75$ (solid line), $S = 1.0$ (short-dashed line), $S = 1.5$ (long-dashed line), $S = 2.0$ (dashed–dotted line), and $S = 2.5$ (short-dashed–long-dashed line), for the case $Y_e = 0.33$ and $Y_v = 0.07$. (a) Γ_s the adiabatic index, with the dotted line showing $\Gamma_s = 4/3$; (b) T (in MeV); (c) P/ρ the total pressure per nucleon; and (d) P_N/ρ the total contribution due to nucleons.

hence the proton fraction depends on W_s exponentially. As we discuss in Section 9.5.1, the final lepton fraction is very sensitive to the value of X_p since this determines the amount of electron capture.

In principle, it is not really correct to vary the bulk symmetry coefficient, W_s, without also varying the surface symmetry contribution. The important measurable parameters are μ_n and μ_p, which are related to the neutron and proton separation energies. The chemical potentials depend on both the bulk and the surface symmetry coefficient, and so we must also vary the surface symmetry coefficient when we vary W_s. With the size energy expression we use in our equation of state, the choice $W_s = 31.5$ MeV gives the best fit to the nuclear mass tables.

Thus, the study that Bruenn (1988a) has performed should be viewed as a parameter study of the effects of varying the proton fraction, rather than one of varying W_s. We should remember that supernova matter is considerably more neutron rich than laboratory nuclei. The extrapolation of liquid drop mass formulas that fit terrestrial nuclei to such conditions contains uncertainties, and the free proton fraction may vary upon the introduction of improved physics.

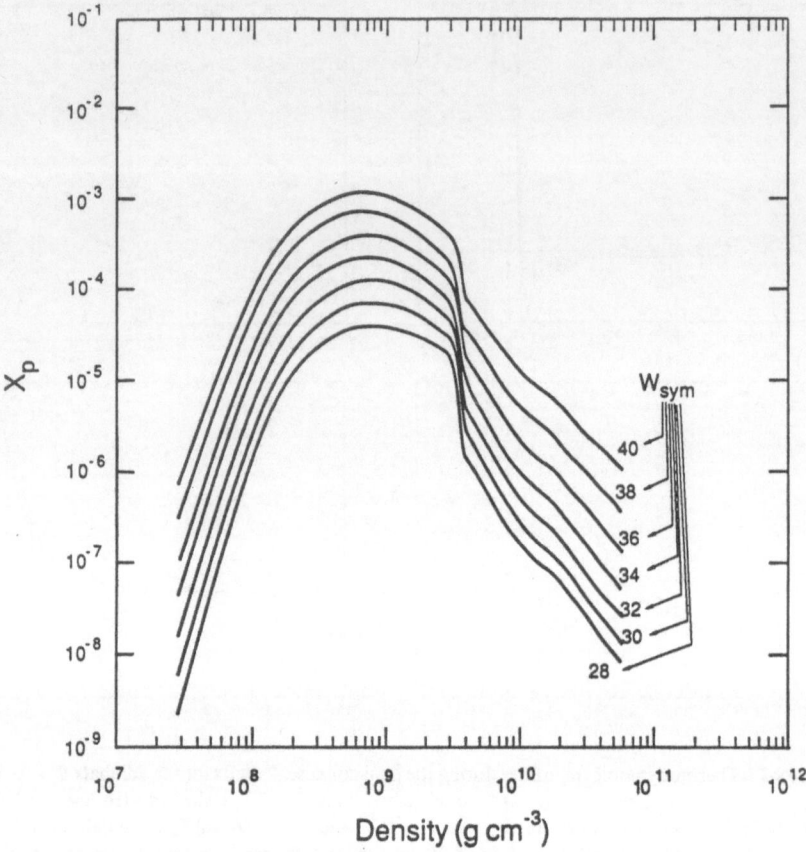

Figure 9.13. Proton fraction for various choices of W_s. The free proton fraction for various choices of the bulk symmetry coefficient W_s for the 15 $M_\odot$ model of Woosley and Weaver (1988) taken from the results of Bruenn (1988a). The numbers refer to the value of W_s in MeV.

9.3.6. The Transition to Nuclear Matter

The transition to uniform nuclear matter from the bubbles (or "casserole") regime is a weakly *first-order* phase transition. In supernova material this transition is muted because of the relativistic leptons which provide the dominant contributions to the energy and pressure, and which do not participate in the transition. Furthermore, as can be seen from Figure 9.11(a), the drip particles tend to get squeezed back into the dense phase as the packing fraction approaches unity. The most natural way to make use of this behavior is to assume that the phase transition is actually *second order* (Cooperstein, 1985), so that Maxwell constructions are avoided which puts much less stress on the supernova hydrodynamical codes. Since the transition is assumed to be second order, quantities such as the energy and their first derivatives, such as the pressure, will be continuous, but the second derivatives, such as the adiabatic index, will be discontinuous.

Of critical importance is the fact that the merger to uniform nuclear matter occurs well below $\rho = \rho_s(Y_e)$, at about 75–80% of that value. It is at this point, not at ρ_s, that the adiabatic index rises and the homology of the collapse is strongly disturbed, even though the nuclear pressure is still negative. The energy of this material is described by the compressible liquid drop E_{bulk} of eq. (9.40), and because we are not far from ρ_s the quadratic approximation in the density deviation from saturation suffices. Above ρ_s, we must apply the formulas of Section 9.4.

The adiabatic index (at $T = 0$) for this subsaturation uniform medium is given by Baron, Cooperstein, and Kahana (1985b)

$$\Gamma = \frac{4}{3}\left(\frac{1 - (K_0/K_1)\theta^{2/3}(2\theta - 3)}{1 - \frac{4}{3}(K_0/K_1)\theta^{2/3}(1 - \theta)}\right), \tag{9.88}$$

where $K_1 \sim 255$ MeV (see eq. (9.92)) and K_0 is the nuclear incompressibility at Y_e. If we take $K_0 = 140$ MeV we find that for $\theta = (0.80, 0.9, 1.0)$, $\Gamma = (1.81, 1.94, 2.06)$. Thus there is a sharp rise in the adiabatic index even before $\theta \to 1$. (If we used a larger value for K_0 this would be even more dramatic.)

The thermal effects on the transition can be seen from Figure 9.12. The phase transition to nuclear matter is signalled by an abrupt increase in the adiabatic index Γ_s displayed in Figure 9.12(a). The temperature, displayed in Figure 9.12(b) is smooth across the transition.

It is very important (as well as convenient) that we have used, for our energy of nuclear matter at densities below saturation, the exact same functional form we employ for the compressible liquid drop nuclear material in the nonuniform system. Knowing that the transition is only weakly first order, this maneuver enables us to assume the transition is second order. If this is not done, new and artificial discontinuities can develop at the transition between the two systems. For example, in our earlier *incompressible* liquid drop model (Bethe *et al.*, 1983), there was an abrupt increase in temperature at the transition. Such behavior (especially discontinuities in pressure) can disrupt the homologous collapse unnecessarily and have a deleterious effect on shock formation.

9.4. Suprasaturation Density Equation of State

The equation of state for matter with $\rho > \rho_0$ controls the bounce of the star. In particular, when the equation of state is relatively soft the matter implodes to a higher density before bouncing at *maximum scrunch* (Cooperstein, 1982). Because it climbs out from deeper within the gravitational well, more energy becomes available to the explosion. Precisely how this occurs is not well understood analytically, especially with general relativistic hydrodynamics, but it is quantitatively known from simulations, and we give results in Section 9.5.2.

Experimentally, only two areas have yielded good data for simulation guidance. The first is the analysis of the breathing mode in the giant monopole resonance, which provides information about the properties of nuclear matter near saturation density. We discussed this matter in Section 9.3.3. The second is the observed masses of neutron stars, which tell us about the properties of neutron matter at the central densities of these objects, $\rho_c \sim 4$–$8\rho_0$ according to the theoretical analysis. We will discuss this constraint in Section 9.4.2.

A third experimental area has been relativistic heavy ion collisions, in which the matter is not neutron rich and may reach densities up to several times ρ_s. Earlier analysis of the data gave results indicative of a stiff equation of state (see Stöck, 1986; and Stöcker and Greiner, 1986, and references therein). But these collisions take place at high momenta (Ainsworth et al., 1987). If we take into account the momentum dependence in the potential, the density dependence of the effective mass, in terms of scattering from the energy-dependent real part of the optical potential, such a stiffening can be accounted for (Ainsworth et al., 1987; Brown, 1988a). Recent calculations including these effects (Aichelin et al., 1987) have shown that a soft but momentum-dependent equation of state can reproduce the observed data in heavy ion collisions. At the present time these experiments have not yielded reliable determinations of the equation of state of dense, cold, uniform material.

The matter found in a collapsing supernova has different properties than that observed in neutron stars, or that measured in the breathing mode experiments. It is neutron rich, but not neutron matter (having $Y_e \sim \frac{1}{3}$), has plenty of relativistic electrons and neutrinos, finite but low temperatures ($T \sim 5$–15 MeV), and densities probably no more than $\rho_c \sim 4\rho_0$. Thus, the equation of state we use for this material, while it should incorporate the known data, must by necessity have extrapolations.

9.4.1. BCK Equation of State at $T = 0$

A simple functional form widely exploited in supernova simulation is the BCK expression for the nucleonic pressure and energy per nucleon

$$P_N = \frac{K_0(x)\rho_0(x)}{9\gamma}(u^\gamma - 1), \tag{9.89}$$

$$E_n = E_{nm} + W_s(1 - 2x)^2 + \frac{K_0(x)}{9\gamma}\left(\frac{u^{\gamma-1} - 1}{\gamma - 1} + \frac{1 - u}{u}\right), \tag{9.90}$$

where $u = \rho/\rho_s(x)$ is the density compression factor. (Note that in the uniform medium the packing fraction is of course unity, and that u here corresponds to θ in the subsaturation material.) The nuclear bulk incompressibility K_0 is defined at ρ_s, the saturation density, Both K_0 and ρ_s are functions of $x = Z/A = Y_e$, as discussed in Section 9.3.2.1. A new important parameter in these formulas is γ which becomes the adiabatic index at high density and which differs from the actual adiabatic index, as we will see in eq. (9.94).

As discussed in Section 9.3.3, the most commonly accepted value from the breathing mode data is $K_0 = 210 \pm 30$ MeV for symmetric matter. BCK have usually used the value $K_0 = 180$ MeV for matter with $x = \frac{1}{2}$, and in the following we will use this value for examining several features of the equation of state.

It is very important that supernova material contains electrons and neutrinos because we will see they dominate the pressure until about $3 \rho_0$ for the BCK equation of state we will use here. Taking $x = \frac{1}{3}$ so that $\rho_s = 0.145$ fm^{-3} we get, for the full equation of state,

$$\frac{P}{\rho_0} = \frac{K_1}{9\gamma_1}u^{4/3} + \frac{K_0}{9\gamma}(u^\gamma - 1), \tag{9.91}$$

where $\gamma_1 = \frac{4}{3}$ and $K_1 = 255$ MeV for $Y_e = 0.33$ and $Y_v = 0.07$, which gives an effective

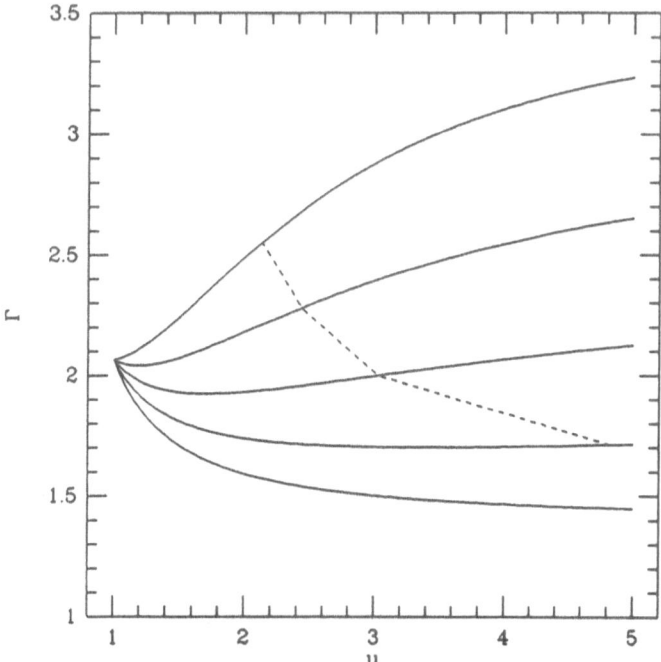

Figure 9.14. Adiabatic index of nuclear matter at $T = 0$. The full adiabatic index of eq. (9.94) is displayed for the BCK equations of state of eq. (9.91), with $\gamma = 2$, 2.5, 3.0, and 4.0. The dashed line shows where the nuclear pressure becomes larger than the leptonic pressure.

lepton number for the pressure of

$$Y_{1,\text{eff}} = (0.33^{4/3} + 2^{1/3}0.07^{4/3})^{3/4} = 0.37. \tag{9.92}$$

This gives for the full incompressibility at ρ_s, with $K_0(\tfrac{1}{3}) = 140$ MeV,

$$K(\rho_0) = K_1 + K_0 \sim 395 \text{ MeV} \tag{9.93}$$

even for a "soft" equation of state, and for the full adiabatic index we obtain

$$\Gamma = \frac{K_1 u^{4/3} + K_0 u^{\gamma}}{(K_1/\gamma_1)u^{4/3} + (K_0/\gamma)(u^{\gamma} - 1)}, \tag{9.94}$$

which is displayed in Figure 9.14. Note that at ρ_0 we have

$$\Gamma = \frac{4}{3}\left(1 + \frac{K_0}{K_e}\right) = 2.07 \tag{9.95}$$

independent of γ. Even at $u \sim 4$ the adiabatic index is nowhere near its asymptotic value of γ, both because of the influence of the leptons and because of the -1 in the pressure due to saturation. The nucleonic pressure surpasses the leptonic pressure only when

$$\frac{K_1}{9\gamma_1}u^{4/3} = \frac{K_0}{9\gamma}(u^{\gamma} - 1) \tag{9.96}$$

and this is shown by the dashed line in Figure 9.14. For $\gamma = 2.5$ this is not accomplished until $u = 3$.

The finite temperature extension of BCK is quite trivial, since we use just the bulk piece of eqs. (9.72)–(9.75), and is just that for a degenerate Fermi gas with $m^*/m = m_s = 0.7$. It introduces a small $\gamma = \frac{5}{3}$ piece in the pressure. This matters right near ρ_s where the $T = 0$ nuclear pressure vanishes, but once we get solidly into nuclear matter the term is not quantitatively important. Since the combined adiabatic index of the electrons plus the $T = 0$ nuclear material is quite close to $\Gamma = 2$, the effect is not that large, as can be seen by examining the adiabats in Figure 9.12(a). But it is another contribution to the pressure and thus it does stiffen the equation of state somewhat.

9.4.2. Neutron Stars and the Equation of State

Objections have been raised that the BCK equation of state described in Section 9.4.1 cannot be used to make neutron stars (e.g., Glendenning, 1986; Takahara and Sato, 1988) of enough mass to satisfy the observations (see Taylor and Weisberg, 1982). These indicate that the maximum neutron star mass is at least $1.55\ M_\odot$, but there is no good data to indicate that it need be larger.

There is a one-to-one correspondence between the stiffness of the neutron matter equation of state and the maximum mass of a neutron star. In fact, in the absence of any other high density data, the maximum mass an equation of state yields is probably the best quantitative definition we could give for stiffness. It is certainly true that if we simply take the BCK equations of state we have used in our supernova calculations, insert them in *unmodified* fashion into the hydrostatic neutron star structure codes, and then find the maximum mass, the resulting masses are too low to satisfy the observations.

However, this is inappropriate, as the BCK equations of state were constructed for only somewhat neutron-rich matter and not for neutron matter (or more precisely, the β-equilibrium, mostly neutron matter) found in the neutron star. While it is true that equations of state, such as that of Glendenning (1986), have specific forms and density dependence for the symmetry energy and thus give the equation of state for symmetric, neutron rich, and neutron matter, the conclusions deduced from them are no more general than the specific model or method used to derive them. Furthermore, the crucial range of density for the equation of state of a neutron star is perhaps 4–$8\ \rho_0$, the density at the center, which lies above the density achieved at maximum scrunch in supernova collapse, as described in Section 9.5.2.

In a recent study, Cooperstein (1988a) again pointed out the fact that equations of state with quite similar behavior near saturation density can give quite different neutron stars, and similarly, equations of state which give similar neutron stars can have quite different behavior at low densities. It depends both on the density-dependence of the symmetry energy, and upon the change in the adiabatic index with density, and the BCK equation is too simple to mock up these additional factors. This question has been taken up recently by Prakash, Ainsworth, and Lattimer (1988), who construct neutron stars with quite high masses with soft symmetric equations of state, utilizing phenomenological Yukawa-like repulsions and attractions, and with a variety of choices for the density dependence of the symmetry energy.

It would be nice to stop here and say that since so little is really known about the connection between neutron star matter at high density and matter near saturation density, the relatively soft equations of state used by BCK are readily permissible. However, a detailed examination of the work of Prakash, Ainsworth, and Lattimer (1988) for $x = \frac{1}{3}$ neutron-rich matter indicates that quite a bit of stiffening already occurs at this point. Furthermore, the symmetry energy of the BCK equation of state definitely rises too slowly with density, and $\hat{\mu}$ can go negative at a density $\rho \sim 4.5 \, \rho_s$, and we do not mean to defend such behavior.

At the present time the reason shocks fail has not much to do with the equation of state of suprasaturation densities. As we discuss in Section 9.5, it is primarily because of too much deleptonization during the infall, and weakening of the shock wave at late times due to radiation of neutrinos of all flavors. Use of very soft equations of state will not make the models which fail at the present time succeed. Thus we have not yet incorporated in our calculations other high-density equations of state because this is not the cutting edge of the problem at the present moment (although with a slight variation in the input physics, such as a reduction of the initial iron core mass, the equation of state could wind up back on the throne as the decisive factor). It is our opinion that the equation of state of neutron-rich matter found at the highest densities of stellar collapse is still not a well-understood problem.

9.5. Numerical Results

Numerical calculations of stellar collapse have become increasingly sophisticated (and hopefully more accurate) in the last few years. Almost all calculations today include general relativistic hydrodynamics, non-LTE neutrino transport with some degree of general relativistic corrections, a host of neutrino absorption, emission, and scattering processes, and varying degrees of sophistication in the equation of state. The increased complexity in the numerical computations has not arisen from a blind "kitchen sink" approach, throwing into the calculation every effect that we can think of. Each one of these complications has been studied and shown to be important in stellar collapse.

Our own calculations incorporate fully general relativistic hydrodynamics and neutrino transport (Baron *et al.*, 1989). A typical calculation includes about 120 Lagrangean mass zones for the iron core. The resolution in the shock propagation region is quite fine, with a maximum $\Delta M = 1/120 \, M_\odot$. Shocks are handled through the standard pseudo-viscosity method (see Baron (1985), for details).

The neutrino transport method is a two-fluid (frequency-integrated) flux-limited diffusion model, where we assume that the distribution function has a Fermi–Dirac form, parametrized by a temperature and chemical potential (van den Horn and Cooperstein 1986; Cooperstein, van den Horn, and Baron, 1986, 1987; Cooperstein, 1988b). Neutrinos are not assumed to be in equilibrium with the matter. The flux limiter is adapted from that of Levermore and Pomraning (1981).

The most important ingredient in the calculation, the equation of state, we have already discussed in some detail.

Throughout this section we will make use of the results of calculations by Bruenn (1989a, b) and by Myra and Bludman (1989). In their most recent work, these authors have adopted our equation of state, but use multi-group flux limited diffusion. As we shall see, the results of both methods are in good agreement, although all three groups differ to some extent on specific details of the results.

9.5.1. Neutrinos and Collapse

Neutrino physics has played an important role in collapse calculations throughout their history. One mechanism for producing the explosion was for neutrinos to deposit their momentum behind the shock wave producing an explosion (Colgate and White, 1966). With the advent of weak neutral currents it was realized that neutrinos are trapped on collapse timescales (Mazurek, 1975; Sato, 1975). Thus large fluxes of neutrinos cannot reach the mantle on collapse timescales. A variant of this model, in which the neutrinos deposit their energy behind the shock on the far longer deleptonization timescale of about 1 s (Bethe and Wilson, 1985; Mayle, 1985; Wilson, 1985), is popular today and is discussed by Mayle in this volume.

During the infall neutrinos are produced mainly by electron capture on free protons, $e + p \rightarrow \nu_e + n$. This is a super-allowed transition which produces an electron-type neutrino ν_e. Electron-type neutrinos can also be produced by the analogous process on nuclei, $e + (A, Z) \rightarrow \nu_e + (A, Z - 1)$. This process tends to be cut off as the matter becomes neutronized since the allowed transition is blocked by a filled neutron shell and the transition must then go by first forbidden reactions (Fuller, 1982; Fuller, Fowler, and Newman, 1982). Thermal effects may mitigate this somewhat (Cooperstein and Wambach, 1984). In any case capture on free protons is probably the main producer of neutrinos.[5]

Other types of neutrinos (including $\bar{\nu}_e$) are not produced in any appreciable quantity on infall since they must be produced by pair emission and the temperatures tend to be too low for this until the shock wave is produced. Electron antineutrinos can also be produced by positron capture upon neutrons, but the high electron degeneracy during infall ensures that there are too few positrons to produce a significant number of antineutrinos.

Until the center reaches trapping density, about 10^{12} g cm^{-3} neutrinos freely escape from the star, reducing the lepton fraction and hence the size of the homologous core and the point where the shock is produced. In Section 9.2.2 we pointed out that the size of the homologous core scales as $\sim Y_l^2$. Equivalently, we note that this reduces the effective adiabatic index, as discussed in Section 9.2.3. At densities above this the neutrinos must diffuse from the core and although some more will escape from the homologous core they do so more slowly. The reduction in the lepton fraction Y_l can be seen in Figure 9.15 taken from the results of Bruenn (1989a). In these calculations the initial Y_l is about 0.42 in the center and rises to reach 0.5

---

[5] The physics of electron capture is, however, not a closed subject. There exist at present no hydrodynamical calculations which include a full treatment of the heavy nucleus electron captures, including thermal unblocking and forbidden captures. Bruenn (1985), however, has parametrized these effects somewhat. See Cooperstein (1988b) for some further analysis of this problem.

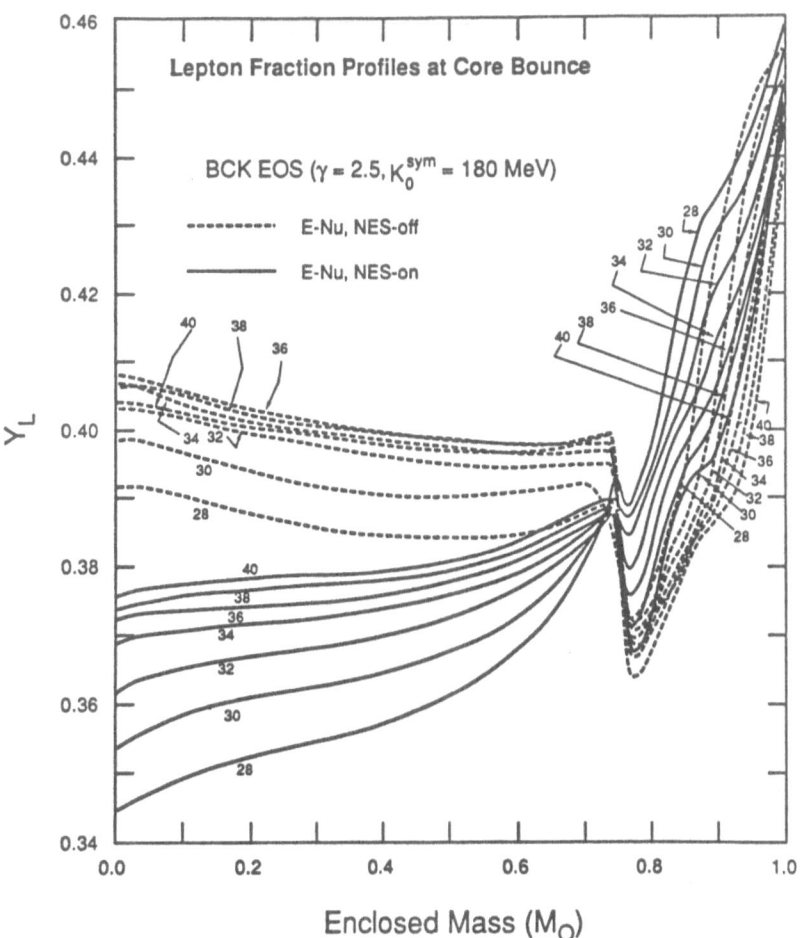

Figure 9.15. Y_l profiles from the results of Bruenn (1988a). The Y_l profiles at bounce from the results of Bruenn (1988a). The initial model is the 15 $M_\odot$ model of Woosley and Weaver (1988). The solid lines refer to calculations with the effects of neutrino–electron scattering (NES) included, and the dashed lines for the case without NES. The numbers refer to the value of the bulk symmetry coefficient W_s which controls the proton fraction, see Section 9.3.5.

at the outside of the iron core. The dashed lines show the value of the lepton fraction for various choices of the amount of free protons (the numbers refer to the value of W_s, see Section 9.3.5). The most realistic case is labeled by the line marked 32 in which case direct and diffusive loses result in a reduction of the central Y_l by about 0.02 to slightly greater than 0.40.

Figure 9.15 also illustrates a much more insidious process that reduces Y_l, that of neutrino electron scattering $v + e \rightarrow v + e$. The effects of this process have been studied in detail by Bruenn (1985, 1989a, 1989b) and by Myra et al. (1987) and Myra and Bludman (1989). Since the electrons tend be extremely degenerate in the iron core, the only way that a neutrino can scatter off an electron is to raise the electron's energy (and lower its own in the process). Thus, this process tends to reduce the average energy of the neutrinos. Because the neutrino opacity goes as the square of the neutrino energy, the matter is less opaque to lower energy neutrinos and the neutrinos get out easier.

We see in Figure 9.15 that for the case $W_s = 32$ MeV the final lepton fraction drops from ~ 0.405 to ~ 0.36 and that the mass where the shock forms, M_{form}, is reduced by $(0.36/0.405)^2 = 0.8$ which is quite significant. The effects of neutrino electron scattering seem to be limited to the infall stage and appear to have little effect on the propagation of the shock.

We should note that all of the above analysis has relied on the results of Bruenn, and our own results differ slightly in the details (Baron, Cooperstein, and Aufderheide, in preparation). When all the effects are included the agreement is quite good; however, we differ on the exact magnitude of the effect of neutrino–electron scattering. Our calculations give significantly lower lepton fractions than do Bruenn's in the case of no scattering. We find $Y_l \sim 0.375$ in that case, as compared to Bruenn's value of 0.405 for the case of $W_s = 32$ MeV. When we turn electron scattering on we get a similar value to Bruenn. While we could speculate on the reason for such a difference it is most important to note that when the best physics is put into both calculations we find a similar result.

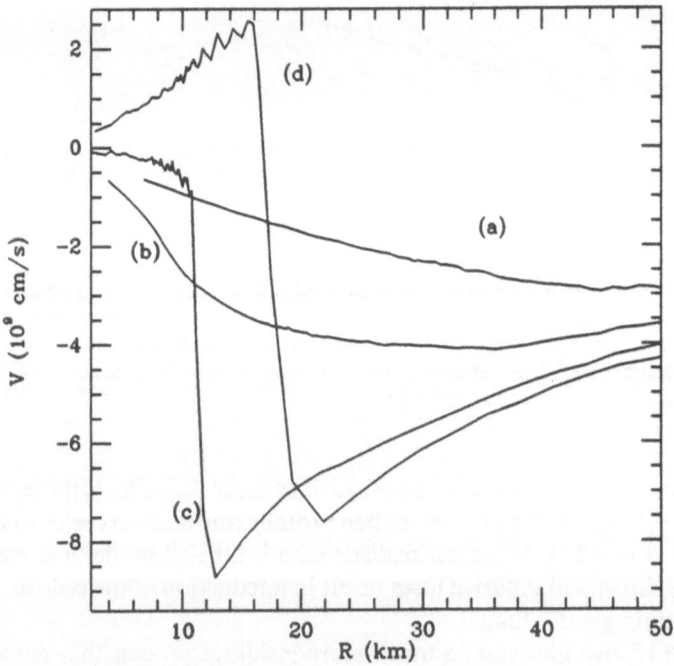

Figure 9.16. Velocity profiles at various times. The velocity profiles taken from the collapse of the 15 $M_\odot$ model of Woosley and Weaver (1988). Time (a) is at last good homology; time (b) is when the center has gone above nuclear matter density and the homology is broken; time (c) is at maximum-scrunch; and time (d) is when the shock wave has been launched.

9.5.2. Bounce and General Relativity

Bounce is the time when core collapse is halted due to the increase in the adiabatic index, Γ_s, since nuclear matter is stiff compared to degenerate electrons. Bounce really begins when the center of the star attains a density greater than about $\frac{2}{3}\rho_s$ (see Sections 9.3.2.5 and 9.3.6), since this is where the equation of state begins to stiffen (Γ_s rises), even though the nuclear pressure is still negative. At this point the central collapse rate slows down and the core begins to be compressed, much like a spring.

Once the equation of state is no longer well described by a single power law, there is a new scale in the problem (the radius at which the adiabatic index rises significantly above $\frac{4}{3}$) and the homology is broken. This can be seen in Figure 9.16 where the velocity is plotted as a function of radius for several times during bounce. At time (a) the flow is homologous. The central density is $\rho_c = 4.8 \times 10^{13}$ g cm^{-3}. This time is known as "last good homology" (Brown, Bethe, and Baym, 1982).

At time (b) in Figure 9.16 the central density has reached $\rho_c = 2.6 \times 10^{14}$ g cm^{-3} and clearly the homology has been broken, in fact, the first hints of a velocity discontinuity (a shock wave) are becoming evident. As the density continues to increase, accoustical waves increase in amplitude, and as more material is brought over ρ_s, further signals are generated. These propagate out towards the "sonic point," the region where the infall velocity becomes supersonic. The sonic point is the point in Figure 9.17 where the sound speed and infall velocity cross. The sound

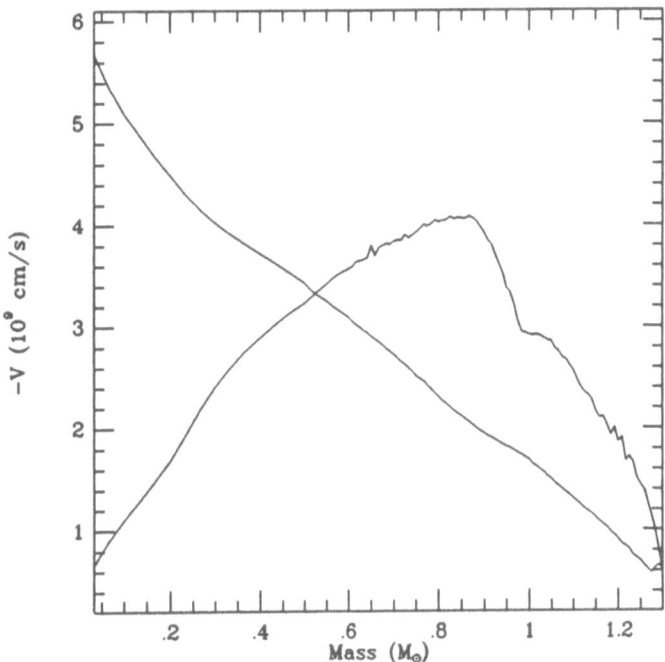

Figure 9.17. Sonic point. The velocity profile of a calculation of the 18 $M_\odot$ star of Weaver and Woosley (1988) plotted with the sound speed (the monotonically decreasing line). The point where the two curves cross is the sonic point and the shock wave will form approximately there.

waves cannot propagate further in space and eventually will crest at the sonic point, add together, and begin to impart enough momentum reversal to the infalling matter to produce discontinuities in velocity, pressure, and density which define the developing shock wave. The sonic point roughly corresponds to M_{form}.

The second phase of bounce is known as "maximum-scrunch" (Cooperstein, 1982) and that is the point where the central density reaches its maximum. At this point the spring is fully compressed and the core is ready to rebound and the shock wave is beginning to form at the sonic point. This can be seen in Figure 9.16. The curve labeled (c) is at maximum-scrunch. The central density is $\rho_c = 9.7 \times 10^{14}$ g cm^{-3}. We can see that the core is completely scrunched with the central 10 km being very nearly at rest. Also note that a shock wave has formed since the velocity is discontinuous. An important dynamical effect of having a relatively soft equation of state is illustrated here, in that the core as one coherent unit is allowed to come to rest by the soft equation of state. If the equation of state is stiff, that never happens and the center starts to move out before the outer parts have stopped completely. This effect seems to lead to an "impedance mismatch" and the core delivers its rebound energy less efficiently to the load, i.e., to the shock wave.

During this phase the general relativistic effects are most important because the compression is at a maximum. The central density reached at maximum-scrunch is important for determining the strength of the shock wave. The higher the density reached the deeper the shock wave digs into the gravitational well, and hence the shock will be launched with a larger energy. Clearly, having a softer equation of state leads to higher maximum central densities.

At time (d) in Figure 9.16 the core has rebounded to a central density of $\rho_c = 6.9 \times 10^{14}$ g cm^{-3}, and the shock is clearly on its way out through the iron core. The last phase of bounce is when the central core rebounds and settles to its equilibrium density, $\rho_c \sim 4 \times 10^{14}$ g cm^{-3}. By this time (roughly 0.5 ms after maximum-scrunch) the shock wave is fully formed (at a radius of about 30 km) and the unshocked core is in hydrostatic equilibrium.

Early calculations of stellar collapse ignored the effects of general relativity, both for simplicity and because it was thought to be only a relatively small correction to the overall dynamics. Van Riper and Arnett (1978) noted that general relativity may be important if high densities are reached in collapse, but cores were large then and it was difficult to explode them. With the advent of smaller cores, and the use of a relatively soft high-density equation of state, the dramatic effects of general relativity became apparent (Baron, 1985; Baron, Cooperstein, and Kahana, 1985a, b).

General relativistic dynamics has two main effects on the collapse and shock formation. The first effect, which is a helpful one, is that general relativity strengthens the gravitational effects over the Newtonian case and hence makes the collapse go to higher densities. This allows the core to dig deeper into the gravitational well and the shock wave is produced with more initial energy than in the Newtonian case. Since the effects of general relativity are amplified as the densities get higher (in a nonlinear way) the coupling of general relativity with a softer equation of state leads to extremely energetic shock waves.

The second effect of general relativity is a harmful one. General relativity alters the critical adiabatic index, γ_{cr}, for a hydrostatic star to be stable against radial

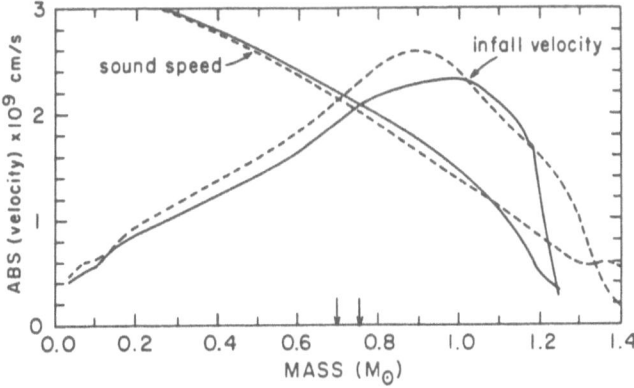

Figure 9.18. Comparison of sonic points for Newtonian and general relativistic calculations. The sonic point is displayed for a Newtonian (solid line) and a general relativistic calculation (dashed line) from Baron (1985). The position of the sonic point is denoted on the mass axis by the arrows. Note the general relativistic effects move the sonic point to a smaller mass.

perturbations. While for an ordinary star that is well described by Newtonian gravity, $\gamma_{cr} = \frac{4}{3}$, and for a general relativistic star, $\gamma_{cr} \sim \frac{4}{3} + 3\sigma$ (Baron, 1985) where $\sigma = P/\rho c^2$ at the center of the star. Thus, γ_{cr} is raised above its Newtonian value. Yahil (1983) has shown, in the Newtonian case, the size of the homologous core goes inversely with $(\gamma_{cr} - \gamma)$ (see Section 9.2.3), so the higher critical adiabatic index leads the shock wave to have a smaller formation mass M_{form}. In early calculations displayed in Figure 9.18 it was found that the shock wave forms at about 0.75 $M_\odot$ in the Newtonian case, and at about 0.66 $M_\odot$ in the general relativistic case. These calculations had a somewhat high value for the lepton fraction, but the difference between them is indicative. In addition to this harmful effect the shock wave must now move out against the stronger gravity.

An additional, somewhat mitigating effect occurs since gravity is stronger and the pressure deficit is greater in the general relativistic case than in the Newtonian case. As can be seen in Figure 9.19, the center falls in faster than the outer parts so the density is lower when the shock reaches the outer parts (just past the homologous core at a mass point of about 0.8 $M_\odot$) which makes it easier for the shock wave to propagate.

It is clear that general relativistic effects cannot be neglected in calculations.

9.5.3. Shock Propagation
The shock's propagation through the iron core is best thought of as a struggle not to lose energy. The shock wave on its propagation through the iron core faces two enormous sinks of energy: dissociation of iron into free nucleons, which we have already discussed in Section 9.2.2, and neutrino losses. If the shock can keep its strength, it can minimize both these losses. Stronger shocks lead to reduced neutrino losses because the time that the shock wave spends in the "mine field" (Bethe, 1981), the region just outside the neutrinosphere where neutrino losses are most copious, is reduced since a stronger shock moves at a higher velocity.

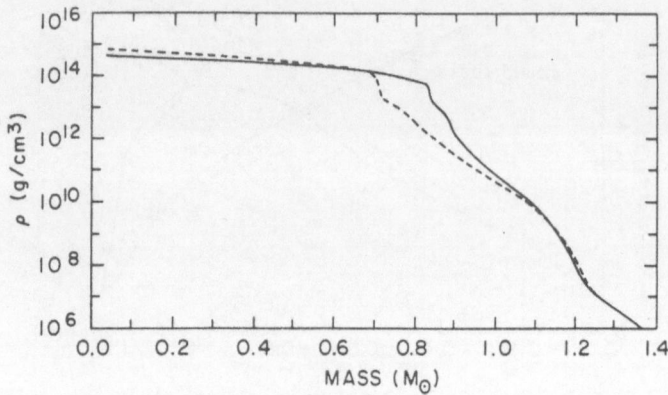

Figure 9.19. Comparison of density profiles for Newtonian and general relativistic calculations. Density profiles at core bounce for a Newtonian (solid line) and a general relativistic (dashed line) calculation from Baron (1985). Not only does the general relativistic calculation go to a higher central density, but the point where the density drops has also moved in. This effect helps to reduce some of the harm caused by the moving in of the sonic point, see Figure 9.18.

Dissociation losses can be reduced because a stronger shock wave will put more energy into the expansion of material, which will reduce the temperature and thus there comes a point where dissociation is no longer complete, but only to α-particles with a much lower energy cost (Cooperstein, Bethe, and Brown, 1984). This lower temperature will also help to lower the neutrino luminosity which is roughly blackbody and so is proportional to T^4.

An important effect on the propagation that was neglected in the early calculations of BCK (1985a, b) is the effect of other types of neutrinos. As has been pointed out by Bruenn (1989a) the effect of μ and τ pairs is only detrimental to the shock wave since they lower the pressure behind the shock wave, weakening it. (Energy is transferred from the less relativistic and hence stiffer dissociated nucleons to the relativistic neutrinos.) They are also an additional source of luminosity behind the shock wave and hence they increase the shock wave's losses. Electron-type antineutrinos are also extremely harmful, in that making a pair of electron–antielectron neutrinos opens an additional channel for electron-type neutrino losses, as well as the effects that antielectron-type neutrinos have in common with μ and τ pairs. Another important effect of including antielectron-type neutrinos is the URCA-like process

$$n + e^+ \rightarrow p + \bar{\nu}_e, \tag{9.97}$$

$$p + e^- \rightarrow n + \nu_e, \tag{9.98}$$

which increases the energy current without increasing the lepton number current (Bruenn 1989a). This may be the largest effect of including additional neutrino types.

Tables 9.3–9.8 illustrate the effect of including all the flavors of neutrinos. They give the history of shock propagation for two calculations. The calculation labeled "All types" has all flavors of neutrinos included, whereas the calculation labeled "Only ν_e" has only electron-type neutrinos included. There is no other difference.

Table 9.3. Quantities at bounce.

	Bounce	
	All types	Only v_e
Time (s)	0.22806	0.223874
ρ_c (g cm^{-3})	9.683×10^{14}	9.710×10^{14}
E_{v_e} (10^{51} ergs)	0.289	0.2839
$E_{\bar{v}_e}$ (10^{51} ergs)	0.0	0.0
E_{pairs} (10^{51} ergs)	0.0	0.0

Table 9.4. Quantities at shock formation.

	Shock formation	
	All types	Only v_e
Time (s)	0.228438	0.224237
t_{ab} (ms)	0.4	0.4
ρ_c (g cm^{-3})	4.664×10^{14}	4.692×10^{14}
R_s (km)	27.84	26.96
M_s ($M_\odot$)	0.9667	0.958
E_{v_e} (10^{51} ergs)	0.303	0.296
$E_{\bar{v}_e}$ (10^{51} ergs)	0.0	0.0
E_{pairs} (10^{51} ergs)	0.0	0.0

Table 9.5. Quantities when shock is near 150 km.

	Shock near 150 km	
	All types	Only v_e
Time (s)	0.231111	0.22662
t_{ab} (ms)	3.1	2.7
ρ_c (g cm^{-3})	4.967×10^{14}	4.942×10^{14}
R_s (km)	142.9	145.8
M_s ($M_\odot$)	1.1916	1.1916
E_s (10^{51} ergs)	1.36	2.94
$Y_{l_{min}}$	0.24	0.29
E_{v_e} (10^{51} ergs)	0.459	0.411
$E_{\bar{v}_e}$ (10^{51} ergs)	0.0	0.0
E_{pairs} (10^{51} ergs)	0.0	0.0

Table 9.6. Quantities when shock is near 300 km.

	Shock near 300 km	
	All types	Only ν_e
Time (s)	0.236099	0.231060
t_{ab} (ms)	8.0	7.2
ρ_c (g cm^{-3})	4.963×10^{14}	4.906×10^{14}
R_s (km)	293.3	315.7
M_s ($M_\odot$)	1.25833	1.25833
E_s (10^{51} ergs)	1.00	2.66
$Y_{l_{min}}$	0.18	0.23
E_{ν_e} (10^{51} ergs)	1.6	0.998
$E_{\bar{\nu}_e}$ (10^{51} ergs)	0.16	0.0
E_{pairs} (10^{51} ergs)	0.4	0.0

Table 9.7. Quantities when shock is near 500 km.

	Shock near 500 km	
	All types	Only ν_e
Time (s)	0.24294	0.23606
t_{ab} (ms)	14.9	12.2
ρ_c (g cm^{-3})	5.066×10^{14}	4.856×10^{14}
R_s (km)	511.5	552.2
M_s ($M_\odot$)	1.2833	1.2833
E_s (10^{51} ergs)	0.72	2.05
$Y_{l_{min}}$	0.17	0.15
E_{ν_e} (10^{51} ergs)	2.86	1.91
$E_{\bar{\nu}_e}$ (10^{51} ergs)	0.547	0.0
E_{pairs} (10^{51} ergs)	1.20	0.0

Table 9.8. Quantities at end of calculation.

	Last snap	
	All types	Only ν_e
Time (s)	0.259311	0.24106
t_{ab} (ms)	31.3	17.2
ρ_c (g cm^{-3})	5.303×10^{14}	4.804×10^{14}
R_s (km)	895.1	747.7
M_s ($M_\odot$)	1.300	1.2917
E_s (10^{51} ergs)	0.8	2.0
$Y_{l_{min}}$	0.12	0.12
E_{ν_e} (10^{51} ergs)	4.15	2.17
$E_{\bar{\nu}_e}$ (10^{51} ergs)	1.8	0.0
E_{pairs} (10^{51} ergs)	2.80	0.0

Both calculations use the 15 $M_\odot$ initial model of Woosley and Weaver (1988), with an iron core mass of ~1.28 $M_\odot$, and have the identical equation of state. The parameters were $\gamma = 2.5$, $K_0(0.5) = 180$ MeV, and $W_s = 36$ MeV. The row labeled time gives the time since the calculation started, t_{ab} gives the time after bounce for each calculation, ρ_c gives the central density, R_s is the radius of the position of the shock wave, and M_s is the mass point corresponding to R_s. E_s gives an estimate of the energy of the shock wave, while $Y_{l_{min}}$ is the minimum value of the lepton fraction behind the shock wave. E_{v_e}, $E_{\bar{v}_e}$, and E_{pairs} list the energy lost from the computational grid in each type of neutrino.

In Table 9.3 we see that there is very little effect on the infall of having the different types of neutrinos since the temperatures are not high enough to produce significant amounts of anything but electron-type neutrinos. The small differences between the two calculations probably represent the idiosyncracies of the printout algorithm rather than actual differences between the calculations.

In Table 9.4 the calculation is 0.4 ms later in both cases. The shock is at almost the exact same place, the difference in mass being only one mass zone. Both calculations have lost about 0.3 of energy in electron-type neutrinos from the computational grid.

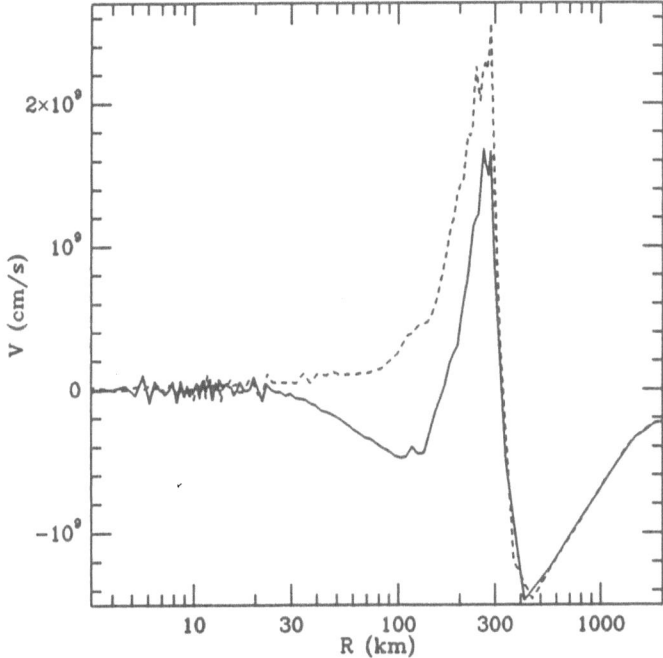

Figure 9.20. Velocity profiles of the two calculations when the shock waves are at about 300 km. The velocity profiles of the two calculations of the 15 $M_\odot$ Woosley and Weaver (1988) start when the shock wave is at about 300 km. The solid line is for the calculation with all types of neutrinos included in the transport, whereas only electron-type neutrinos are included in the case displayed by the dashed line. Note that not only is the size of the velocity peak greater in the case with only electron-type neutrinos, but the width is also bigger. Also matter is already failing back in the case with all types of neutrinos.

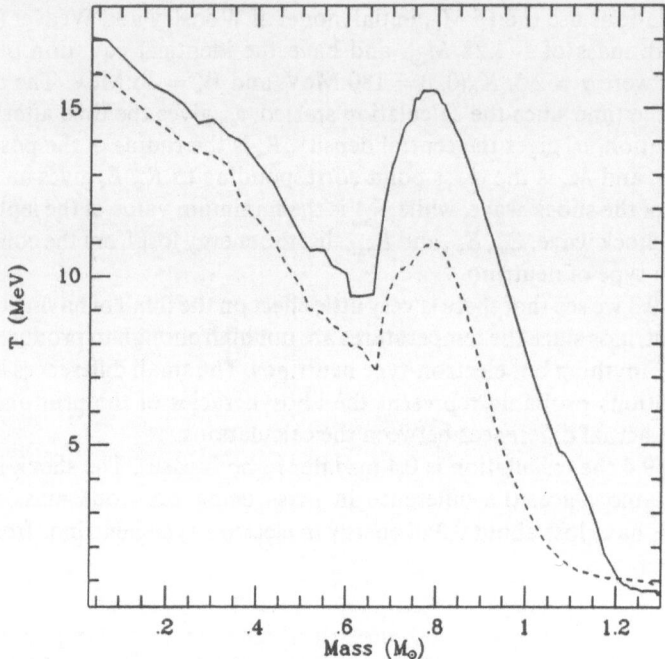

Figure 9.21. Temperature profiles at the last snapshot. A comparison of the temperature profiles of the two calculations described in Figure 9.20. Note the temperature is everywhere higher in the case with all types of neutrinos, which corresponds to a weaker shock wave.

Table 9.5 shows the calculations when the shock has reached approximately 150 km. Note that the calculation with all types of neutrinos has taken longer to get to the same point. The shock wave is already weaker and hence slower. The shock waves are well past the neutrinosphere which is at about 60 km. The neutrino energy losses are the total losses from the computational grid and hence underestimate the losses since there are many neutrinos that are still on the grid, but are already lost from behind the shock. Note that the approximate energy of the shock, E_s, for the calculation with only electron-type neutrinos is more than twice that of the calculation with all types of neutrinos. Also the lepton fraction is lower for the all-type calculation than for the other calculation. This last feature is not found by Bruenn (1989a), but appears to be due to the fact that the shock wave is weaker in the all-type case and hence there has been more time for electron captures.

In Table 9.6 the shock wave has propagated to 300 km. Again note the difference in time. By this time there have been significant losses of the other types of neutrinos from the computational grid. The greater energy of the shock is also illustrated in Figure 9.20 which compares the velocity profiles in the two calculations at about the same time. Note that not only is the maximum positive velocity larger, but more mass is moving out (the peak is both higher and wider). In the weaker case there is a large amount of material already falling back onto the hydrostatic core.

In Table 9.7 we show the shock at 500 km and finally, in Table 9.8, we compare the two calculations at the last snapshot. While the *only* v_e calculation is at 750 km in 17.2 ms it has taken 31.2 ms for the other calculation to get to 900 km. The

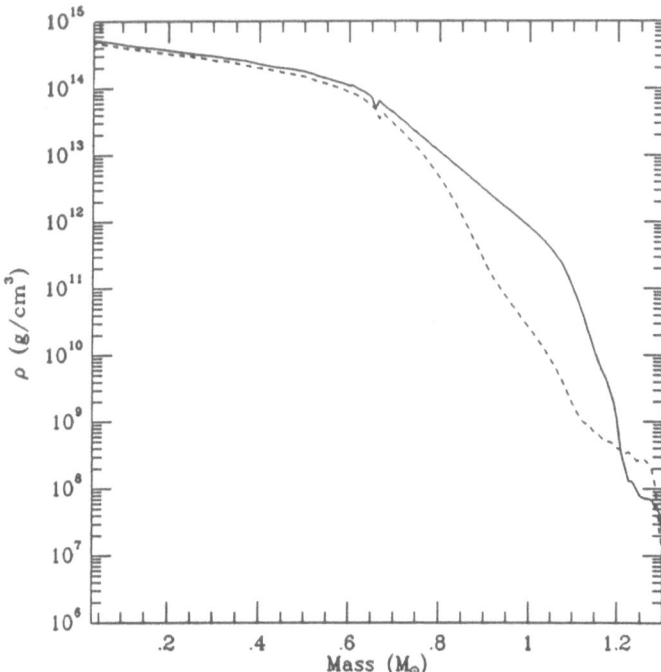

Figure 9.22. Density profiles at the last snapshot. A comparison of the density profiles at the same time as in Figure 9.21.

difference in shock energies has been maintained, with the weak calculation getting weaker since it has had more time to radiate and the temperatures are higher as illustrated in Figure 9.21. Clearly, the stronger shock wave has lower temperatures. In the inside this is due to the fact that more material has fallen back on the hydrostatic core compressing the material in the weaker case, on the outside it is due to the fact that more energy goes into expanding the material, so the temperature stays lower. These higher temperatures can be directly associated with differences in the density profiles shown in Figure 9.22. We note that even though the calculation with all types is quite weak it is not a complete failure. That is because both of these calculations neglected the effect of electron scattering on the infall, and the symmetry energy $W_s = 36$ MeV, so that free protons were probably underabundant on the infall. Even without these deleterious effects the all-type shock wave would not make it to the edge of the computational grid (nearly 2500 km) before stalling.

9.6. Rotation

So far we have only discussed calculations that are spherically symmetric and have therefore avoided the effects of rotation and higher-dimensional effects such as convection. Part of the reason is that two-dimensional and three-dimensional hydrodynamics are extremely complicated and the codes can therefore not incor-

porate all the physics that has been shown to be important, although progress has
been made. The two-dimensional calculations that have been done to date show
that there may be two possible complementary effects of rotation.

In one model the extra centrifugal acceleration helps to hold up the outer
equatorial region and matter is blown out the equator (Müller, Różyczka, and
Hillebrandt, 1980; Müller and Hillebrandt, 1981; Mönchmeyer and Müller, 1989).
In these calculations, hot spots develop along the axis, which would tend to enhance
neutrino emission from these regions. So far, these calculations seem to show that
quite a bit of rotation is required to produce a large effect (they do not actually
find explosions). The amount of rotation necessary would leave collapsed remnants
spinning with periods of $P \lesssim 10$ ms. While some neutron stars may be born with
such fast periods, statistical analyses indicate that the bulk of pulsars are injected
rotating rather slowly with initial periods $P \gtrsim 100$–500 ms (Vivekanand and
Narayan, 1981; Chevalier and Emmering, 1986; Narayan, 1987; Dewey *et al.*, 1988).
The period of the pulsar that is presumably lurking in SN 1987A, will provide much
information on the periods of pulsars at birth.

The second effect (which may be complementary to the first) envisions the effects
of explosive oxygen burning along the poles as providing 10^{51} ergs of energy
(Bodenheimer and Woosley, 1983). While the matter is being held up at the equator,
material from the oxygen shell has had a chance to fall into the shock wave. These
calculations seemed to require large amounts of oxygen which may not be available
in the smaller progenitor stars (see Nomoto and Hashimoto, 1988), and particular
assumptions about the initial distribution of angular momentum in the collapsing
star, although the model needs to be studied in greater detail. Thus, at the moment,
it does not appear that rotation as such can provide a general mechanism for
explosions, but more work remains to be done.

9.7. Discussion

The results of numerical simulations have varied back and forth between success
and failure over the past few years. Some recent numerical calculations had shown
that with the combined effects of smaller initial models, the use of general relativistic
hydrodynamics, and a relatively soft equation of state for densities above nuclear
matter density, that strong explosions could be obtained (Baron, 1985; Baron *et al.*,
1985b; Baron *et al.*, 1987). These conclusions seem inescapable and have been
confirmed by Bruenn (1989a, b) and Myra and Bludman (1989). These results
depend strongly on the equation of state used at all densities (both below and above
saturation density) and both of the above studies used our equation of state.

However, these same authors have also pointed out that when the detailed effects
of neutrino transport are taken into account, currently available initial models do
not seem to be able to explode promptly, a conclusion that our own results also
bear out (Baron and Cooperstein, 1990).

In our earlier work (BCK, 1985a, b; Baron *et al.*, 1987) the neutrino transport
was modeled by a simple leakage scheme that only included the effects of electron-
type neutrinos, and ignored the effects of $\bar{\nu}_e$, ν_μ, $\bar{\nu}_\mu$, ν_τ, $\bar{\nu}_\tau$. As seen in Section 9.5.1,

this is a good assumption during the infall epoch of the collapse, but the other types of neutrinos are extremely important to the propagation of the shock wave.

The leakage scheme we used consists of two simple equations that depend on two adjustable parameters which are functions of time, and local and global conditions. The details of this procedure are specified in Baron (1985) and Cooperstein (1988b). The important thing is that any neutrino transport calculation can be characterized by an effective trapping density. Whereas the trapping density of a full transport calculation is an output, for the case of a leakage scheme it is an input. That is, we tune the parameters in order to obtain a desired trapping density.[6]

In early calculations we used a trapping density of about 3×10^{11} g cm^{-3}, whereas the correct trapping density is more like 1×10^{12} g cm^{-3}. The amount of lepton loss during infall, and thus the energy costs to the shock wave were therefore underestimated. In later work (Baron et al., 1987) the parameters of the leakage scheme were adjusted to increase the trapping density.

The importance of the effects of neutrino transport cannot be underestimated, and can turn a successful shock wave into a failed one. This is not to say that it is the only significant effect in stellar collapse. While neutrino transport is extremely important, there are other crucial factors. The most important question is arguably that of the presupernova evolutionary calculations.

The effects of convection remain a major uncertainty in presupernova evolution. The two groups actively working on evolving massive stars to the collapse phase use different prescriptions for convection, Woosley and coworkers using the Ledoux criteria, invoking semiconvection and overshoot, while Nomoto and coworkers use the Schwarzschild criterion and neglect semiconvection and overshoot. In both calculations convection is modeled using standard mixing-length theory.

In order to fit the observed blue–red–blue evolution of the progenitor SK-69 202 to SN 1987A (see Kirshner, 1988) Woosley, Pinto, and Ensman (1988) found that it was necessary to turn off the effects of semiconvection and overshoot in their calculations. Saio et al. (1988a, b) were able to accomplish the same task by artificially mixing significant amounts of helium into the hydrogen shell. Thus, it is clear that our knowledge about convection remains rather uncertain and its effects on initial models as we have discussed are extremely important.

Another uncertainty in the presupernova evolution is the rather imprecise knowledge of important nuclear reaction rates at astrophysical energies. The preferred value of the $^{12}C(\alpha, \gamma)^{16}O$ rate went up and then went back down by a factor of 3 in the last few years (see Caughlan et al., 1985; Caughlan and Fowler, 1989). This variation simply represents our uncertainty in the rate, which determines the amount of carbon left after helium burning. A lower rate leads to a smaller iron core. The more time there is for carbon burning the lower the entropy because plasma neutrinos become an important energy sink at that time. The carbon burning rate is not the only one that is uncertain by factors of 2 or so.

---

[6] By trapping density we mean the density at which the total lepton number ceases to decrease on dynamical time scales during the gravitational collapse. This is not quite the same as other definitions often used, such as when the neutrinos are trapped in space rather than mass, or are equilibrated. See Cooperstein (1988b) for a full discussion.

Silicon burning, the ashes of which are the iron core, is an incredibly complex nonequilibrium process. For example, currently used values of neutron reaction rates have recently been questioned (Thielemann and Arnett, 1985; Malaney, 1988; Petrov and Shlyakhter, 1988) and this could have important consequences on nuclear burning. All in all, presupernova evolution calculations remain rather uncertain, and we should temper conclusions based upon these calculations.

Convective processes may play a role in the hydrodynamical simulation of supernova explosions. The effects of convection can be accurately modeled only with the use of two-dimensional and even three-dimensional hydrodynamical codes. Mixing-length studies in one-dimension show it may have effects on the neutrino luminosities (Burrows and Lattimer, 1986; Mayle, 1985, and this volume), but the detailed hydrodynamic effects are not yet known. As far as the direct mechanism is concerned, the shock wave moves too rapidly for convection to have time to play much of a role before the shock has exited the iron core. It may, of course, have an effect on the neutrino emission over the deleptonization timescale of seconds.

Self-consistent calculations which follow the supernova process from the initial instability, through collapse, bounce, shock propagation, and neutron star formation still remain to be done. (The exception is the calculations of Wilson and Mayle which follow the evolution until about 1 s after bounce in the delayed scenario; see Mayle's chapter in this volume.) It is important to do this for several reasons. Such calculations can obtain the true energy of the shock wave as it exits the star, which is difficult to ascertain unambiguously from the shorter-time hydrodynamical simulations. They can also determine the mass of the resulting neutron star, and under which conditions a black hole is the final result. This is a crucial question which has not yet been addressed in detail. Finally, such self-consistent calculations can be used to predict the nucleosynthetic yield of the explosion, the input for galactic evolution models. We are presently pursuing these goals.

Acknowledgments
We would like to thank Steve Bruenn, Eric Myra, Stan Woosley, and Ken Nomoto for generously providing us with the results of their calculations prior to publication and for many invaluable discussions. We would also like to thank our collaborators, Morry Aufderheide, Hans Bethe, Gerry Brown, Sid Kahana, Friedel Thielemann, and Leo van den Horn, for kindly allowing to use the results of our joint work. The calculations reported were performed at the National Magnetic Fusion Computer Center under the auspices of the U.S. Department of Energy, Nuclear Physics Division. This work has been supported by the U.S. Department of Energy under Contract No. DE-AC02-76CH00016 and grant DE-FG02-88ER40388.

References
Aichelin, J., Rosenhauer, A., Peilert, G., Stöcker, H., and Greiner, W. 1987, *Phys. Rev. Lett.*, **58**, 1926.
Ainsworth, T., Baron, E.A., Brown, G.E., Prakash, M., and Cooperstein, J. 1987, *Nucl. Phys.*, **464**, 740.
Baron, E. 1985, Ph.D. thesis, Stony Brook.
Baron, E., Bethe, H.A., Brown, G.E., Cooperstein, J., and Kahana, S. 1987 *Phys. Rev. Lett.*, **59**, 736.
Baron E. and Cooperstein, J. 1990, *Ap. J.*, **352**, in press.
Baron, E., Cooperstein, J., and Kahana, S. 1985a, *Phys. Rev. Lett.*, **55**, 126.
Baron, E., Cooperstein, J., and Kahana, S. 1985b. *Nucl. Phys.*, **A440**, 744.
Baron, E., Myra, E., Cooperstein, J., and van den Horn, L. J. 1989, *Ap. J.*, **339**.

Baym, G., Bethe, H.A., and Pethick, C.J. 1971, *Nucl. Phys.*, **A175**, 225.

Bethe, H.A. 1981, in *Supernovae: A Summary of Current Research*, eds. M.J. Rees and R.J. Stoneham (Dordrecht: Reidel).

Bethe, H.A., Brown, G.E., Applegate, J.H., and Lattimer, J. 1979, *Nucl. Phys.*, **A324**, 487.

Bethe, H.A., Brown, G.E., Cooperstein, J., and Wilson, J.R. 1983, *Nucl. Phys.*, **A403**, 507.

Bethe, H.A. and Wilson, J.R. 1985, *Ap. J.*, **295**, 11.

Blaizot, J.P. 1980, *Phys. Rep.*, **64**, 171.

Blaizot, J.P., Gogny, D., and Grammaticos, B. 1976, *Nucl. Phys.*, **A265**, 315.

Bodenheimer, P. and Woosley, S.E. 1983, *Ap. J.*, **269**, 381.

Brown, G.E. 1988a, *Phys. Rep.*, **163**, 167.

Brown, G.E. 1988b, *Nucl. Phys*, **A488**, 689c.

Brown, G.E., Bethe, H.A., and Baym, G. 1982, *Nucl. Phys.*, **A375**, 481.

Brown, G.E. and Osnes, E. 1985, *Phys. Lett.*, **B154**, 223.

Bruenn, S. 1985, *Ap. J. Suppl.* **58**, 771.

Bruenn, S. 1989a, *Ap. J.*, **340**, 955.

Bruenn, S. 1989b, *Ap. J.*, **341**, 385.

Brunish, W.M. and Truran, J. 1982, *Ap. J. Suppl.*, **49**, 447.

Burrows, A. and Lattimer, J.M. 1983, *Ap. J.*, **270**, 735.

Burrows, A. and Lattimer, J.M. 1984, *Ap. J.*, **285**, 294.

Burrows, A. and Lattimer, J.M. 1986, *Ap. J.* (*Letters*), **307**, L178.

Caughlan, G.R. and Fowler, W.A. 1989, *Atomic Data Nucl. Tables*, **40**, 283.

Caughlan, G.R., Fowler, W.A., Harris, M., and Zimmerman, B.A. 1985, *Atomic Data Nucl. Tables*, **32**, 197.

Chevalier, R. and Emmering, R.T. 1986, *Ap. J.*, **304**, 140.

Colgate, S. and White, R.H. 1966, *Ap. J.*, **143**, 626.

Cooperstein, J. 1982, Ph.D. thesis, Stony Brook.

Cooperstein, J. 1985, *Nucl. Phys.*, **A438**, 722.

Cooperstein, J. 1988a, *Phys. Rev.*, **C37**, 786.

Cooperstein, J. 1988b, *Phys. Rep.*, **163**, 95.

Cooperstein, J., Bethe, H.A., and Brown, G.E. 1984, *Nucl. Phys.*, **A429**, 527.

Cooperstein, J. and Brown, G.E. 1985, in *Relativistic Astrophysics*, eds. J. Centrella, J. LeBlanc, and R. Bowers (Boston: Jones and Bartlett).

Cooperstein, J., van den Horn, L.J., and Baron, E. 1986, *Ap. J.*, **309**, 653.

Cooperstein, J., van den Horn, L.J., and Baron, E. 1987, *Ap. J.* (*Letters*), **321**, L129.

Cooperstein, J., and Wambach, J. 1984, *Nucl. Phys.*, **A420**, 591.

Dewey, R.J., Taylor, J.H., Maguire, C.M., and Stokes, G.H. 1988, *Ap. J.*, **332**, 762.

El Eid, M.F. and Hillebrandt, W. 1980, *Astron. Astrophys. Suppl. Ser.*, **42**, 215.

Epstein, R.I. and Arnett, W.D. 1975, *Ap. J.*, **201**, 202.

Fuller, G.M. 1982, *Ap. J.*, **252**, 741.

Fuller, G.M., Fowler, W.A., and Newman, M.J. 1982, *Ap. J.*, **252**, 715.

Glendenning, N. 1986, *Phys. Rev. Lett.*, **57**, 1120.

Goldreich, P. and Weber, 1980, *Ap. J.*, **238**, 991.

Kirshner, R.P. 1988, in *SN 1987A in the Large Magellanic Cloud*, eds. M. Kafatos and A. Michalitsianos (Cambridge: Cambridge University Press).

Kolehmainen, K., Prakash, M., Lattimer, J., and Treiner, J. 1985, *Nucl. Phys.*, **A439**, 535.

Hillebrandt, W., Hoflich, P., Truran, J.W., and Weiss, A. 1987, *Nature*, **327**, 597.

Lamb, D.Q., Lattimer, J.M., Pethick, C.J., and Ravenhall, D.G. 1978, *Phys. Rev. Lett.*, **41**, 1623.

Lamb, D.Q., Lattimer, J.M., Pethick, C.J., and Ravenhall, D.G. 1981, *Nucl. Phys.*, **A360**, 459.

Lamb, D.Q., Lattimer, J.M., Pethick, C.J., and Ravenhall, D.G. 1983, *Nucl. Phys.*, **A411**, 449.

Lamb, D.Q., Lattimer, J.M., Pethick, C.J., and Ravenhall, D.G. 1985, *Nucl. Phys.*, **A432**, 646.

Lattimer, J.M. 1981, *Ann. Rev. Nucl. Part. Sci.*, **31**, 337.

Lattimer, J.M. and Ravenhall, D.G. 1978, *Ap. J.*, **223**, 223.

Levermore, C.D. and Pomraning, G.C. 1981, *Ap. J.*, **248**, 321.

Malaney, R.A. 1988, *M.N.R.A.S.*, **231**, 657.

Mayle, R.W. 1985, Ph.D. thesis, Livermore Report UCRL-53713.

Mazurek, T. 1975, *Ap. Sp. Sci.*, **35**, 117.

Mönchmeyer, R. and Müller, E. 1989, in *Timing in Neutron Stars*, eds. H. Ögelmann and E. van den Heuvel NATO ASI Series (Dordrecht: Kluwer), p. 549.

Müller, E. and Hillebrandt, W. 1981, *Astr. Ap.*, **103**, 358.

Müller, E., Różyczka, M., and Hillebrandt, W. 1980, *Astr. Ap.*, **81**, 288.

Myra, E.S. and Bludman, S.A. 1989, *Ap. J.*, **340**, 384.

Myra, E.S., Bludman, S.A., Hoffman, Y., Lichtenstadt, I., Sack, N., and Van Riper, K.A. 1987, *Ap. J.*, **318**, 744.

Narayan, R. 1987, *Ap. J.*, **319**, 169.

Nomoto, N. and Hashimoto, M. 1988, *Phys. Rep.*, **163**, 13.

Pethick, C.J., Ravenhall, D.G., and Lattimer, J.M. 1984, *Nucl. Phys.*, **A414**, 513.

Pethick, C.J., Ravenhall, D.G., and Lattimer, J.M. 1983, *Phys. Lett.*, **128B**, 137.

Petrov, Yu. V. and Shlyakhtev, A.I. 1988, *Ap. J.*, **327**, 294.

Pines, D., Quader, K.F., and Wambach, J. 1988, *Nucl. Phys.*, **A477**, 365.

Prakash, M., Ainsworth, T.L., and Lattimer, J.M. 1988, *Phys. Rev. Lett.*, **61**, 2518.

Ravenhall, D.G., Bennett, C.D., and Pethick, C.J., 1972, *Phys. Rev. Lett.*, **28**, 978.

Ravenhall, D.G., Pethick, C.J., and Lattimer J.M. 1983, *Nucl. Phys.*, **A407**, 571.

Ravenhall, D., Pethick, C.J., and Wilson, J.R. 1983, *Phys. Rev. Lett.*, **150**, 2006.

Saio, H., Kato, M., and Nomoto, K. 1988, *Ap. J.*, **331**, 388.

Saio, H., Nomoto, K., and Kato, M. 1988, *Nature*, **334**, 508.

Sato, K. 1975, *Prog. Theor. Phys.*, **54**, 1352.

Sharma, M.M., Borghols, W.T.A., Brandenburg, S., Crona, S., van der Woude, A., and Harakeh, M.N. 1988, *Phys. Rev.*, **C38**, 2562.

Stöck, R. 1986, *Phys. Rep.*, **135**, 259.

Stöcker, II. and Greiner, W. 1986, *Phys. Rep.*, **137**, 277.

Takahara and Sato, K. 1988, *Ap. J.*, **335**, 301.

Taylor, J. II. and Weisberg, J.M. 1982, *Ap. J.*, **253**, 908.

Thielemann, F.K. and Arnett, W.D. 1985, *Ap. J.*, **295**, 604.

Truran, J.W. and Weiss, A. 1988, in *SN 1987A in the Large Magellanic Cloud*, eds. M. Kafatos and A. Michalitsianos (Cambridge: Cambridge University Press), p. 331.

van den Horn, L.J., and Cooperstein, J. 1986, *Ap. J.*, **300**, 142.

Van Riper, K. and Arnett, W.D. 1978, *Ap. J. (Letters)*, **225**, L129.

Vivekanand, M. and Narayan, R. 1981, *Ap. Astr.*, **2**, 315.

Weaver, T.A. and Woosley, S.E. 1988, private communication.

Williams, K. and Koonin, S.E. 1985, *Nucl. Phys.*, **A435**, 844.

Wilson, J.R. 1985, in *Relativistic Astrophysics*, eds. J. Centrella, J. LeBlanc, and R. Bowers, (Boston: Jones and Bartlett).

Woosley, S.E., Pinto, P., and Ensman, L. 1988, *Ap. J.*, **324**, 466.

Woosley, S.E. and Weaver, T.A. 1984, *Bull. AAS*, **16**, 971.

Woosley, S.E. and Weaver, T.A. 1986, *Ann. Rev. Astr. Ap.*, **24**, 205.

Woosley, S.E. and Weaver, T.A. 1988, *Phys. Rep.*, **163**, 79.

Yahil, A. 1983, *Ap. J.*, **265**, 1047.

10. Neutrino Heating Supernovae

Ronald W. Mayle

10.1. Introduction

Exploding stars were suggested by Baade and Zwicky (1934) as the source of bright optical displays (10^{49} ergs emitted in light energy, 10^{51} ergs in kinetic energy of particle motion) abruptly appearing inside galaxies and then fading away on time-scales of hundreds of days. Most astrophysicists today agree that this conjecture is correct, but most would not agree on the detailed physical scenario that leads to the destruction of a star.

There is probably more than one way to blow up a star; this is reflected in the many attempts at providing theoretical explanations since the 1930s. Early work includes Hoyle's paper (1946) in which he showed that a highly evolved star, whose core had fused light elements into iron group elements, is gravitationally unstable to collapse. He proposed that such a star would become rotationally unstable (if it were initially rotating) and material would be expelled as soon as the centrifugal force overcame gravity. Burbidge *et al.* (1957) argued that the dynamic implosion, triggered by the gravitational instability earlier proposed by Hoyle, caused rapid compression and heating of light elements surrounding the iron core and initiated a thermonuclear explosion and disruption of the star.

Present-day workers in the field of supernova theory agree that some supernovae are triggered by a gravitationally unstable iron core while others explode due to the energy released by the burning of light elements into iron group nuclei (see Woosley and Weaver, 1986, and references therein). The progenitors of the latter group of exploding stars are white dwarfs composed mainly of helium, carbon, oxygen, and neon. Continuous mass accretion onto a white dwarf will eventually drive its mass above the Chandrasekhar mass limit for stability; implosive burning of the light elements into iron creates more than enough energy to explode the star. However, exploding white dwarfs produce negligible emission of neutrinos. It is the supernovae triggered by unstable iron cores that produce in excess of 10^{53} ergs of energy in neutrinos (the binding energy of a neutron star). The energy delivered to the extended hydrogen–helium envelope surrounding the collapsing iron core is around 1% of the energy emitted in neutrinos; the exact mechanism for coupling the released gravitational binding energy to the envelope is not completely certain as of this day. This chapter is concerned with the role neutrinos play in the explosion of a highly evolved massive star triggered by a collapsing iron core.

10.2. Brief Review of Numerical Work

Colgate and White (1966) published results of a numerical calculation that obtained an explosion of a "collapse driven supernova" (one that is triggered by the gravitational energy released by a collapsing iron core). Neutrinos produced by the capture of electrons on protons in the dense inner core region deposited energy in the outer part of the core by rare interactions with matter (neutrino capture on neutrons). The envelope gained an escape velocity by the absorption of neutrinos and an explosion resulted. The time duration of the burst of electron capture neutrinos was around 0.01 s. Colgate and White did not use radiation transport to follow the neutrinos; they used an entropy source term in the matter internal energy equation to model the effects of the neutrino interactions. Wilson (1971) published results of numerical work that included neutrino transport in a fully consistent general relativistic calculation modeling a collapse driven supernova. He showed that the electron capture neutrino burst was not strong enough to eject material as seen by Colgate and White. He also saw that the rebound of the collapsing iron core upon reaching nuclear densities started a shock wave that could eject the envelope of stars with low mass iron cores.

In the later 1960s and into the 1970s the understanding of the weak interaction was changing. The Weinberg–Salam model of the electroweak interaction became the standard theory and showed new possibilities for neutrino interactions with matter (neutral current interactions). For the supernova problem, coherent scattering of neutrinos through neutral current interactions increases the neutrino coupling with light elements surrounding the iron core over that calculated assuming only charged current interactions (see Freedman, 1974). It was hoped that this increased opacity would revive the Colgate and White (1966) scenario for supernova explosions. Many workers (e.g., Wilson, 1974; Bruenn, 1975; Schramm and Arnett, 1975) included the new weak interaction cross sections in their calculations but found that explosions still did not result from neutrinos (produced by electron capture on protons in the inner core) interacting with matter farther out in the star.

Attention was also being paid to the possibility that the hydrodynamic bounce of the iron core would form a shock wave strong enough to produce an explosion. Bethe made an important contribution in pointing out that during the collapse phase the large number of excited nuclear states available to the heavy nuclei in the core would thermostat the temperature. Thus the collapse would continue until the matter reached supranuclear densities, stopping only because of the nearly incompressible nature of uniform nuclear matter. This in turn meant that more gravitational energy was transformed into kinetic energy and possibly into an energetic shock wave. The details of the rebound of the iron core are sensitive to the properties of matter near and above normal nuclear matter density; much research into the supranuclear density equation of state was motivated by the supernova problem (see Bethe et al., 1979). Numerical modelers (e.g., Bowers and Wilson, 1982; Cooperstein, 1982) using the new work on the equation of state found that the shock wave stalled on its exit from the core. This was mainly due to the shock dissociation of the infalling iron to free baryons (or helium) thus robbing the shocked material of thermal pressure. More recent work by Baron, Cooperstein, and Kahana (1985) found that a soft enough supranuclear density equation of state along with a small enough iron core (less than around 1.20 $M_\odot$) forms a shock energetic enough to produce a prompt hydrodynamic bounce supernova on timescales of 20 ms. However, Wilson (1985) found that the shock wave, if stalled, could be revived at late times (about 1 s after bounce) by interaction of neutrinos with the shocked material (mostly in the form of free baryons). In later sections of this chapter the late-time heating mechanism discovered by Wilson (1985) for supernova explosion will be discussed in detail, as well as the current state of affairs regarding the viability of this mechanism as the true way nature explodes stars.

10.3. Collapse-Driven Supernova

10.3.1. Precollapse Physical Conditions

The evolution of massive stars (mass between 10–50 $M_\odot$) produces iron cores around 1–2 $M_\odot$ in size (radius about 1000 km) at the center of an extended hydrogen–helium envelope (see Woosley and Weaver, 1986). The thermal energy which supplies the pressure needed to support the mass of the star against the inward pull of gravity leaks out, largely in photons (neutrino emission becomes the dominant energy loss mechanism shortly before collapse) and is replenished by nuclear reactions. As heavier and heavier elements are produced from the original hydrogen–helium composition in the core of the star, the energy released per nucleon decreases. Since iron is the most tightly bound nucleus, when the core has become predominantly iron group nuclei no further fusion energy is available. The main source of pressure support for the iron core is electron degeneracy pressure. From energy considerations (assuming nuclear energy generated during the formation of the iron core has been radiated away):

$$\frac{3\mu}{4}\frac{\rho Y_e}{m_b} = \frac{\rho GM}{R},$$
(10.1)

where μ is the electron chemical potential (the average energy per electron being $3/4\mu$ for a relativistic degenerate electron gas), Y_e is the number of matter electrons per baryon, ρ is the matter density, m_b is the mass of a nucleon, R is the radius of the iron core, M is the core mass, and G is Newton's gravitational constant. Using the expression relating electron chemical potential to the number density of electrons in a relativistic degenerate electron gas

$$\mu = 11(\rho_{10} Y_e)^{1/3} \text{ MeV}$$
(10.2)

(ρ_{10} is density in units of 10^{10} g cc^{-1}) gives a value of 5 MeV for μ and 2×10^9 g cc^{-1} for ρ when $R = 10^8$ cm, $M = 1.3$ $M_\odot$, and $Y_e = 0.46$ (the value for iron) is used. The temperature of the core is near the temperature for the onset of silicon burning (about 0.5 MeV), so the electrons are highly degenerate in the precollapse iron core.

10.3.2. Collapse

10.3.2.1. *Iron Disintegration*

A collapse is triggered by loss of pressure support. Two mechanisms contribute to this loss, iron thermal disintegration and electron capture on protons. Iron thermal disintegration refers to the breaking up of heavy nuclei to lighter ones in response to an increase in density and temperature. In the iron core this occurs as the matter remains in nuclear statistical equilibrium so entropy is unchanged. The breaking up of the iron requires energy to overcome the nuclear binding energy; this energy is lost to pressure support. In order to compensate for the pressure deficit, the star begins to contract, increasing density and temperature. This leads to more iron disintegration and thus the nature of the instability is seen (this is Hoyle's (1946) gravitational instability). The internal kinetic energy in the iron core at the onset of collapse is dominated by the degenerate electron gas part of the composition, and can be found using typical numbers from evolutionary calculations ($\rho \approx 2.6 \times 10^9$ g cc^{-1}, $T \approx 0.6$ MeV, and $Y_e \approx 0.42$, see Chandrasekhar (1967) for the relevant formula) to be about 2×10^{18} ergs g^{-1}. Only 4×10^{17} ergs g^{-1} is due to the finite temperature of the core, the remainder is the energy density of a zero temperature degenerate electron gas. Since 1.65×10^{18} ergs g^{-1} is needed to convert 1 g of iron to 1 g of helium, a rough estimate of the amount of iron breakup needed to completely exhaust the thermal energy in the core is 25%. The larger the size of the iron core, the more important is the iron disintegration in initiating collapse; the electron capture instability, to be discussed next, competes with the iron disintegration for small cores.

10.3.2.2. *Electron Capture Instability*

Electrons can be captured on protons in nuclei, or on free protons. Unlike iron thermal disintegration, this is not an equilibrium process if the neutrino escapes the star and lowers the entropy; however, since capture rates lag behind equilibrium, the entropy can increase (bulk viscosity). In practice, the entropy remains almost unchanged.

The reaction rate for electron capture on free protons can be written, using Fermi's golden rule (assuming a given initial electron state), as

$$dΓ = \frac{4\pi^2}{h^4} H(1 - f) \, d^3p \, \delta(E + Q - \varepsilon), \qquad (10.3)$$

where H is the square of the matrix element for interaction (averaged over the initial electron spin), f is the Fermi–Dirac neutrino distribution function (the factor of $(1 - f)$ being a blocking factor if the final neutrino state is occupied), p is the neutrino momentum, E is the neutrino energy, ε is the electron energy, Q is the difference in the energies of the final and initial nuclear states (in this case the difference between the neutron and proton rest mass), and h is Planck's constant. At the temperatures seen in a supernova (1–100 MeV), the weak interaction can be considered a contact interaction and the protons never become relativistic, so the use of Fermi's golden rule in the nonrelativistic limit is justified. Assuming the dipole approximation in evaluating H, we see that H depends on weak interaction constants (no energy dependence) and (10.3) can be rewritten as

$$d\Gamma = \frac{\sigma_0}{2}\left(\frac{1 + 3\alpha^2}{4}\right)c(1 - f)\left(\frac{E}{m_e c^2}\right)^2 dE\,\delta(E + Q - \varepsilon), \qquad (10.4)$$

where $\sigma_0 = 1.7 \times 10^{-44}\,\text{cm}^2, \alpha = 1.2$ (see Tubbs and Schramm (1975) for σ_0 in terms of more fundamental constants and for a definition of α), $Q = 1.3$ MeV (rest mass energy difference between the neutron and proton), and c is the speed of light.

To find the luminosity from emission of neutrinos by electron capture on free protons integrate (10.4) over the initial electron states after first multiplying by E (the neutrino energy) and the number of protons. Since the electron chemical potential is high in the iron core (about ten times the electron rest mass) and more than twice the Q value for the reaction, the integration simplifies to give

$$F = (4.9 \times 10^{13})X_p \rho \left(\frac{\mu}{m_e c^2}\right)^6 \frac{\text{ergs}}{\text{s cm}^3}, \qquad (10.5)$$

where X_p is the mass fraction of free protons. The average energy of the emitted neutrino can be found using (10.4). The expression is

$$E_{av} = \frac{\int_Q^\mu d\varepsilon\, \varepsilon^2(\varepsilon - Q)^3}{\int_Q^\mu d\varepsilon\, \varepsilon^2(\varepsilon - Q)^2} \qquad (10.6)$$

giving a value of $\frac{5}{6}\mu$. The difference in the chemical potentials of the neutron and proton is given by the difference between μ and E_{av}, so the ratio of the number of free protons to free neutrons is given by

$$\frac{n_p}{n_n} = \exp\left(\frac{E_{av} - \mu}{kT}\right), \qquad (10.7)$$

where k is the Boltzmann constant and T is the temperature. With the above (10.5) becomes

$$\frac{F}{\rho} = (4.9 \times 10^{13})X_n \exp\left(\frac{\mu}{6kT}\right)\left(\frac{\mu}{m_e c^2}\right)^6 \frac{\text{ergs}}{\text{g s}}. \qquad (10.8)$$

The vast majority of protons are in iron (heavy nuclei) and helium (formed from iron thermal disintegration). This is true since the free proton fraction is about one-fifth the free neutron fraction (use (10.7) with $\mu/kT = 10$), the free neutron fraction is about one-thirteenth the helium mass fraction (assume the average iron nucleus breaks up into thirteen helium nuclei plus four neutrons), and the helium mass fraction stays less than about 0.25 during collapse. A free proton mass fraction of about 0.003 can be inferred from the above (this justifies the assumption that iron decomposes mainly to helium and neutrons).

An estimate for the neutrino energy production rate from electron capture on free protons during collapse is found from (10.8) using $\mu/kT = 10$ (for a free relativistic electron gas undergoing an adiabatic compression, this ratio is constant) and $X_n = 1/(4 \times 13)$ to be

$$\frac{F}{\rho} \le (2 \times 10^{11})\left(\frac{\mu}{m_e c^2}\right)^6 \frac{\text{ergs}}{\text{g s}}. \qquad (10.9)$$

Putting in a value of 5 MeV for μ in (10.9) gives 2×10^{17} ergs g^{-1} s^{-1} for the neutrino flux. An estimate for the time in which energy loss to neutrinos from electron capture on free protons would (if the energy production rate remained constant, which of course it does not) affect the dynamics of the core (a quasi-static configuration going into collapse due to loss of pressure support) can be found by equating the 2×10^{18} ergs g^{-1} of internal energy in the electron gas at the start of the collapse to the above energy production rate times time. Electron capture on free protons would exhaust this energy store in about 10 s. Even though the free proton mass fraction is small (0.003), the loss of energy to neutrino production by capture on free protons is appreciable. Neutrino capture on bound protons, discussed next, turns out to be a more important sink of energy.

Electron capture on the protons bound in heavy nuclei is complicated by the nuclear physics. The neutron produced must find an available state in the nucleus. Since heavy nuclei in a supernova collapse have more neutrons than protons, part of the energy of the captured electron must go into boosting the proton to a higher energy unoccupied neutron state. Bethe et al. (1979) argue that the daughter nucleus formed after electron capture finds itself in an excited nuclear state about 3 MeV above the ground state of the parent nucleus. A Fermi gas model of a nucleus can be used (along with the fact that part of the initial electron energy, the 3 MeV mentioned above, goes into promoting the proton to an available state) to find an approximate rate for electron capture on heavy nuclei and the associated neutrino luminosity (see Bethe et al. (1979) for this type of calculation or Fuller, Fowler, and Newman, (1982a, b) for detailed calculations using more realistic nuclear models). For our purpose we will estimate that 10% of the protons in each nucleus (those near the Fermi energy in a statistical nuclear model) are energetic enough to capture electrons and we rewrite equation (10.5) as

$$\frac{F}{\rho} = (4.9 \times 10^{13}) \left(\frac{0.1Z}{A} \right) \left(\frac{\mu}{m_e c^2} \right)^6 \frac{\text{ergs}}{\text{g s}}, \tag{10.10}$$

where Z is the number of protons and A is the number of nucleons in a nucleus. Using a value of 0.46 for the ratio of Z to A and 5 MeV for μ in (10.10) gives a luminosity of 2×10^{18} ergs g^{-1} s^{-1}. The timescale for this energy loss rate to exhaust the internal energy in the cores is about 1 s. The contributions to neutrino production from bound and free protons are additive. However, the contribution of bound protons is more important than that due to free protons even though bound protons contend with final nuclear state blocking in neutrino capture, while the free protons do not. It is only because there are so few free protons that the bound protons are the major source of neutrinos from electron capture during the collapse.

A total luminosity in electron capture neutrinos can be roughly estimated from (10.10) using a mass of 1×10^{33} g (the mass of the inner core) to be 2×10^{51} ergs s^{-1}. The neutrino Eddington luminosity L_{EDD} is the luminosity in neutrinos required so that momentum deposition by neutrinos balances the force of gravity. The derivation of the expression for L_{EDD} can be found in many textbooks (e.g., see Lightman et al. (1975), but replace the word "photon" by "neutrino") and is given by

$$L_{\text{EDD}} = \frac{4\pi GMc}{\kappa}, \tag{10.11}$$

where G is the gravitational constant, M is the mass of the core, c is the speed of light, and κ is the Planckian average opacity. It is the coherent neutral current neutrino scattering reactions with the light elements surrounding the core that are important in momentum transfer. The correct opacity to be used in (10.11) can be found from an expression similar to (10.4) along with the fact that the neutrino energy distribution is close to a neutrino blackbody at the matter temperature. A rough estimate is fine for our purposes, so using (10.4) as a guide an expression for the opacity is

$$\kappa \approx \frac{A\sigma_0}{m_b} \left(\frac{E_{av}}{m_e c^2} \right)^2, \tag{10.12}$$

where E_{av} is an average neutrino energy and A is the number of nucleons in the scatterer (the factor A appears because of the coherent effect). Using a value of 4 MeV for the average neutrino energy during the collapse phase (about $\frac{5}{6}\mu$) and an A of 40, gives κ around 3×10^{-17} cm^2 g^{-1}. With these numbers a value for L_{EDD} is found to be around 10^{54} erg s^{-1}. It is clear that momentum transfer by neutrinos is not responsible for expelling the envelope during the collapse phase in a supernova explosion (or even later when the peak neutrino luminosity reaches $\approx 10^{53}$ ergs s^{-1}, see Section 10.4 of this chapter).

10.3.3. Bounce and Shock Formation

When pressure support is lost through iron thermal disintegration and electron capture on protons, the core collapses on freefall timescales (about 0.1 s). The neutrinos produced by the electron capture cease escaping freely from the star as densities become large enough that the neutrino mean free path becomes on the order of the core radius. The neutrino mean free path between emission and absorption interactions can be estimated using (10.12) (without the factor A, since charged current interactions are incoherent). The mean free path is about 10^6 cm when $E_{av} \approx 10$ MeV and $\rho \approx 2.5 \times 10^{11}$ g cc^{-1}. After a neutrinosphere (taken to be the radius where the neutrino "optical" depth to infinity is $\frac{2}{3}$) forms, neutrinos diffuse outward from the inner core regions.

Figure 10.1 shows a sequence of infall velocity curves taken from a numerical calculation, parametrized by the central density, and covering a 0.001 s span in time. The last curve, with $\rho = 1.3 \times 10^{14}$ g cc^{-1} is just before bounce and shock formation. Numerical calculations of massive star evolution give a value of the entropy per baryon (in units of the Boltzmann constant) around one in the iron core at the onset of collapse. Bethe et al. (1979) point out that this value will not appreciably change during collapse due to the large statistical weight associated with the many possible excited states of heavy nuclei. Most of the nucleons will stay inside the heavy nuclei during the collapse phase, and the nuclear contribution to the pressure will be much smaller than the pressure of the degenerate electrons. However, when the core shrinks to the point when the nuclei begin to touch each other, a uniform sea of nuclear matter begins to form and the repulsive part of the nuclear force comes into play. The near incompressible nature of uniform nuclear matter is responsible for the halting of the collapse.

Since the major pressure support during collapse is degenerate electrons, the adiabatic equation of state for the matter in the core is given approximately

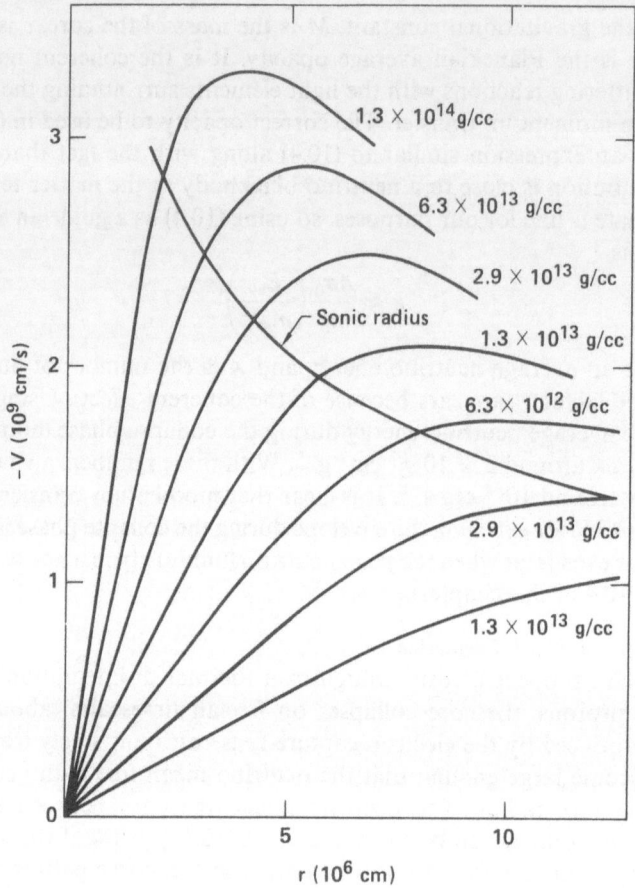

Figure 10.1. Infall velocity versus radius curves during the final 0.001 s before bounce, parametrized by the value of the central density.

by

$$P = K\rho^\gamma, \tag{10.13}$$

where P is pressure, K is a constant (see Chandrasekhar, 1967), and γ is $\frac{4}{3}$. It is well known that a $\gamma = \frac{4}{3}$ gas is neutrally stable to gravitational collapse. One way to see this is in the fact that no length scale can be made from the important dimensional parameters K, G, and M (the core mass). Material inside the sonic point (the point in the star where the sound speed equals the magnitude of the infall velocity) stays in communication and collapses homologously (velocity roughly proportional to radius) in what are successive $n = 3$ polytrope configurations (see Chandrasekhar (1967) for a discussion of polytropes). The material outside the sonic point falls in quasi-free fall with velocity proportional to the inverse square root of the radius. This structure is clearly seen in the curves drawn in Figure 10.1. Considerable analytic work has been done on this phase of collapse by Goldreich and Weber (1980) and Yahil and Lattimer (1982), who predict that the size of the homologous core is roughly the Chandrasekhar mass ($M_{Ch} = 5.76 Y_e^2\ M_\odot$). Due to electron capture, the value of Y_e in the iron core falls from near 0.46 (iron) prior to collapse to around 0.33. This occurs on collapse timescales (around 0.1 s); the value of Y_e then continues to decrease slowly below 0.33 on neutrino diffusion timescales (around 10 s). (The matter in the core becomes opaque to neutrinos as soon as the density exceeds 10^{12} g cc^{-1}.) For a Y_e of 0.33, $M_{Ch} = 0.63\ M_\odot$; in realistic numerical calculations, the mass of the homologous core (taken when the central density reaches 1×10^{11} g cc^{-1}) is about 0.55 $M_\odot$. The size of the homologous core is smaller than 0.63 $M_\odot$ due mainly to the effects of general relativity.

As the central parts of the core reach and exceed nuclear matter density, γ begins to rise above $\frac{4}{3}$ due to the near incompressible nature of nuclear matter above nuclear matter density and the collapse halts. Brown, Bethe, and Baym (1982) describe the process of shock formation as the build up of pressure waves (carrying information of the impending bounce to the still infalling core) at the sonic point (the waves cannot pass this point as the infall velocity equals the wave speed there). The homologous core bounces as a unit; the subsequent rebound launches the shock wave into the outer core.

10.3.4. Late-Time Neutrino Heating Supernovae

The shock wave starts at the radius which encloses about 0.70 $M_\odot$ and may promptly (about 20 ms after formation) exit the outer core and travel into the hydrogen–helium envelope. For this to happen the shock must be energetic enough to overcome loss of energy to the thermal decomposition of the iron in the outer core as it passes through the shock wave. To be quantitative, numerical calculations show that the kinetic energy developed during collapse is about 1×10^{52} ergs. It takes 1.6×10^{52} ergs per $M_\odot$ to completely decompose iron to free neutrons and protons. This implies that 0.63 $M_\odot$ of iron can be decomposed by the transformation of the kinetic energy to nuclear unbinding energy. Thus the shock stalls for iron cores of mass greater than about 1.33 $M_\odot$.

In the following discussion the initial iron core mass is assumed greater than that which would allow a prompt explosion. The gravitational binding energy of a cold neutron star is about 0.1 Mc^2. Using $M = 1.6\ M_\odot$ gives about 3×10^{53} ergs of energy that must be radiated away in neutrinos of all types. The nucleons outside the neutrinosphere may absorb enough energy by neutrino interactions to heat the post-shocked material and revive the shock; this is the scenario for "late-time neutrino heating supernovae" (see Bethe and Wilson, 1985; Wilson, 1985).

An estimate of the composition of the post-shocked material can be made using energy-conservation arguments. Assuming iron is dissociated to neutrons and protons (taking 8 MeV per nucleon of the shock energy) gives the following equation:

$$\frac{GM}{R} m_b \approx 8 \text{ MeV} + \tfrac{3}{2} T \tag{10.14}$$

and using $M = 1.6\ M_\odot$ gives

$$\frac{22}{R_7} \approx 8 + \tfrac{3}{2} T_{MeV}, \tag{10.15}$$

where $R_7 = 10^{-7} \times R$ with R in centimeters. Assuming iron dissociates to helium

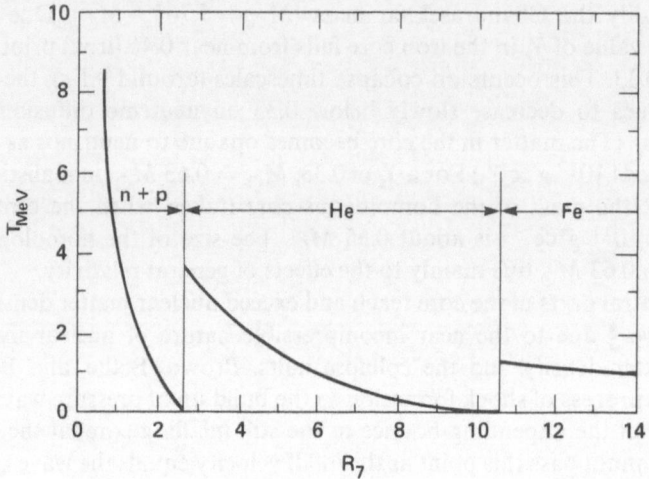

Figure 10.2. Expected post-shocked material temperature and composition versus radius.

(taking about 2 MeV per nucleon) we find

$$\frac{22}{R_7} \approx 2 + \tfrac{3}{2}T_{\text{MeV}}. \tag{10.16}$$

Finally, assuming that no decomposition takes place gives

$$\frac{22}{R_7} \approx \tfrac{3}{2}T_{\text{MeV}}. \tag{10.17}$$

In Figure 10.2, eqs. (10.15), (10.16), and (10.17) are graphed and the regions where the shock can dissociate the material into baryons, helium, or no dissociation can be seen. The actual temperature would be an interpolation between these disjoint segments. It can be seen from Figure 10.2 that the shock wave would be expected to stall somewhere around $(2–8) \times 10^7$ cm from the center of the star (the neutrino-sphere is located at about $(2–6) \times 10^6$ cm). The presence of free neutrons and protons outside the neutrinosphere is important because electron neutrinos and antineutrinos are freely emitted and absorbed by these baryons. The cross section for emission or absorption of neutrinos on heavy elements is much less due to the effects nuclear structure. The first excited state of the helium nucleus is about 20 MeV above the ground state while the average energy of the electron neutrino or antineutrino is about 10 MeV (at this point in the supernova evolution); thus the opacity for neutrino interaction with helium is also much less than for free baryons.

Inelastic scattering of neutrinos by heavy nuclei also occurs. In calculations we have done (Wilson and Mayle, 1988) we model this effect using a cross section for interaction per nucleon similar to the elastic scattering from the free baryons. The number of nucleons participating are those near enough to the Fermi level in a statistical model of the nucleus to be excited above the Fermi level by the neutrino. The energy lost by the neutrino is taken to be the energy gained by the nucleon. (However, Haxton (1988) argues that giant resonant states of nuclei are excited by inelastic scattering of neutrinos and that modeling the reaction as above is not adequate.)

10.3.4.1. *First Heating*

An estimate of the net heating rate for free baryons can easily be done. In the region near $R = 10^7$ cm the density of the shocked material is around 10^8 g cc^{-1} and from Figure 10.2 the temperature is around 2 MeV. In this region electrons are relativistic and quasi-degenerate. The important neutrino reaction is

$$b + v_l \;\Rightarrow\; b + l, \tag{10.18}$$

where b represents a free neutron or a proton and l the electron or positron with v_l being either an electron neutrino if the outgoing lepton is an electron or an electron antineutrino if the outgoing lepton is a positron. Assume that the number densities of electrons and positrons are nearly the same and ignore the mass difference between the neutron and proton. With these assumptions the identity of the baryon (neutron or proton), lepton (electron or positron), or neutrino (neutrino or anti-neutrino) can be ignored (and will be in the following calculation). The net heating rate is found by subtracting from the total heating rate the total cooling rate (the reverse reaction to (10.18)).

The matter heating rate per baryon can be written as

$$\dot{E}_H = \int E f_v \frac{E^2\, dE}{(hc)^3} \frac{d\tilde{\Gamma}}{dE}\, d\Omega, \tag{10.19}$$

where E is the neutrino energy, Ω is the solid angle, and the reaction rate is given by

$$\frac{d\tilde{\Gamma}}{dE} = \sigma_0 \left(\frac{1 + 3\alpha^2}{4} \right) c \left(\frac{\varepsilon}{m_e c^2} \right)^2 \delta(E - \varepsilon), \tag{10.20}$$

with ε the lepton energy. Since the region of interest is outside the neutrinosphere, the neutrino distribution function f_v at R has roughly the value of the neutrino distribution function at the neutrinosphere if $\hat{\Omega}$ points from a position on the neutrinosphere to the observation point, or zero if $\hat{\Omega}$ points in any other direction. We take f_v at the neutrinosphere to be

$$f_v = \left[\exp\left(\frac{E}{kT_v} \right) + 1 \right]^{-1}, \tag{10.21}$$

with T_v the temperature of the neutrinos at the neutrinosphere (this will be close to the matter temperature at that radius). Using the above the expressions in (10.19) gives the heating rate as

$$\dot{E}_H = 2010 \sigma_0 c \left(\frac{m_e c}{h} \right)^3 m_e c^2 \left(\frac{R_v}{2R} \right)^2 \left(\frac{kT_v}{m_e c^2} \right)^6, \tag{10.22}$$

where R_v is the position of the neutrinosphere.

The matter at R is cooled by emitting neutrinos principally by the reaction reverse to (10.18). The cooling rate per baryon is given by

$$\dot{E}_C = 8\pi \int E f_l \frac{\varepsilon^2\, d\varepsilon}{(hc)^3} \frac{d\Gamma}{d\varepsilon}, \tag{10.23}$$

with similar notation as above. $d\Gamma$ is found from (10.4) and f_l (lepton distribution function) is taken as

$$f_1 = \left[\exp\left(\frac{\varepsilon}{kT}\right) + 1 \right]^{-1}. \qquad (10.24)$$

The factor of 8π comes from the solid angle (the neutrinos are emitted isotropically) and the the fact that electrons have twice as many spin states as neutrinos. The integral (10.23) can be done to find

$$\dot{E}_C = 2010 \sigma_0 c \left(\frac{m_e c}{h}\right)^3 m_e c^2 \left(\frac{kT}{m_e c^2}\right)^6. \qquad (10.25)$$

Measuring temperatures in MeV and evaluating the constants gives an expression for the net heating rate per baryon as

$$\dot{E}_{NET} = 3.3 \times 10^{-6} \left[\left(\frac{R_\nu}{2R}\right)^2 T_\nu^6 - T^6 \right] \frac{\text{ergs}}{\text{baryon s}}. \qquad (10.26)$$

This equation is also given in Bethe and Wilson (1985) (they do not derive it, so the above discussion can supplement the reading of their paper). The condition for (10.26) to be positive is that T is less than $(R_\nu/2R)^{1/3} T_\nu$. The fact that the shock decomposed the iron means the post-shock temperature is low (1–2 MeV as compared to about 10 MeV if no decomposition occurred) and the neutrinosphere temperature is characteristic of the hotter regions closer to the center of the star ($T_\nu \approx 5$–6 MeV and $R_\nu \approx 3 \times 10^6$ cm). Using these values for the temperatures and neutrinosphere radius as well as a realistic dependence of T on R, the region where (10.26) is positive is for $R > 2 \times 10^7$ cm. The shocked material outside this radius can gain energy and provide a push on the back side of the shock wave enabling it to begin moving outward in radius (numerical calculations show the shock reaches a radius around $(3$–$4) \times 10^7$ cm before becoming an accretion shock).

10.3.4.2. Second Heating

As the shock wave moves outward a "bubble" of low density (ρ less than about 10^5 g cc^{-1}) material forms between the surface of the protoneutron star and the material moving outward with the shock. The density at the surface of the protoneutron star goes as

$$\rho \approx \rho_s \exp\left(-\frac{g_s m_b R}{kT_s}\right), \qquad (10.27)$$

where T_s is the surface temperature, g_s is the surface gravitational acceleration, and ρ_s is the surface density. This is an isothermal atmosphere solution, and is a good approximation as the temperature gradient is much smaller than the density gradient at this time (due to neutrino diffusion smoothing out temperature differences). Putting in typical numbers from a numerical calculation gives a density scale height ($kT_s/g_s m_b$) of about 1 km. A sharp density gradient means that neutrinos leaving the neutrinosphere can annihilate into electron–positron pairs in a region of such low density that $e^+ e^-$ annihilation into neutrino–antineutrino pairs (or electron and positron capture on free baryons) occurs at a negligible rate. Thus a second type of neutrino heating is possible and can boost the energy of explosion provided the first heating moves the shock outward and allows the large density gradient at the protoneutron star surface to form. The above scenario was suggested by Goodman, Dar and Nussinov (1987); numerical calculations (Wilson and Mayle, 1988) show that it does occur and is important in getting an explosion if the first heating phase does not provide enough energy for the outward moving shock to overcome the ram pressure (ρv^2) of the infalling material (second shock stalling).

The heating rate for $\nu\bar{\nu}$ annihilation can be written as (see Cooperstein, van den Horn, and Baron, 1987)

$$q = \frac{1}{9}\frac{\sigma_0 c}{(m_e c^2)^2} D \int d\tau (E_n + E_a) F_n F_a (1 - \hat{\Omega}_n \cdot \hat{\Omega}_a)^2, \tag{10.28}$$

where the subscripts n and a refer to neutrino and antineutrino respectively, E is neutrino energy, and $D = 1 + 4 \sin^2 \Theta + 8 \sin^4 \Theta$ for $\nu_e \bar{\nu}_e$ annihilation or $D = 1 - 4 \sin^2 \Theta + 8 \sin^4 \Theta$ for $\nu_\mu \bar{\nu}_\mu$ or $\nu_\tau \bar{\nu}_\tau$ annihilation, Θ being the Weinberg angle ($\sin^2 \Theta \approx 0.21$). The two expressions for D take into account the fact that charged and neutral currents participate in $\nu_e \bar{\nu}_e$ anninilation while only neutral currents are involved in $\nu_\mu \bar{\nu}_\mu$ or $\nu_\tau \bar{\nu}_\tau$ annihilation to $e^+ e^-$ pairs. F is defined by

$$F = \frac{E^3}{(hc)^3} f, \tag{10.29}$$

where f is the Fermi–Dirac neutrino distribution function (in thermal equilibrium f is given by (10.21)). The integration volume $d\tau$ is $dE_a \, dE_n \, d\Omega_a \, d\Omega_n / (4\pi)^2$. One factor of $(1 - \hat{\Omega}_n \cdot \hat{\Omega}_a)$ comes from the relative velocity of the colliding neutrinos; the other from the matrix element for the reaction. In the limiting case, where neutrinos are free streaming from a sharply defined surface emitting isotropically in the outward direction, the angular and energy integrals in (10.28) can be done separately. Using the same arguments about the angular distribution of the neutrinos used in obtaining (10.22), (10.28) can be rewritten as

$$q = \frac{1}{9}\frac{\sigma_0 c}{(m_e c^2)^2} D\chi \int dE_a \, dE_n (E_n + E_a) F_n F_a, \tag{10.30}$$

where χ is the integrated angular factor given by

$$\chi = \tfrac{1}{8}(1 - x)^2 (x^2 + 4x + 5), \tag{10.31}$$

with $x = (1 - (R_\nu/R)^2)^{1/2}$, R_ν being the radius of the sharply defined surface. I have been following the development given in Cooperstein, van den Horn, and Baron (1987); they further write (10.30) as

$$q = \frac{1}{9}\frac{\sigma_0 c}{(m_e c^2)^2} \frac{L_a L_n}{4\pi^2 c^2 R_\nu^4}(\omega_a + \omega_n)(1 - x)^2 \chi, \tag{10.32}$$

where L is the neutrino luminosity (energy per second), ω is an average neutrino energy, and the extra angular factor of $(1 - x)^2$ comes from the $(1/R)^2$ behavior of F outside the sharply defined surface. Equation (10.32) is a heating rate per unit volume and can be integrated over the exterior volume (from R_ν to ∞) which results in

$$4\pi \int_{R_\nu}^{\infty} qR^2 \, dR = \frac{1}{27}\frac{\sigma_0 c}{(m_e c^2)^2} D\frac{L_a L_n}{c^2 R_\nu}\frac{(\omega_a + \omega_n)}{8}. \tag{10.33}$$

Putting in values for L of 5×10^{52} ergs s^{-1}, $R_v \approx 20$ km and ω about 15 MeV gives a energy deposition rate on the order of 10^{51} ergs s^{-1}. Thus the $v\bar{v}$ annihilation reactions are capable of delivering an amount of energy to the bubble material that is of the same order as that observed in supernova explosions.

Cooperstein, van den Horn, and Baron (1987) wrote their paper to discuss the proposition made by Goodman, Dar, and Nussinov (1987) ("second heating"). They felt that the energy delivered by $v\bar{v}$ annihilation would be lost in neutrino cooling reactions, correctly pointing out that the neutrinosphere is not a sharp surface and that about half of the energy delivered to the matter is in a region from R_v to $1.1R_v$ (a region about 3 km wide). Thus the turning on of the second heating phase depends critically on the size of the density gradient at the surface of the protoneutron star.

10.3.5. Summary
Supernovae are produced by collapsing iron cores when the gravitational binding energy released (the core compresses to a volume one thousand million times smaller that its original size) couples to the hydrogen–helium envelope surrounding the massive star. Approximately 1% of this energy must couple to the envelope in order to match observations (see Woosley, 1988). Gravitational energy is released in neutrinos of all types (electron, muon, and tauon) and the mode of coupling of this energy to the envelope may depend on neutrino interactions as first suggested in the calculations done by Colgate and White (1966) over 20 years ago.

10.4. Calculational Neutrino Spectra and Luminosity from Collapsed Stellar Cores

10.4.1. Introduction
Neutrinos from SN 1987A were detected by the Kamiokande-II detector in Japan (Hirata *et al.*, 1987) and the IMB detector in the United States (Bionta *et al.*, 1987). In this section the observed neutrino signal will be compared with that predicted by numerical calculations. The possibility that the observed signal contains evidence for or against the late-time neutrino heating mechanism is discussed.

10.4.2. Early-Time Neutrino Signal
Before SN 1987A was observed numerical calculations of the expected early-time (about 1 s after collapse, bounce, and shock wave formation) neutrino signal from core collapse supernovae were done by Mayle (1985), Woosley, Wilson, and Mayle (1985), and Mayle, Wilson, and Schramm (1987). The remainder of this section summarizes their results.

Prior to the inner 0.55 $M_\odot$ of the collapsing core reaching and exceeding nuclear matter density (the rebound launching the shock wave into the outer core) the neutrino signal is dominated by electron neutrinos. These electron neutrinos are produced by electron capture on protons, both free and bound in heavy nuclei. The energy emitted in v_e prior to bounce (products of nuclear fusion reactions) is around 0.1% of the gravitational binding energy of a neutron star. The v_e luminosity reaches a peak just after the shock wave has moved outside the neutrinosphere, as the free protons, produced by shock dissociation of the iron, capture electrons, and emit v_e.

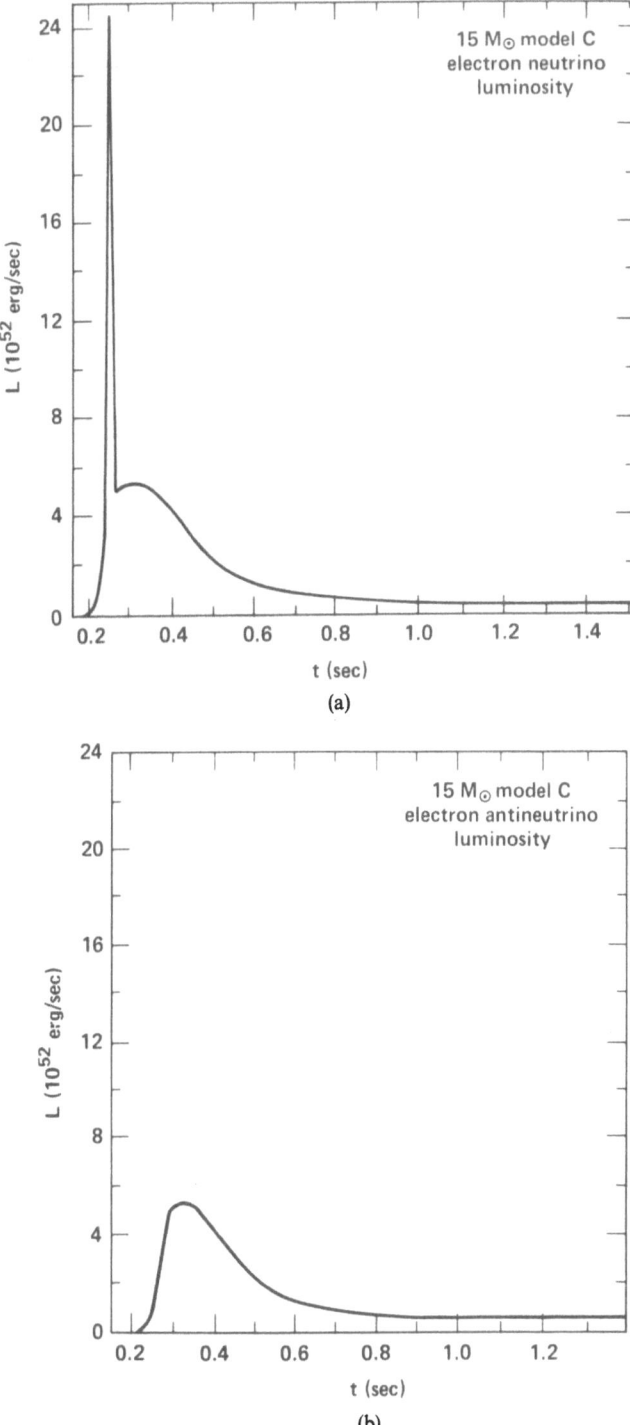

Figure 10.3. Electron–neutrino, electron–antineutrino, and muon–neutrino luminosities versus time are shown in (a), (b), and (c), respectively. These results are from the calculation of a 15 $M_\odot$ star (model C in Wilson *et al.*, 1985).

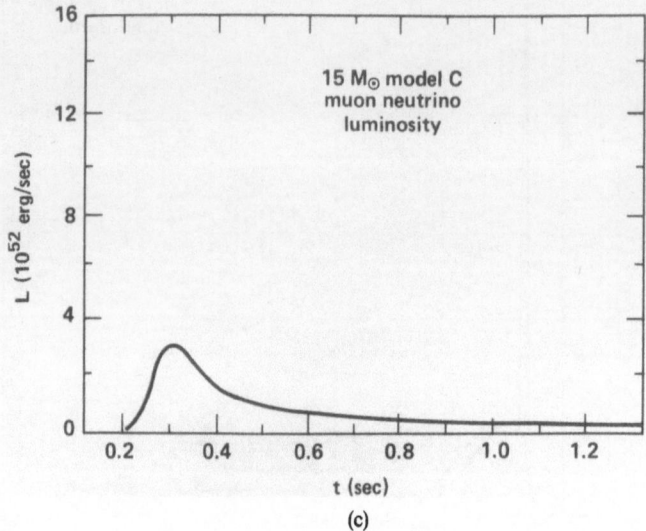

Figure 10.3 (*continued*)

The width in time of this deleptonization pulse is about 5 ms and accounts for about 1% of the total energy that will be emitted in neutrinos as the protoneutron star cools.

Deleptonization continues throughout the evolution of the protoneutron star; however, only about 5% of the neutron star binding energy is associated with the process of electron capture on protons. Neutrinos of all flavors are thermally produced by e^+e^- annihilation reactions after the outer part of the core is heated by the passage of the shock wave. The bulk of the binding energy is emitted in neutrinos of all flavors on neutrino-diffusion timescales (seconds). For the models studied by Mayle (1985), Woosley, Wilson, and Mayle (1985), and Mayle, Wilson and Schramm (1987), the supernovae exploded by the late-time neutrino heating mechanism (see Section 10.3.4 of this chapter) following a period of mass accretion onto the protoneutron star (while the shock is an accretion shock).

Figure 10.3(a), (b), (c) displays, respectively, the ν_e, $\bar{\nu}_e$, and ν_μ luminosity taken from a numerical calculation (see Wilson *et al.*, 1985). The electron capture burst is clearly seen in Figure 10.3(a) as the spike in luminosity; the rapid decline signals the formation of a neutrino sphere. Figure 10.3(b), (c) show similar features as Figure 10.3(a), the main difference being the absence of a burst structure. The thermal nature of the emission from the shock-heated outer core can be seen in the almost equal magnitudes of the ν_e, $\bar{\nu}_e$, and ν_μ luminosities (neglecting the spike in the ν_e curve). In the numerical calculation, $\bar{\nu}_\mu$, ν_τ, and $\bar{\nu}_\tau$ luminosities are assumed equal to the ν_μ since physical conditions found in supernovae do not favor the formation of free muons or tauons in large numbers (they are too massive), and their cross section for interaction with the matter is almost independent of neutrino flavor (ν_μ, $\bar{\nu}_\mu$, ν_τ, or $\bar{\nu}_\tau$). The shock stalled but revived at about 0.4 s after bounce, ending the mass accretion phase. This is seen in Figure 10.3(a), (b), (c) as the neutrino luminosity drops by about a factor of about 2–3 at this time. Mass accretion keeps the neutrino luminosity high, since kinetic energy developed by the infalling material is transformed to internal energy (at the protoneutron star surface) and finally into neutrinos (material cooling).

During the accretion phase, the average energy of the electron antineutrinos is about two-thirds the value attained after the envelope has separated from the core, and the protoneutron star has entered a slow (relative to collapse timescales) contraction and cooling phase (Kelvin–Helmholtz cooling, seen in Figure 10.3(a), (b), (c) as the slow decline in luminosity after 0.5 s). The reason for the rise in the average $\bar{v}_e$ energy is that $\bar{v}_e$ are emitted and absorbed by protons, and mass accretion brings material with proton fraction near one-half close to the neutrinosphere. After mass accretion is over, the opacity for $\bar{v}_e$ emission and absorption reaction drops as the matter at the neutrinosphere is neutron rich and remains so (no additional protons are being supplied by accretion). The mean free path of the $\bar{v}_e$ is increased allowing them to escape from hotter regions closer to the center of the core.

The muon and tauon neutrinos and their antiparticles only scatter by neutral current interactions from baryons and electrons and positrons, so their mean free path is always longer than that for v_e or $\bar{v}_e$. Consequently, as they become free streaming, v_μ, $\bar{v}_\mu$, v_τ, and $\bar{v}_\tau$ neutrinos are more energetic than free streaming v_e or $\bar{v}_e$. We find, after the envelope has separated from the core, that the v_e average energy is 10 MeV, the $\bar{v}_e$ average energy is 15 MeV, and the muon–tauon neutrinos and antineutrinos average energy is 24 MeV. These energies are those measured by an earthly observer, redshifted by the gravitational force of the core.

10.4.3. Kelvin–Helmholtz Cooling

In the remaining part of this section calculations of the Kelvin–Helmholtz phase of neutron star evolution are described (more detail can be found in Mayle and Wilson, 1987). The initial data for the neutron stars was artificially constructed from the cores of presupernova stars provided by S. Woosley. A 1.64 $M_\odot$ protoneutron star was made by following the inner three solar masses of the 25 model C (Wilson et. al., 1985) for 50 ms after bounce, and then removing all material just outside the shock position (the 1.64 $M_\odot$ point). This simulated a prompt explosion, leaving the hot protoneutron star to evolve into the Kelvin–Helmholtz contraction phase. A 1.27 $M_\odot$ neutron star was also constructed by evolving the core of model 15AB (Woosley, Pinto, and Ensman, 1988) well past bounce and removing the material outside the shock.

We evolved three models using the two cores described above and two different supranuclear density equations of state (one stiff, the other much less so). The results of the calculations are shown in Figure 10.4 which graphs total energy emitted in $\bar{v}_e$ versus time. We display the emitted energy in $\bar{v}_e$, since $\bar{v}_e$ capture on hydrogen protons in H_2O molecules is the most likely interaction to occur in the two earthly neutrino detectors (K II and IMB). The evolution was continued until the total luminosity in all neutrino types had fallen below 5×10^{51} ergs s^{-1} and was rapidly declining. At this time, roughly 80% of the binding energy had been radiated away (the binding energy of the 1.27 $M_\odot$ soft core, the 1.64 $M_\odot$ hard core, and the 1.64 $M_\odot$ soft core was 2.3×10^{53} ergs, 2.7×10^{53} ergs and 4.1×10^{53} ergs, respectively).

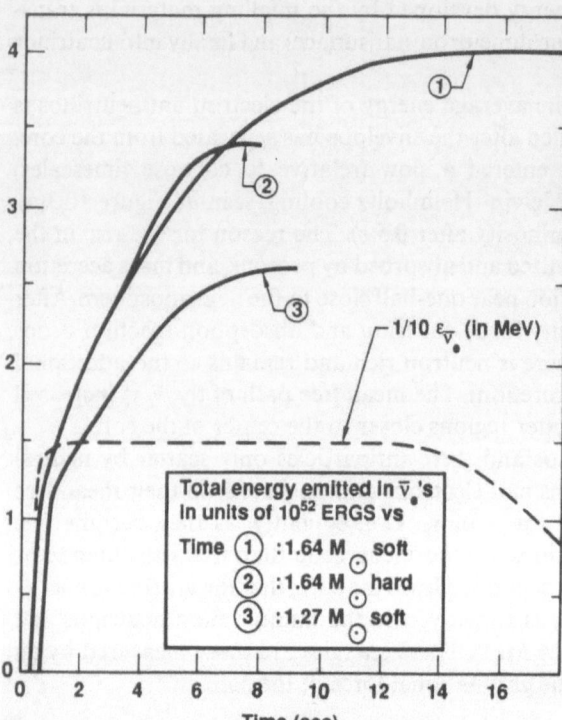

Figure 10.4. Total energy emitted in electron antineutrinos versus time for the Kelvin–Helmholtz phase of neutron star cooling for three model neutron star cores. Also shown is the average antineutrino energy.

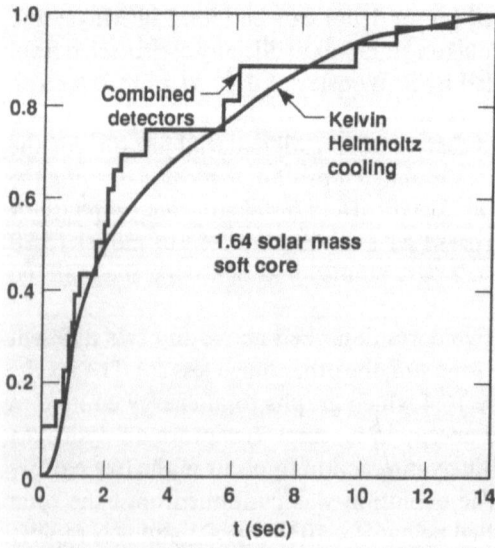

Figure 10.5. Energy emitted in electron antineutrinos for the 1.64 $M_\odot$ soft core (normalized to unity at 14 s) along with the time integrated number of detections from the K II and IMB detectors (also normalized to unity).

Figure 10.6. Detector efficiencies versus energy for the K II and IMB detectors.

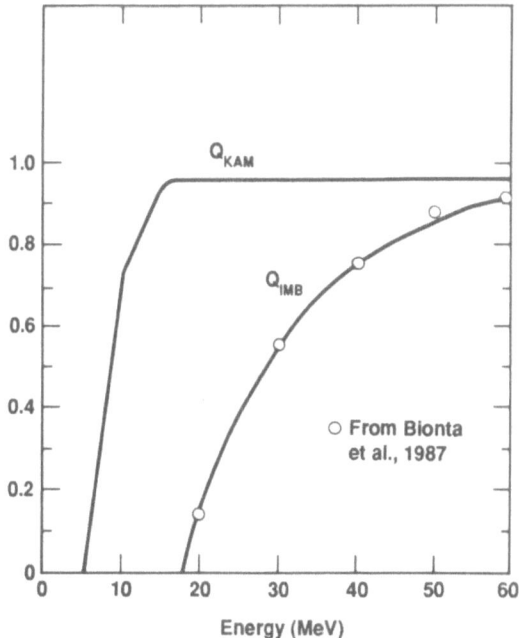

In Figure 10.5 the $\bar{\nu}_e$ luminosity for the 1.64 $M_\odot$ soft core is shown normalized to unity (at 14 s) along with the time-integrated counts from the K II and IMB detection events (also normalized to unity). Since K II saw eleven events and IMB saw eight, the stair step jumps in Figure 10.5 for the K II events are rather arbitrarily taken to be eight-elevenths the size of the IMB events. It is seen that the 1.64 $M_\odot$ soft core $\bar{\nu}_e$ luminosity is similar in shape to the curve of the detection events. (Since appreciable $\bar{\nu}_e$ emission ceased after about 8 s in the 1.64 $M_\odot$ hard core or the 1.27 $M_\odot$ soft core, these two models do not fit well the curve of the detection events.) The K II and IMB detectors also measured the energies of the neutrinos, which we discuss next.

Figure 10.6 shows the K II and IMB detector efficiencies as a function of energy. It is seen that the IMB detector is most sensitive to high-energy neutrinos. This helps in understanding why the 5000 t IMB water Cherenkov detector saw only eight events while the 2000 t K II water Cherenkov detector, with an efficiency near unity for almost all the energy range of interest in a supernova explosion, saw eleven. The $\bar{\nu}_e$ spectrum from the 1.64 $M_\odot$ soft core calculation along with the predicted $\bar{\nu}_e$ detection spectrum (both normalized so that their maximum is unity) is graphed in Figure 10.7(a), (b) for the K II and IMB detectors. Also shown is the detection histogram, with the energy of the detected neutrinos binned in energy bins of width 10 MeV (for the actual detection energies, see Bionta *et al.* (1987) and Hirata *et al.* (1987)). The energy of the detected event is 1.3 MeV less than the energy of the incoming $\bar{\nu}_e$ (the proton is transformed into the more massive neutron). Taking this into account the expected average energy of the detection events can be

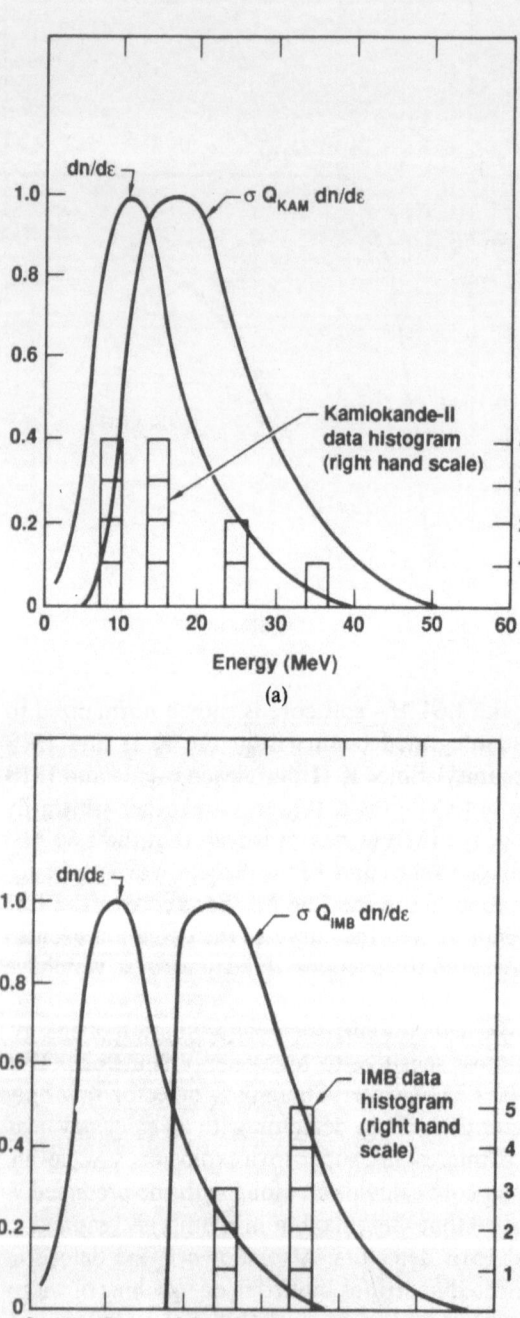

Figure 10.7. Energy spectrum of emitted electron antineutrinos (from the 1.64 $M_\odot$ soft core), as well as the expected detected spectrum, are shown for the K II and IMB detectors in (a) and (b), respectively. Also seen are the data histograms with the energy of the detected neutrinos placed in bins of width 10 MeV.

written as

$$E_{\mathrm{DET}} = \frac{\int (E - 1.3 \text{ MeV}) \sigma Q \dfrac{dn}{dE} \, dE}{\int \sigma Q \dfrac{dn}{dE} \, dE}, \qquad (10.34)$$

where dn/dE is the electron–antineutrino distribution function (the integral over dE gives the total number of ν_e emitted), Q is the detector efficiency, and σ is the cross section for positron capture on protons. For the $\bar{\nu}_e$ luminosity of the 1.64 $M_\odot$ soft core we find an E_{DET} for the K II detector of 20.3 MeV and 28.9 MeV for the IMB detector. The actual detected average energies were 15.4 MeV and 32.5 MeV for the K II and IMB detectors, respectively.

The total number of detected neutrinos is given by the following expression:

$$N_{\mathrm{DET}} = N_{\mathrm{p}} \int \frac{\sigma}{4\pi R^2} Q \frac{dn}{dE} \, dE, \qquad (10.35)$$

where N_{p} is the total number of hydrogen protons in the detector, R is the distance to the exploding star, and the rest of the notation is as above. Using the distance to the Large Magellanic Cloud, which contained SN 1987A, of 52 kpc (about 1.5×10^{23} cm), and the the $\bar{\nu}_e$ luminosity for the 1.64 $M_\odot$ soft core gives 14.8 and 8.9 detected events for the K II and IMB detectors, respectively. Given the nature of the quasi-Poisson process involved in detecting events, the predicted number of detections is in excellent agreement with the eleven events seen at K II and the eight events seen at IMB.

More sophisticated tests can be made (Kolomogorov–Smirnov goodness of fit) as to the likehood that SN 1987A is well modeled by the 1.64 $M_\odot$ soft core evolution (see Mayle and Wilson, 1987). It is found that, due the small number of detected events, the above results for 1.64 $M_\odot$ soft core evolution are in quite good agreement with the detected events. Unfortunately, the question of the explosion mechanism remains unanswered, as the frequency of the detected events was too small to find a numerical derivative (to get a luminosity). A phase of mass accretion lasting about 1 s prior to the shock exiting the core would give (as described above) a luminosity that remained high before declining as the protoneutron star entered the Kelvin–Helmholtz cooling phase. This luminosity structure would rule out a prompt explosion (the core quickly enters the Kelvin–Helmholtz phase). For other numerical calculations modeling the neutrino signal for SN 1987A the reader can consult Bludman and Schinder (1988), Bruenn (1987), Burrows and Lattimer (1987), or the chapter by Adam Burrows in this volume.

10.5. The Neutrino Heating Mechanism for Supernova Explosions: A Critique

Only one group of supernova modelers (the Livermore group) has seen numerical supernova explode by the mechanism of neutrino heating. A computer program that has any chance of being able to model a supernova must include an extremely complex network of detailed physics. It must have an equation of state that covers densities and temperatures in the range $10^2 < \rho < 10^{15}$ g cc^{-1} and $0.5 < T < 100$

MeV, electron fractions (Y_e) between 0.05 (neutron stars) and 0.5 (symmetric matter). For matter not in NSE (nuclear statistical equilibrium) a nuclear burn network should be included (granted, it can be a simple network of a few important elements as a detailed nucleosynthesis calculation is not the concern). The six neutrino fields need to be modeled by Boltzmann-type equations and all important neutrino–matter interactions must be included. This is best done using Monte Carlo, but Monte Carlo has not yet been applied to neutrino transport in a dynamic calculation. All the equations used (hydrodynamics, neutrino–evolution, etc.) must be consistent with the theory of general relativity. (This last point is very important for neutrino heating supernovae and we will return to it later.) Finally, the nonlinear partial differential evolution equations must be converted to a finite difference scheme before the computer can integrate the equations in time. There is also freedom in the choice of numerical scheme or even the physical approximations used by the modeler (what is the equation of state of supranuclear density matter?). It is thus apparent that there will be some controversy as to the meaning of numerical calculations, and to the confidence placed in the numerical output.

The numerical supernova explosions reported in Wilson et al. (1985) were the results of a quasi-Newtonian calculation (an effective static general relativistic gravitational potential was used in the matter evolution equations but the neutrinos were not redshifted). A rewritten version of the supernova code that includes a consistent general relativistic treatment of matter and neutrinos has since been used to study supernovae (for a description of the new code see Mayle and Wilson (1990)). In a recent calculation (reported in Wilson and Mayle, 1988) of a collapsed 1.3 $M_\odot$ iron core we find that the redshifted neutrino energy in the post-shock heating region (called first heating in Section 10.3.2) is about 98% (z of 0.02) the value if no redshift occurred. Since luminosity scales with redshift as $(1 + z)^{-2}$, the free baryon heating rate (proportional to neutrino energy squared times the luminosity) scales as $(1 + z)^{-4}$. A z of 0.02 translates to a 8% decrease in neutrino heating rate (first heating). This decrease in heating translates into a decreased explosion energy over a similar quasi-Newtonian calculation (about 5×10^{50} ergs in the calculation of Wilson and Mayle (1988) versus a value of 1×10^{51} ergs for a quasi-Newtonian calculation. Note that in Wilson and Mayle (1988) the value of the explosion energy reported was a factor of 2 too large due an error in an outer boundary condition). Woosley (1988) finds that the smallest allowable explosion energy consistent with the observation of SN 1987A is 7×10^{50} ergs. Neutrino heating supernovae tend to produce explosion energies that are less than observed. However, since a change in the heating rate of 8% can change the explosion energy by 50% (nonlinear process), it is clear that we must have a physical model and a numerical algorithm that represents the physics to better than a few percent. We are continuing to improve the physical models and the numerical methods used in our supernova computer code, and believe that future results will confirm that the neutrino-heating mechanism for supernovae is, in fact, the way nature explodes stars.

Acknowledgments

This work was performed under the auspices of the U.S. Department of Energy through the Lawrence Livermore National Laboratory under contract number W-7405-ENG-48.

References

Baade, W. and Zwicky, F. 1934, *Phys. Rev.*, **45**, 138.

Baron, E., Cooperstein, G., and Kahana, S. 1985, *Phys. Rev. Lett.*, **55**, 126.

Bethe, H.A., Brown, G.E., Applegate, J., and Lattimer, J. 1979, *Nucl. Phys.*, **A324**, 487.

Bethe, H.A. and Wilson, J.R. 1985, *Ap. J.*, **263**, 366.

Bionta *et al.* 1987, *Phys. Rev. Lett.*, **58**, 1494.

Bludman, S.A. and Schinder, P.J. 1988, *Ap. J.*, **326**, 265.

Bowers, R.L. and Wilson, J.R. 1982, *Ap. J.*, **263**, 366.

Brown, G.E., Bethe, H.A., and Baym, G. 1982, *Nucl. Phys.* **A265**, 315.

Bruenn, S.W. 1975, *Ann. N.Y. Acad. Sci.*, **262**, 80.

Bruenn, S.W. 1987, *Phys. Rev. Lett*, **59**, 938.

Burbidge, E.M., Burbidge, G.R., Fowler, W.A., and Hoyle, F. 1957, *Rev. Mod. Phys.*, **29**, 547.

Burrows, A. and Lattimer, J. 1987, *Ap. J.*, **318**, L63.

Chandrasekhar, S. 1967, *An Introduction to the Study of Stellar Structure* (New York: Dover).

Colgate, S.A. and White, R.H. 1966, *Ap. J.*, **143**, 626.

Cooperstein, G. 1982, Ph.D. thesis, SUNY, Stony Brook.

Cooperstein, G., van den Horn, L.J., and Baron, E. 1987, *Ap. J.*, **321**, L129.

Freedman, D.Z. 1974, *Phys. Rev. D.*, **9**, 1389.

Fuller, G.M., Fowler, W.A., and Newman, M.J. 1982a, *Ap. J. Suppl.*, **48**, 279.

Fuller, G.M., Fowler, W.A., and Newman, M.J. 1982b, *Ap. J.*, **252**, 715.

Goldreich, P. and Weber, S.V. 1980, *Ap. J.*, **238**, 991.

Goodman, J., Dar, A and Nussinov, S. 1987, *Ap. J.*, **314**, L7.

Haxton, W.C. 1988, *Phys. Rev. Lett.*, **60**, 1999.

Hirata *et al.* 1987, *Phys. Rev. Lett.*, **58**, 1490.

Hoyle, F. 1946, *M.N.R.A.S.*, **106**, 343.

Lightman, A.P., Press, W.H., Price, R.H., and Teukolsky, S.A. 1975, *Problem Book in Relativity and Gravitation* (Princeton: Princeton University Press).

Mayle, R.W. 1985, Ph.D. thesis, U.C. Berkeley, Lawrence Livermore Laboratory report UCRL-53713.

Mayle, R.W. and Wilson, J.R. 1987, *Ap. J.*, preprint.

Mayle, R.W. and Wilson, J.R. 1990, *Phys. Rep.*, submitted.

Mayle, R.W., Wilson, J.R., and Schramm, D.N. 1987, *Ap. J.*, **318**, 288.

Schramm, D.N. and Arnett, W.D. 1975, *Ap. J.*, **198**, 629.

Tubbs, D.L. and Schramm, D.N. 1975, *Ap. J.*, **201**, 467.

Wilson, J.R. 1971, *Ap. J.*, **163**, 209.

Wilson, J.R. 1974, *Phys. Rev. Lett.*, **32**, 849.

Wilson, J.R. 1985, in *Numerical Astrophysics*, eds. J. Centrella, J. Leblanc and R.L. Bowers (Boston: Jones and Bartlett), p. 422.

Wilson, J.R. and Mayle, R.W. 1988, *Supernova Modeling and SN 1987A*, *Proceedings of the Fifth Marcel Grossman Conference on General Relativity*, in preparation.

Wilson, J.R., Mayle, R.W., Woosley, S.E., ad Weaver, T.A. 1985, *Ann. N.Y. Acad. Sci.*, **470**, 267.

Woosley, S.E. 1988, *Ap. J.*, **330**, 218.

Woosley, S.E., Pinto, P.A., and Ensman, L. 1988, *Ap. J.*, **324**, 466.

Woosley, S.E. and Weaver, T.A. 1986, *Ann. Rev. Astron. Astrophys.*, **24**, 205.

Woosley, S.E., Wilson, J.R., and Mayle, R.W. 1985, *Ap. J.*, **302**, 19.

Yahil, A. and Lattimer, J.M. 1982, in *Supernova: A Survey of Current Research*, eds. M.J. Rees and R.J. Stoneham (Dordrecht: Reidel).

Index

In this index, f following a page number refers to a figure, t, to a table, and more extensive or longer discussions are given in boldface. The table of contents should also be consulted since it was not practical to enter into the index the frequent references in each chapter to the chapter topic.